Between Tradition and Innovation

Jesuit Studies

MODERNITY THROUGH THE PRISM OF JESUIT HISTORY

Editor

Robert A. Maryks (*Independent Scholar*)

Editorial Board

James Bernauer, S.J. (*Boston College, emeritus*)
Louis Caruana, S.J. (*Pontificia Università Gregoriana, Rome*)
Emanuele Colombo (*DePaul University*)
Paul Grendler (*University of Toronto, emeritus*)
Yasmin Haskell (*University of Western Australia*)
Ronnie Po-chia Hsia (*Pennsylvania State University*)
Thomas M. McCoog, S.J. (*Loyola University Maryland*)
Mia Mochizuki (*Independent Scholar*)
Sabina Pavone (*Università degli Studi di Macerata*)
Moshe Sluhovsky (*The Hebrew University of Jerusalem*)
Jeffrey Chipps Smith (*The University of Texas at Austin*)

VOLUME 32

The titles published in this series are listed at *brill.com/js*

Between Tradition and Innovation

Gregorio a San Vicente and the Flemish Jesuit Mathematics School

By

Ad J. Meskens

With Contributions by

Herman van Looy

BRILL

LEIDEN | BOSTON

Cover Illustration: Thesis print of Jan-Antoon Tucher, *Antverpia Surrounded by Allegorical Representations* (Antwerp: n.p., n.d. [1641?]). © Rijksmuseum Amsterdam, RP-P-OB-67.969A.

Library of Congress Cataloging-in-Publication Data

Names: Meskens, Ad, author. | Looy, Herman van, author.
Title: Between tradition and innovation : Gregorio a San Vicente and the
   Flemish Jesuit mathematics school / by Ad J. Meskens with contributions
   by Herman van Looy.
Description: Leiden ; Boston : Brill, [2021] | Series: Jesuit studies :
   modernity through the prism of Jesuit history, 2214-3289 ; volume 32 |
   Includes bibliographical references and index.
Identifiers: LCCN 2020058155 (print) | LCCN 2020058156 (ebook) | ISBN
   9789004414990 (hardback) | ISBN 9789004447905 (ebook)
Subjects: LCSH: Saint-Vincent, Grégoire de, 1584-1667. |
   Jesuits—Education—Belgium—History—17th century. |
   Jesuits—Belgium—Intellectual life. |
   Mathematics—Belgium—History—17th century. |
   Mathematicians—Belgium—Biography. | Mathematics
   teachers—Belgium—Biography. | Belgium—Intellectual life—17th
   century.
Classification: LCC QA27.B4 M47 2021 (print) | LCC QA27.B4 (ebook) | DDC
   510.71/1493—dc23
LC record available at https://lccn.loc.gov/2020058155
LC ebook record available at https://lccn.loc.gov/2020058156

Typeface for the Latin, Greek, and Cyrillic scripts: "Brill". See and download: brill.com/brill-typeface.

ISSN 2214-3289
ISBN 978-90-04-41499-0 (hardback)
ISBN 978-90-04-44790-5 (e-book)

Copyright 2021 by Koninklijke Brill NV, Leiden, The Netherlands.
Koninklijke Brill NV incorporates the imprints Brill, Brill Hes & De Graaf, Brill Nijhoff, Brill Rodopi, Brill Sense, Hotei Publishing, mentis Verlag, Verlag Ferdinand Schöningh and Wilhelm Fink Verlag.
All rights reserved. No part of this publication may be reproduced, translated, stored in a retrieval system, or transmitted in any form or by any means, electronic, mechanical, photocopying, recording or otherwise, without prior written permission from the publisher. Requests for re-use and/or translations must be addressed to Koninklijke Brill NV via brill.com or copyright.com.

This book is printed on acid-free paper and produced in a sustainable manner.

Contents

Preface IX
Abbreviations XIII

Introduction: The Low Countries, Spain, and Europe 1
   1   The Jesuits in the Netherlands 6

1   The College and Its School of Mathematics 11
   1   Schools in Antwerp 11
   2   Jesuit Educational Policy 12
   3   Mathematics in the Jesuit Curriculum 14
   4   The Academy of Mathematics at the Collegio Romano 20
   5   The College of Leuven 22
   6   The Antwerp College in the Sixteenth Century 23
   7   The Antwerp College in the Seventeenth Century 26
   8   The School of Mathematics 30
   9   Michiel Coignet and the Jesuits 35

2   The Seventeenth Century: The Dawn of a New Era 42
   1   Conic Sections 42
   2   Squaring the Circle the Archimedean Way 47
   3   The Humble Beginnings of Infinitesimal Calculus 49
   4   Infinitesimals: The Keplerian Revolution 53
   5   Cavalieri's Indivisibles 57
   6   The Jesuits and Indivisibles 58

3   Francisco de Aguilón and Mathematical Optics 62
   1   *Opticorum libri sex* 64
   2   Aguilón's *Catoptrica* Manuscript 68

4   Gregorio a San Vicente: An Ignored Genius 81
   1   A Tragic Life 81
   2   Mathematical *Oeuvre* 87
   3   The Mechanics Theses 91

5   The Creative Antwerp–Leuven Period 94
   1   Trisection of an Angle 94
   2   Mean Proportionals 98
   3   Properties of Conic Sections 103

6   **Exhaustion: The Road to Infinitesimals**   117
    1   Sequences and Series   117
    2   The Exhaustion Method   121
    3   San Vicente's Use of Infinitesimals   124
    4   The Cavalieri Dispute   128

7   **Infinitesimal Calculus at Work**   132
    1   The Hyperbola   132
    2   Calculation of the Volume of Ductus Figures   140
    3   Lateral Area of the *Ungula cylindrica* and Relations between Ductus Figures   147

8   **Rome and Prague, the Final Discoveries**   153
    1   The Missives to Rome   153
    2   The *Chartae Romanae*   155
    3   San Vicente's Legacy   158
    4   Conclusion   160

9   **The Erroneous Circle Quadrature**   162

10 **Joannes della Faille and the Beginning of Projective Geometry**   165
    1   An Itinerant Life   165
    2   Conic Sections   173
    3   *De centro gravitatis*   182

11 **The Antwerp Students**   184
    1   Philip Nuyts   184
    2   Ignatius Derkennis   189
    3   Other Students   190

12 **The Leuven Students**   192
    1   Theodorus Moretus   192
    2   Jan Ciermans   198
    3   Willem Boelmans   206
    4   Willem Hesius   208
    5   Other Students   209

13 The Later Disciples 211
    1   Andreas Tacquet 211
    2   Gilles-François de Gottignies 218
    3   Alphonse Antonius de Sarasa 220

14 The Jesuit Architects 222
    1   Ad maiorem Dei gloriam 222
    2   *Descensus ad inferos* 238

15 The Influence of the School of Mathematics 244

Appendix 1: Chronology of San Vicente's Manuscripts 251
Appendix 2: Students of the School of Mathematics after 1625 255
Bibliography 257
Index 285

Preface

In his masterful book, *Infinitesimal: How a Dangerous Mathematical Theory Shaped the Modern World* (2014), Amir Alexander writes that the French courtier Samuel Sobrière's (1615–70) network included "some of the greatest luminaries in France, and philosophers and scientists in Italy, the Dutch Republic and England." For a Fleming, like myself, it is striking that Alexander omits Spain and, more importantly, the Spanish Netherlands, thus reinforcing, albeit unwittingly, the idea that the Southern Netherlands was a scientifically barren nation after the re-conquest by the Spanish. It is certainly true that the Northern Netherlands gained a great deal from the exodus from the South. Not in the least did Amsterdam flourish because of the Antwerp merchants, with their capital and commercial network, who had found refuge there. In their wake came numerous painters and scientists, Gillis Coignet (c.1542–99) and Simon Stevin (c.1548–1620) among them. But this cultural and scientific hemorrhage did not suck the lifeblood out of the South. The South was crippled, but not on its knees; on the contrary, this was the age of Peter Paul Rubens (1577–1640) and Gregorio a San Vicente (1584–1667).

On the back cover of Alexander's book, we find:

> On August 10, 1632 five Jesuits met in an austere Roman palace to pass judgment on the veracity of an apparently simple idea: that a continuous line is made up of distinct and limitless small parts. On that fateful day the judges ruled that this notion wasn't only improbable but actually repugnant—with the stroke of a pen it was banished.

A number of mathematicians within the Jesuits' own ranks would have vehemently disagreed with this judgment. Indeed, it was one of their own who had invented the concept of summating an infinity of terms to find the area of a plane figure or of a solid; it was one of their own who, although intuitively, defined the limit concept of a series; it was one of their own who pushed the exhaustion method to its limits. And it was some of their own, who despite the judgment, would be at the cradle of integral calculus. These Jesuits were San Vicente and his students.

The idea for this book arose when I was asked to write a contribution on mathematics for a book on the development of Antwerp's intellectual life. Writing the chapter brought to light how little attention had been paid to these Jesuit-mathematicians, despite their remarkable and ground-breaking contributions to mathematics.

This book is about the Flemish Jesuit school of mathematics, in which San Vicente played a major role. This school would have a profound influence on the course of mathematics in the seventeenth century, an influence that has been hugely underscored up to now. Not only did San Vicente arrive at his most influential insights when working in Antwerp but he also taught students who would become renowned mathematicians in their own right.

It all began with a simple Jesuit house and college, which became a focus of a triumphant Catholicism during the Counter-Reformation. The book will show that the Flemish school of mathematics was a major force of change within mathematics and mechanics during the first part of the seventeenth century. However, although San Vicente may have been the most important Flemish Jesuit mathematician, he was by no means the only one with any influence. Joannes della Faille's (1597–1652) theorem, for instance, is still compulsory knowledge for any mechanics student. The failure of San Vicente to bring his circle quadrature to a good end still casts a dark and undeserved shadow over his reputation and that of his students. To quote William Shakespeare (c.1564–1616): "The evil that men do lives after them; The good is oft interred with their bones" (Julius Caesar, act 3, scene 4), or in this case—"The error a mathematician makes blots his reputation, his grand theorems are oft forgotten to be his."

This book deals with mathematics and mathematicians, but not entirely as we now know them. At the turn of the century in 1600, mathematics would have included subjects we know as accounting, astronomy, physics, engineering, wine-gauging, cryptography, and, to a lesser degree, architecture. For court mathematicians, it would, in many cases, also have included astrology and perhaps even numerology. On the other hand, some mathematical subjects, such as logic, were part of the philosophy course, while astrology was studied in conjunction with medicine. By 1700, pure mathematics was more clearly defined as an abstract subject and began to resemble our concept of mathematics. This is in no small part due to the development of infinitesimal calculus, which, in a sense, can be viewed as the fusion of algebra and geometry. Under the influence of Johannes Kepler (1571–1630) and Galileo Galilei (1564–1642), other sciences became mathematized, not only physics but also geography, stonecutting or stereotomy, and even music. The distinction between a mathematician and an engineer or a scientist remained hard to make. Some even ventured into what we would call the occult sciences, the best-known example being Isaac Newton's (1643–1727) (al)chemical interests. Fortunately for us, our subjects conform to our image of a mathematician, their prime research interest being geometry, and with it a nascent calculus. The book will, however, also point out whether and to what extent they ventured into other sciences.

PREFACE   XI

In writing this book, I sometimes had to make difficult decisions. Keeping San Vicente's style would deter any modern reader, and I have consequently chosen to "translate" his prose into a more modern mathematical style, even though many historians of mathematics would disagree. Not doing so would result in a very hard to read and lengthy text. Moreover, San Vicente was a genius when it came to finding particular cases and proving all of them, usually without drawing a generalization. Yet every aspect of this generalization is present in his work.

For geographical names, the book uses the names that are presently in use, even though they may have been different in the seventeenth century. An exception was made for names that have become standard in English, such as Prague or Ypres, instead of Praha or Ieper.

The lack of an appropriate historical-geographical terminology for the Low Countries in English for this specific period may create some confusion. The book uses the term "Seventeen Provinces" or "the Netherlands" (de Nederlanden) for the country ruled by Charles V (1500–58, r.1515–55) and Philip II (1527–97, r.1555–97) that was created by the Pragmatic Sanction (1548), as opposed to the nation we now call the Netherlands (Nederland). The Low Countries also includes the independent prince-bishopric of Liège. In general, Flanders is used to indicate the northern Dutch-speaking part of the Southern Netherlands; however, in citations of a seventeenth-century document, Flanders refers to the then County of Flanders. The religious–civil war at the end of the sixteenth century created two nations in the Low Countries: the Southern or Spanish Netherlands on the one hand and the Northern Netherlands or the Republic of the United Provinces (the republic for short) on the other.

The people studied in this book are referred to by the name under which they were known when they were alive or by their baptismal name. An exception is made for Gregorius a Sancto Vincentio who, in this monograph, is called Gregorio a San Vicente. His students are referred to by their proper names.

Writing the book also involved navigating the difficulty of how best to arrange the material. This could have been done chronologically, in which case the theater would have had to change every few pages, going from Prague to Madrid to Stockholm, without any coherence in the mathematical subject matter. The material could have been arranged thematically, but in that case, some of the individuals' work would had to have been torn apart, making it hard to see a line of thought in their work. It could have been arranged by person, but this would have made it difficult to see the connections between the individuals' work. As none of the options was ideal, it was ultimately this last option that was chosen, although architecture has been kept as a separate subject. This allows the narrative to focus in particular on San Vicente.

Much of the material presented here has been published before, but it is dispersed over many articles and sometimes hard to find books. San Vicente has already been studied by Herman van Looy, Jean Dhombres, and Patricia Radelet-de Grave; Francisco de Aguilón (1567–1617) has been studied by August Ziggelaar, S.J.; and della Faille has been the subject of my own research, not to mention the earlier studies by the Jesuits Henri Bosmans (1852–1928) and Omer van de Vyver (1915–97). The same remark goes for the archival material, which is dispersed over many archives. Nevertheless, I feel I have been able to bring together the core of this material.

It is fortunate that Van Looy allowed me to use his material, as the description of San Vicente's manuscripts is based to a significant extent on his doctoral thesis. These parts have been duly indicated.

I hope I have brought all of this material, and new archival sources, together in a coherent narrative, bringing a new view on this remarkable group of men. This book would not have been possible without the help of numerous people. At the risk of forgetting someone, we mention Fr. Raúl González, S.J. (ARSI), Van Looy, Stephen Hargreaves, Philippe Martin, and Miguel Robert (University of Barcelona).

The staff of Erfgoedbibliotheek Hendrik Conscience Antwerpen, Museum Plantin Moretus (UNESCO World Heritage), Felixarchief Antwerpen, KaDoc Leuven, the State Archives of Belgium at Beveren, and the libraries of the University of Antwerp (especially Ruusbroecgenootschap) and Artesis Plantijn University College Campus Noord were, as usual, very helpful. The cooperation with Paul Tytgat, who provided advice on the drawings, proved as invaluable as ever.

Last, but certainly not least, I have to thank my wife Nicole, without whom this book would never have materialized. Her support genuinely is a labor of love.

Ad Meskens

Abbreviations

ABML	Archives de la Province belge méridionale et du Luxembourg, Kadoc, Leuven
ABSE	Archief Vlaamse Jezuïeten, Kadoc, Leuven
AFL	Archief della Faille de Leverghem
AP	Artesis Plantijn Hogeschool Antwerp
APUG	Archives of the Pontifical Gregorian University, Rome
ARA	Algemeen Rijksarchief, Brussels
ARAA	Algemeen Rijksarchief in de Provinciën, Antwerp
ARAL	Algemeen Rijksarchief in de Provinciën, Leuven
ARSI	Archivum Romanum Societatis Iesu, Rome
BPM	Biblioteca de Palacio de Madrid
BRAHM	Biblioteca de la Real Academia de la Historia, Madrid
EHC	Erfgoedbibliotheek Hendrik Conscience, Antwerp
FA	Felixarchief, Antwerp
Kadoc	Kadoc, Katholieke Universiteit Leuven, Leuven
KBDH	Koninklijke Bibliotheek, The Hague
KBR	Koninklijk Bibliotheek Albert I, Brussels
KIK-IRPA	Royal Institute for Cultural Heritage, Brussels
KUL	Katholieke Universiteit Leuven
MPM	Museum Plantin-Moretus/Prentenkabinet, Antwerp—Unesco-World Heritage
UA	Universiteit Antwerpen
UG	Universiteit Gent

INTRODUCTION

The Low Countries, Spain, and Europe

The origins of the Flemish colleges go back half a century before their school of mathematics was erected, to a time when the Jesuits found themselves in the midst of a religious and political struggle that tore the Low Countries apart.[1]

In 1555, together with the crown of Spain, Philip II inherited the Seventeen Provinces, one of the most prosperous countries of Europe. During his reign, Protestantism swept throughout the country, despite the imposition of an Inquisition. Beginning on August 10, 1566 in the Flemish town of Steenvoorde, the Beeldenstorm (Iconoclastic revolt) spread across the Dutch-speaking provinces, marking the beginning of the Eighty Years' War (1566–1648).

In July 1581, the States General proclaimed the *Plakkaat van Verlatinghe* (Act of abjuration), renouncing the king of Spain. In a counteroffensive, the Spanish, under Alexander Farnese, duke of Parma (1545–92), succeeded in taking town after town. Invariably, this prompted an exodus to the north.

Parma's troops arrived at the gates of Antwerp in July 1584. After a siege of barely a year, the city that had become the symbol of the revolt in the south fell to the Spanish. This victory seemed to have opened the road to the north, but the fledgling Dutch Republic was saved by Philip II's decision to order Parma and his troops to Dunkirk to prepare for an invasion of England. Parma's Dutch offensive thus ground to a halt on the banks of the great rivers.

---

1 As the region's religious troubles have been thoroughly documented and studied elsewhere, the historical material in this introduction is intended simply as an outline to anchor the reader in the prevailing context of the time. For a detailed account of this period in the history of the Low Countries see, among other works, Michiel A. G. de Jong, *Staat van Oorlog (1585–1621)* (Hilversum: Verloren, 2005); Jonathan I. Israel, *The Dutch Republic* (Oxford: Clarendon Press, 1998); Paul Janssens, ed., *België in de 17de Eeuw: De Spaanse Nederlanden en het Prinsbisdom Luik; Band I; Politiek* (Brussels: Dexia Bank-Snoeck, 2006); Janssens, ed., *België in de 17de Eeuw: De Spaanse Nederlanden en het Prinsbisdom Luik; Band II; Cultuur en leefwereld* (Brussels: Dexia Bank-Snoeck, 2006); Geoffrey Parker, *The Army of Flanders and the Spanish Road* (Cambridge: Cambridge University Press, 1972); Parker, *Spain and the Netherlands: Ten Studies* (Glasgow: Fontana Press, 1990); Ben H. Roosens, "Habsburgse defensiepolitiek en vestingbouw in de Nederlanden (1520–1560)" (PhD diss., Universiteit Leiden, 2005); Ivo Schöffer, Herman van der Wee, and Johannes A. Bornewasser, *De Lage Landen 1500–1780* (Amsterdam: Agon, 1988); Werner Thomas, ed., *De Val van Het Nieuwe Troje: Het Beleg van Oostende, 1601–1604* (Leuven: Davidsfonds, 2004); Willem J. M. van Eysinga, *De wording van het Twaalfjarig Bestand van 9 april 1609* (Amsterdam: Noord-Hollandsche Uitgeversmij, 1959).

After 1585, the more or less periodic shifting of the frontline northward and southward drove the economies of Flanders and Brabant into a deep recession. Many of Antwerp's merchants now emigrated, especially to Rotterdam and Amsterdam, which had become boomtowns.[2] In the wake of these merchants came artists and scientists.[3]

Shortly before his death, Philip II ceded the Low Countries to his daughter Isabella (1566–1633). Together with her husband Albert (1559–1621), the archduchess would rule the Netherlands as a sovereign in her own right.[4]

In 1609, after an armistice and two years of negotiations, the war-weary parties concluded the Twelve Years' Truce. An undeniable truth became plain for all to see: the *Leo Belgicus* (Belgian Lion) no longer existed and had been superseded by two independent nations, *Belgica regia* (Royal Belgium) and *Belgica*

---

2 These merchants kept their overseas network intact and, in many cases, did not cut all ties with their hometown. Antwerp merchants also acted as go-betweens for Dutch traders who, more often than not, were their emigrated next of kin. See Clé Lesger, *Handel in Amsterdam Ten Tijde van de Opstand* (Hilversum: Verloren, 2001) on the port of Amsterdam in the second half of the sixteenth century; Ad Meskens and Godelieve van Hemeldonck, "Quignet, Quingetti, Cognget, Coignet: An Antwerp Family of Goldsmiths, Some Painters, One Mathematician, and a Lot of Merchants," *Antwerp Royal Museum Annual 2015–2016* (2018): 75–138, for an example of the diaspora of the Antwerp merchant family Quinget(ti), who had offices, run by members of the family, in many important commercial centers such as Amsterdam, Rotterdam, Rouen, Venice, Cologne, and Hamburg; and James R. Sadler, "Family in Revolt: The Van Der Meulen and della Faille Families in the Dutch Revolt" (PhD diss., University of California at Los Angeles, 2015) for the relations within the della Faille–Van der Meulen merchant family.

3 The best-known example of course being Simon Stevin. But there are many others, for example the playwright and schoolmaster Peter Heyns (1537–98), whose odyssey, following the merchant Antonio Anselmo (1536–1611), took him to Frankfurt, Stade, and finally Haarlem, or the painter Gillis Coignet, who fled Antwerp for Amsterdam and later moved on to Hamburg. On Heyns, see Ad Meskens, "Peter Heyns: Leaving the Market Place of the World," in *Rechenmeister und Mathematiker der frühen Neuzeit*, ed. Rainer Gebhardt (Annaberg-Buchholz: Adam-Ries-Bundes, 2017), 287–96; Hubert Meeus, "Peeter Heyns, a 'French Schoolmaster,'" in *Grammaire et enseignement du Français 1500–1700*, ed. Jean de Clercq, Nico Lioce, and Pierre Swiggers (Leuven: Peeters, 2000), 301–16; Meeus, "De *Spieghel der Werelt* als spiegel van Peeter Heyns," in *Ortelius's* Spieghel der Werelt: *A Facsimile for Francine de Nave*, ed. Elly Cockx-Indestege et al. (Antwerp: Vereniging van Antwerpse Bibliofielen, 2009), 31–47. On Coignet, see Meskens and Van Hemeldonck, "Quignet, Quingetti, Cognget, Coignet," and Barbara Uppenkamp, "Gilles Coignet: A Migrant Painter from Antwerp and His Hamburg Career," *De Zeventiende Eeuw* 31 (2015): 55–77.

4 On the reign of the archdukes, see, e.g., Werner Thomas and Luc Duerloo, eds., *Albert and Isabella 1598–1621: Essays* (Turnhout: Brepols, 1998); Luc Duerloo, *Dynasty and Piety: Archduke Albert (1598–1621) and Habsburg Political Culture in an Age of Religious Wars* (London: Routledge, 2012); Dries Raeymaekers, *One Foot in the Palace: The Habsburg Court of Brussels and the Politics of Access in the Reign of Albert and Isabella, 1598–1621* (Leuven: Leuven University Press, 2013).

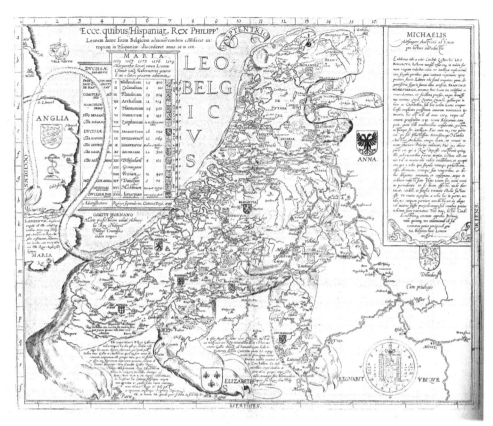

FIGURE 0.1 The *Leo Belgicus*—an allegorical map of the Low Countries—at the time of Philip II. Frans Hogenberg (1535–90) in Michael Eytzinger, *De Leone Belgico, ejusque topographica atque historica description liber* (Cologne: Gerardus Campensis, 1583)
© EHC K 55537

foederate (Federal Belgium). In the north, the Dutch Republic would develop into a leading mercantile power. In the south, the truce was the period in which the economic recovery gained momentum. The south could breathe again, after having been the theater of hostilities for forty years. Fiscal policies were relaxed, and cultural expenditure rose to levels never seen before. The archdukes supported the Counter-Reformation with unprecedented vehemence.[5] This was by no means coincidental, as from 1590 onward scions of the major families of all Habsburg territories began to abandon the reformed churches in favor of Catholicism.

---

5 Werner Thomas, "Andromeda Unbound: The Reign of Albert & Isabella in the Southern Netherlands, 1598–1621," in Thomas and Duerloo, *Albert and Isabella*, 1–14, here 9–10.

For the Southern Netherlands, this was a silver age and a cultural heyday. The entire visual fabric of the church needed reconstruction and renewal. Flemish churches once again exhibited high-quality artwork. New and magnificent churches and convents were built on a scale unparalleled outside of Italy.[6] From 1585 onward, the Spanish government tried to resurrect the Catholic faith, which resulted in a thriving artistic scene. The destruction wrought by the Iconoclastic Revolt was gradually repaired, and artwork was replaced. The guilds were expected to re-erect their chapels in the churches, thereby becoming pillars of Catholicism once again. The image that was created of Protestants was one of villains and iconoclasts who had been responsible for the famines of the late 1580s.[7]

In cities such as Antwerp, international trade had been replaced by a luxury industry. The South was not an impoverished country, as it had high-yield agriculture and a wide range of industries with products of an impressive quality. It was hoped, in vain, that the River Scheldt would be reopened.[8]

The Twelve Years' Truce was not renewed, possibly because of the damage the Dutch were inflicting on the trade fleet to the Indies and the Americas. Moreover, the Spanish were reluctant to renew the truce for political reasons. If Spain failed to intervene in the Germanic lands, in what was to become the Thirty Years' War (1616–48), the Holy Roman Empire would be under the control of Protestants, which would also undermine Spain's position in the Netherlands and Italy.[9] The Dutch–Spanish war now merged into the Thirty Years' War. Initially, the Thirty Years' War was a war between various Protestant and Catholic states in the fragmented Holy Roman Empire. However, it gradually developed into a more general conflict involving most of the great powers

---

6 See, e.g., Jeffrey Muller, "Jesuit Uses of Art in the Province of Flanders," in *The Jesuits II: Cultures, Sciences, and the Arts 1540–1773*, ed. John W. O'Malley, S.J., et al. (Toronto: University of Toronto Press, 2006), 113–56; Anna C. Knaap, "Meditation, Ministry, and Visual Rhetoric in Peter Paul Rubens's Program for the Jesuit Church in Antwerp," in O'Malley et al., *Jesuits II*, 157–81.

7 See, for instance, the book by Frans de Costere, *Bewiis der ouder catholiicker leeringhe met andtwoorde op sommighe teghenstellinghen* (Antwerp: Ioachim Trognesius, 1601). See Guido Marnef, "Protestant Conversions in an Age of Catholic Reformation: The Case of Sixteenth-Century Antwerp," in *The Low Countries as a Crossroads of Religious Belief*, ed. Arie J. Gelderblom, Jan L. de Jong, and Marc van Vaeck (Leiden: Brill, 2004), 33–48, here 44.

8 Roland Baetens, *De Nazomer van Antwerpens welvaart: De diaspora en het handelshuis De Groote tijdens de eerste helft der Zeventiende Eeuw* (Brussels: Gemeentekrediet, 1976); Israel, *Dutch Republic*, 410–20.

9 Geoffrey Parker, *The Thirty Years' War* (London: Routledge, 1993), 51.

and would become the longest war in European history. Indeed, when one takes into account that battles were also waged on the high seas and in the colonies, this was the first war on a truly global scale. Unfortunately, the Southern Netherlands became the preferred theater of the war, where Spain confronted France and the republic to restore its hegemony in Europe.

Albert and Isabella's marriage remained childless, and after Albert's death the Southern Netherlands reverted to Spanish control, with Isabella as their governess.[10] In 1625, Ambrogio Spinola (1569–1630) succeeded in reducing the Dutch fortress of Breda, and the Spanish were once again in a position to enforce a tight economic blockade, not only of the continental river trade but also of ports and fisheries. For most of the decade, Spain held the initiative in the North Sea, and the effect on Dutch commerce was devastating. The total embargo on trade with the Dutch never materialized, however, and was actually undermined in practice by central government itself.[11] By 1629, the fortunes of the war had reversed, and with the Dutch on the offensive, there was growing unrest among the Flemish and Walloon nobility together with unmistakable signs of popular agitation. The States General decided to open direct negotiations with its Dutch counterpart.[12]

In 1648, the Thirty Years' War drew to a close, as the Peace of Westphalia redrew the map of Europe, with the Dutch Republic being recognized as an independent state. Spain's worries were not over, however. Revolts continued in Portugal and Catalonia, thanks to French support; in 1647, a major rebellion broke out in the city of Naples, and shortly afterward there were uprisings in Sicily and Andalusia. Moreover, the whole of Spain was afflicted by a virulent outbreak of plague. On the other hand, the peace with the Dutch Republic gave Spain the opportunity to turn the Army of Flanders against France.[13]

---

10   The seminal study on the history of the Southern Netherlands in the seventeenth century is René Vermeir, *In Staat van Oorlog: Filips IV en de Zuidelijke Nederlanden 1629–1648* (Maastricht: Shaker, 2001).
11   Robert A. Stradling, *Europe and the Decline of Spain: A Study of the Spanish System, 1580–1720* (London: Allen & Unwin, 1981), 57–58, 63.
12   Della Faille was kept informed about these events by Balthasar II Moretus (1615–74), who regularly wrote to him with news of the negotiations and hoped they would lead to a permanent peace. However, by the middle of 1633, Moretus had given up hope. MPM Arch 144, 322, 338, 341–42, 345, 356, 360. Also, René Vermeir, "'Oorloghsvloeck en Vredens Zegen': Madrid, Brussel en de Zuid-Nederlandse Staten over oorlog en vrede met de Republiek, 1621–1648," *Bijdragen en Mededelingen betreffende de Geschiedenis der Nederlanden* 115 (2000): 1–32; Vermeir, *In Staat van Oorlog*, 63–79.
13   Parker, *Thirty Years' War*, 167.

1 The Jesuits in the Netherlands

From 1542 onward, barely two years after the Society of Jesus's foundation, Jesuits arrived in the Netherlands.[14] Some would reside in a semi-official residence in Leuven, while a residency for traveling Jesuits was later established in Tournai.[15]

Charles V, and his regent Mary of Hungary (1505–58, r.1531–55), were reluctant to subsidize the Jesuit order or give it any financial benefits. However, shortly after Charles's abdication, Ignatius of Loyola's (c.1491–1556) special envoy Pedro de Ribadeneyra (1527–1611) succeeded in obtaining Philip II's recognition of the order in the Netherlands.[16] The idea was to erect several colleges in which to instruct the young.[17] Perhaps unsurprisingly, the founding of colleges in the French-speaking parts of the country was easier than in the Dutch-speaking part, which leaned heavily toward Protestantism.[18]

In 1564, the Low Countries, which had previously been part of the Lower German province (including the Rhineland), became a province in their own right, known as the Belgian province, at a time when Catholicism in these regions was near its nadir. In 1612, the Jesuits separated the Belgian province along the linguistic border into a Flandro-Belgian province (with Dunkirk, Bruges, Antwerp, and Brussels, among others) and a Gallo-Belgian province (with Dinant, Tournai, Lille, etc.).[19]

The first college the Jesuits opened in the Low Countries was the one in Tournai (1562), soon to be followed by Dinant (1563), Cambrai (1563), Leuven

---

14   Lodewijk Brouwers, S.J., "Aperçu historique sur la province belge de la Compagnie de Jésus" (MS, Archief Jezuïeten Heverlee [Kadoc] 12429, n.d.), handwritten, 1. Eight Spanish Jesuits had been expelled from France and came to Leuven, Pedro de Ribadeneyra among them. They immediately attracted youths to join the Society. As early as 1544, eight new members were admitted.

15   Herman van Goethem et al., *Antwerpen en de Jezuïeten, 1562–2002* (Antwerp: UFSIA, 2002), 12; Jan Roegiers, "Awkward Neighbours: The Leuven Faculty of Theology and the Jesuit College (1542–1773)," in *The Jesuits of the Low Countries: Identity and Impact*, ed. Rob Faesen and Leo Kenis (Leuven: Peeters, 2012), 153–76, here 154.

16   ARAA T14/034 123, 124, 126, 3701; FA KK590, 1584, copy of the admission of the Jesuits in the Netherlands; Kadoc ABSE 2.

17   Luce Giard, "Les collèges jésuites des anciens Pays-Bas et l'élaboration de la *Ratio studiorum*," in Faesen and Kenis, *Jesuits of the Low Countries*, 83–108, here 92–93.

18   "In sola Flandria et Brabantia frigeat et ne locum quidem hospiti inveniat"; Laínez to Mercurian, Epist. Gen., September 5, 1559, cited in Poncelet, *Histoire de la Compagnie de Jésus*, 1:121; Van Goethem et al., *Antwerpen en de Jezuïeten*, 12.

19   Van Goethem et al., *Antwerpen en de Jezuïeten*, 12.

(1565), and Saint-Omer (1566).[20] With the exception of Leuven, it would take another decade before colleges in the Dutch-speaking part were erected: Antwerp, Bruges, and Maastricht all opened their doors in 1575.[21] These were short-lived, however: during the revolt, the Jesuits were banished, and the colleges were only re-erected after the re-conquest by Parma.[22]

When the Spanish colony in Antwerp applied for a chaplain who spoke their native tongue to cater for the needs of the Spanish faithful, the Jesuits obliged in 1562 by sending one of their own, soon to be followed by other Iberian Jesuits.[23] Antwerp thus became the third house of the Jesuits in the Low Countries.[24] Yet their success was short-lived, as a wave of Protestantism swept

---

20 Alfred Poncelet, S.J., *Histoire de la Compagnie de Jésus dans les anciens Pays-Bas: Établissement de la Compagnie de Jésus en Belgique et ses développements jusqu'à la fin du règne d'Albert et d'Isabelle* (Brussels: Marcel Hayez, 1927–28), 1:153ff.

21 Poncelet, *Histoire de la Compagnie de Jésus*, 1:213ff. Sometimes, 1562 is mentioned for the founding of the Antwerp college, but this is the founding date of the house, not the college. Although it has been argued that the implantation of Jesuit colleges followed a grand scheme as part of a defense against Protestantism, this has been adequately disproved (see Marc Venard, "Y-a-t-il une 'Stratégie scolaire' des jésuites en France au XVIe siècle?," in *L'Université de Pont-à-Mousson et les problèmes de son temps* [Nancy: Université de Nancy 2, 1974], 67–85). However, Philip II did urge the duke of Alba, who was unsympathetic toward the Jesuits, to erect colleges in the northern provinces. "A loccasion de quoy avois pense sil ne seroit convenable et prouffictable que pardela se procurast que lesdects Jesuites se colloquassent en aulcunes provinces ou Ilz ne sont point et mesmes en celles que sont plus eslongnes de la Court et ou les Evesques nont point encoires de Channoynes pour leurs assistents, sicomme Frise, Groeningen, Gheldres, Overissel et semblables" (It is thought that it would be convenient and profitable for the Jesuits to settle in any province where they are not yet present and in those farthest away from court and those where the bishops do not even have canons as assistants, such as Frisia, Groningen, Guelders, Overijssel, and the like). April 22, 1570, cited in Jozef Andriessen, *De Jezuïeten en het samenhorigheidsbesef der Nederlanden 1585–1648* (Antwerp: De Nederlandsche Boekhandel, 1957), 3–4n23. The image of the Jesuit colleges as the spearhead of the Counter-Reformation is also reinforced by a letter sent by the duke of Parma to Philip II in which he writes: "Your Majesty desired me to build a citadel at Maastricht; I thought that a College of the Jesuits would be a fortress more likely to protect the inhabitants against the enemies of the Altar and the Throne. I have built it." Cited in John Patrick Donnelly, S.J., "Padua, Louvain, and Paris: Three Case Studies of University—Jesuit Confrontation," *Louvain Studies* 15 (1990): 38–52, here 42.

22 Alfons K. L. Thijs, *Van Geuzenstad tot katholiek bolwerk: Antwerpen en de Contrareformatie* (Turnhout: Brepols, 1990), 67–69.

23 Poncelet, *Histoire de la Compagnie de Jésus*, 1:215–17.

24 Van Goethem et al., *Antwerpen en de Jezuïeten*, 12; Andriessen, *De Jezuïeten*, 9–11. When, during the Spanish Fury in 1576, citizens begged the clergy for help, the Jesuits were expressly shunned and kept out. Later, the legend surfaced that the Jesuits had cooperated with the Spanish soldiers in ransacking the city.

FIGURE 0.2 The expulsion of the Jesuits from Antwerp. Michael Eytzinger, *De Leone Belgico, ejusque topographica atque historica description liber* (Cologne: Gerardus Campensis, 1583)
© MPM R25.8

the Dutch-speaking provinces.[25] In April 1578, the city's government expected the clergy to take an oath of allegiance to Matthias (1557–1619, r.1612–19), the emperor's son, whom the insurgents had chosen as their new sovereign. The Jesuits' refusal to do so led to them being expelled from the city.[26]

At the end of August 1585, days after Antwerp's reconciliation with Spain, three Jesuits came to the city to resurrect their community. Upon their return, they found a population that had experienced the economic, moral, and religious ravages of two decades of civil war and a year-long siege. Within four years, the population would be halved. The Jesuits were helped in no small means by Parma, who in 1587 granted them a subsidy of three thousand

---

25 For a detailed account of this period at Antwerp, see the outstanding work by Guido Marnef, *Antwerp in the Age of Reformation: Underground Protestantism in a Commercial Metropolis* (Baltimore, MD: Johns Hopkins University Press, 1996).

26 FA KK 589, letter, 1657; ARAA T14/O34 3666.

guilders a year from the city treasury.[27] The Jesuits were also granted the same rights as they had in Spain, Portugal, and Italy, such as tax exemptions.[28] By 1603, the Antwerp Jesuits counted thirty-one members, and by 1622 they could open the most beautiful church of the city.

The violent early period of their existence made the Belgian Jesuits different from their confrères in other European countries. For one, they would not become the target of toxic anti-Jesuit propaganda: their members were pious, austere, and had an impeccable reputation. Even their enemies recognized that they were indulgent to human frailty. Perhaps nowhere was this as clear as in Antwerp. During the Counter-Reformation, Antwerp was undoubtedly one of the foci of the Jesuits' activities, which stretched across a whole spectrum from religious education, the teaching and study of philosophy, humanities, and science, to taking care of the poor and sick. Their work among plague victims resulted in a large number of Jesuits succumbing to the infection.[29] Through their work, they came to be seen as the spiritual leaders of the people in the struggle against Protestantism.

For the Jesuits, the Low Countries became an important focus. On the one hand, they supported the Counter-Reformation in the South, while on the other they were doing missionary work in the North, the so-called Holland mission,[30] and, to a lesser extent, in England and Wales.

In 1585, Farnese founded a Jesuit army almoner's service. Two years later, the provincial officialized this service in a *missio castrensis* (military mission). The chaplains were not only expected to read Mass and do pastoral work but also had to act as apothecaries and nurses and as garrison clerks as far as orphans and widows of casualties were concerned. Initially, this corps consisted of twenty-four Jesuits, but this was later reduced. Jesuit teachers at the local college often doubled as almoner for the garrison stationed in their town.[31]

---

27 ARAA T14/034 1917; FA KK590, September 3, 1591, copy of Alexander Farnese, duke of Parma.

28 FA KK590, 1584, copy of admission by Philip II. Also, ARAA T14/034 142–47; 156, 1918.

29 ARAA T14/034 1933 and 2041. Rudi Mannaerts, *Sint-Carolus Borromeus: De Antwerpse Jezuïetenkerk, Een Openbaring* (Antwerp: Vzw Maria-Elisabeth Belpaire-vzw Toerismepastoraal Antwerpen, 2011), 28.

30 On the Jesuit contribution to the Holland mission, see Gerrit vanden Bosch, "Saving Souls in the Dutch Vineyard: The *Missio Hollandica* of the Jesuits (1592–1708)," in Faesen and Kenis, *Jesuits of the Low Countries*, 139–52, and Lodewijk Brouwers, *Carolus Scribani 1561–1629* (Antwerp: Ruusbroecgenootschap, 1961), 234–45. See also ARAA T14/034 2826; 2827, and Paul Arblaster, "The Archdukes and the Northern Counter-Reformation," in Duerloo and Thomas, *Albert and Isabella*, 87–92.

31 Janssens, *België in de 17de Eeuw*, 2:68–69; Willem Audenaert, S.J., Prosopographia iesuitica Belgica antiqua: *A Biographical Dictionary of the Jesuits in the Low Countries 1542–1773*:

In the first third of the seventeenth century, the Jesuits received large amounts in alms, subsidies, and gifts.[32] In most years, the total amount exceeded twenty-thousand guilders. The most important gifts came from Antwerp's foremost families, such as the Van der Cruyce, De Smidt, Van Hove, and Boot families,[33] and later in the century from the Houtappel sisters.[34] Gifts also came from members themselves, with Aguilón donating 5,416 guilders on April 1, 1601,[35] for example, while over the course of the period from 1616 to 1625 Philip Nuyts (Nutius [1597–1661]) donated no less than 7,967 guilders.[36] Furthermore, there were also donations by or legacies from families of the Society's members.[37]

The Jesuits' erudition and learning earned them the respect of the political and economic elite, many of whom sent their sons to the Jesuit college, which saw student numbers double from about three hundred in 1591 to more than six hundred in 1613. The archdukes wanted to restore the educational infrastructure, but they were particularly selective, and their attention was focused on Jesuit education. In Antwerp, for example, the archdukes financed the construction of an additional building to the college.[38]

PIBA (4 Vols.) (Heverlee: Filosofisch en Theologisch College van de Sociëteit van Jezus, 2000), appendix 9, for a list of all Jesuits involved, and Brouwers, *Carolus Scribani 1561–1629*, 231–34.

32 ARAA T14/034 142–47; 1918; 1920.

33 Marie-Juliette Marinus, "De financiering van de contrareformatie te Antwerpen (1585–1700)," in *Geloven in het Verleden: Studies over het godsdienstig leven in de Vroegmoderne Tijd, aangeboden aan Michel Cloet*, ed. Eddy Put, Marie Juliette Marinus, and Hans Storme (Leuven: Universitaire Pers, 1996), 239–52, 246. See also ARAA T14/034 1920; 1938; 1939.

34 ARAA T14/015.01 185. Bert Timmermans, "The Chapel of the Houtappel Family and the Privatisation of the Church in Seventeenth-Century Antwerp," in *Innovation and Experience in the Early Baroque in the Southern Netherlands: The Case of the Jesuit Church in Antwerp*, ed. Piet Lombaerde (Turnhout: Brepols, 2008), 175–86. In ARAA T14/034 1938, we find that the foundation of Maria Houtappel was valued at sixty thousand guilders (fol. 19^{r-v}) and that of Anna Houtappel at 120,000 guilders (fol. 20^{r-v}).

35 ARAA T14/034 1938, fol. 1^v.

36 ARAA T14/034 1938, fol. 4^v.

37 ARAA 14/034 1370 (Nuyts), 1371 (Jacob van Rasseghem), 1373 (Thomas Aynscombe), and 1374 (Aynscombe and Nuyts).

38 See Werner Thomas, "Andromeda Unbound," 10, on the influence of the Jesuits and their colleges on Protestant Europe; see also Arblaster, "Archdukes," and FA KK 590–91.

CHAPTER 1

The College and Its School of Mathematics

1   Schools in Antwerp

In its sixteenth-century heyday, Antwerp may have boasted as many as 150 schools.[1] Although it is difficult to find comparable figures for other towns,[2] it is safe to conclude that Antwerp indeed enjoyed a very high concentration of teachers and schools. For Leuven, on the other hand, there is no trustworthy evidence for the sixteenth century. By the end of the seventeenth century, Leuven seems to have had about ten schools for boys and fifty for girls (but these were schools for handicrafts).[3]

In June 1570, the synod of Mechelen decided to establish Sunday schools. These institutions were supposed to cater for thousands of children who were unable to attend regular schools because they had to work. Bishop Laevinus Torrentius (Lieven van der Beke [1525–95]) founded the first such school in Antwerp in 1592, and his commitment to establishing as many schools as possible would be continued by his successors. While the primary focus of the Sunday schools was on religious instruction, children were also taught reading, writing, and arithmetic at a very basic level.[4]

---

1   This number is mentioned in the 1612 edition of Lodovico Guicciardini's *Beschryvinghe van alle de Nederlanden*, where it appears as an addition by Cornelis Kiliaan (1528–1607) based on a remark by Carlo Scribani, the rector of the Jesuit college: "150 schools in which youths were taught all sciences and languages" (Lodovico Guicciardini, *Antwerpen, Mechelen, Lier e. a.* [*vertaald door Kiliaan*] [Deurne: Soethoudt, 1979], 123). If the estimate takes into account schools where girls were taught needlework, it may well be quite accurate.
2   Nearby Mechelen and Tournai had about twenty-five thousand inhabitants (Jan de Vries, *European Urbanization 1500–1800* [London: Methuen & Co, 1984], 272) and eleven teachers each Jan G. C. A. Briels, "Zuidnederlandse Onderwijskrachten in Noord-Nederland 1570–1630," *Archief voor de Geschiedenis van de Katholieke Kerk in Nederland* 14 [1972]: 89–169, here 92). A similar proportion is found for the Germanic town of Annaberg (the then second-largest city in Saxony), with its population of around twelve thousand inhabitants and six teachers in 1540 (Peter Rochhaus, "Adam Ries und die Annaberger Rechenmeister zwischen 1500 und 1604," in *Rechenmeister und Cossisten der frühen Neuzeit*, ed. Rainer Gebhardt, Freiberger Forschungshefte 201 [Annaberg: Technische Universität Bergakademie, 1996], 95–96).
3   Eddy Put, *De Cleijne Scholen: Het volksonderwijs in het Hertogdom Brabant tussen Katholieke Reformatie en Verlichting (eind 16de eeuw–1795)* (Leuven: Universitaire Pers, 1990), 91.
4   Marie Juliette Marinus, *Laevinus Torrentius als tweede bisschop van Antwerpen (1587–1595)* (Brussels: Paleis der Academiën, 1989); Alfons K. L. Thijs, "De Contrareformatie en het

After 1585, the parish schools were reinstated, but by this time the colleges established by the religious orders were gaining in importance.[5] The Jesuit college had been founded as early as 1575 and would reach its zenith at the beginning of the seventeenth century. In 1605, the Dominicans founded a college, while the Augustinians, at the behest of the city council, opened their school in 1607. Toward the end of 1617, after certain difficulties had been overcome, a school of mathematics was opened as part of the Jesuit college.

2 Jesuit Educational Policy

The Society's founders did not have the intention of erecting schools. In fact, they more or less glided into the decision of taking up the ministry of education. Yet once this decision had been made, the Jesuits took up the challenge with extraordinary zeal, creating the most extensive educational system of its age.[6] Although initially intended to cater for the needs of the Society itself, the Jesuit colleges soon opened their doors, accepting "everybody, poor or rich, free of charge and for charity's sake, without accepting any remuneration."[7]

In 1541, the document *Fundación de collegio* (Foundation of a school) sketched an academic program in which it was insisted that members should know those sciences needed to defend the faith and to evangelize effectively.[8]

---

economisch transformatieproces te Antwerpen na 1585," *Bijdragen tot de Geschiedenis* 70 (1987): 97–124, here 110–11; Put, *De Cleijne Scholen*, 104–5.

5 On the founding of colleges in the Southern Netherlands, see Dirk Leyder, "L'éclosion scolaire," *Paedagogica historica* 36, no. 3 (2000): 1003–51; Put and Wynants, *De Jezuïeten in de Nederlanden*.

6 O'Malley, *First Jesuits*, 200ff.; John W. O'Malley, S.J., "How the First Jesuits Became Involved in Education," in *The Jesuit* Ratio studiorum *of 1599: 400th Anniversary Perspectives*, ed. Vincent J. Duminuco (New York: Fordham University Press, 2000), 56–74; Antonella Romano, *La Contre-Réforme mathématique: Constitution et diffusion d'une culture mathématique jésuite à la Renaissance (1540–1640)* (Rome: École française de Rome, 1999), 36–39.

7 Letter on the subject of education from Juan Alfonso de Polanco to all members of the Society, December 1, 1551, as cited in O'Malley, *First Jesuits*, 206. As is often the case in Jesuit history, we find conflicting opinions. Some scholars would have it that the Jesuits' motivation lay in gaining power in society by influencing future policymakers. Undoubtedly, to keep the colleges financially afloat, good relations with the elites were necessary. On the other hand, it also has to be admitted that through free education the Jesuits supported upward social mobility. In a sense, they were social evolutionaries. See Steven J. Harris, "Transposing the Merton Thesis: Apostolic Spirituality and the Establishment of the Jesuit Scientific Tradition," *Science in Context* 3 (1989): 29–65, here 53ff.; Judi Loach, "Revolutionary Pedagogues? How Jesuits Used Education to Change Society," in O'Malley et al., *Jesuits II*, 66–85.

8 László Lukács, S.J., "A History of the Jesuit *Ratio studiorum*," in *Church, Culture, and Curriculum: Theology and Mathematics in the Jesuit* Ratio studiorum, ed. László Lukács, S.J., and Guiseppe Cosentino, S.J. (Philadelphia: Saint Joseph's University Press, 1999), 1–46, here 18; Cristiano

Ignatius's *Constitutions*, which sets out the order's governing principles, contains seventeen chapters on studies. As such, it became the order's first unofficial curriculum.[9] In the *Constitutions*, Ignatius had given the sciences (physics and astronomy, and therefore also mathematics) a subordinate position, their role being solely to support the study of theology.[10] For the natural sciences, he recommended that Aristotle (384–322 BCE) be followed.[11] Ignatius more or less followed the *modus Parisiensis*, in which mathematics featured less than in the curricula of Italian and German universities.[12]

When Ignatius died, his successor Diego Laínez (1512–65, in office 1558–65) was confronted with a severe shortage of teachers. His solution was simultaneously pragmatic, surprising, and revolutionary: as the ministry of schools was as important as all other ministries combined, every Jesuit would from now on be expected to teach at some point in their career.[13]

The Jesuits founded schools for what we would call secondary education and called them colleges. A pupil's education would start at the age of nine or ten and, if he enrolled for the entire cycle, would not end until his twentieth birthday.[14] Pupils could even enroll for a preparatory course from the age of six, to learn to read and write. Jesuit colleges were organized in two sections,

---

Casalini and Claude Pavur, S.J., eds., *Jesuit Pedagogy, 1540–1616: A Reader* (Chestnut Hill, MA: Institute of Jesuit Sources, 2016), 7.

9 Oskar Garstein, *Rome and the Counter-Reformation in Scandinavia: Jesuit Educational Strategy* (Leiden: Brill, 1992), 41.

10 "And also mathematical topics, in so far as they are in accord with the end proposed to us […]." Translation by Dennis Chester Smolarski, S.J., "The Jesuit *Ratio studiorum*, Christopher Clavius, and the Study of Mathematical Sciences in Universities," *Science in Context* 15, no. 3 (2002): 447–57, here 453; *Constitutiones*, part 4, cap. 12, [451]. See also François de Dainville, *L'éducation des jésuites (XVIe–XVIIIe siècles)* (Paris: Les éditions de minuit, 1976), 324.

11 Giuseppe Cosentino, S.J., "Mathematics in the Jesuit *Ratio studiorum*," in Lukács and Cosentino, *Church, Culture, and Curriculum*, 47–80, here 54.

12 Gabriel Codina Mir, S.J., *Aux sources de la pédagogie des jésuites: Le "Modus Parisiensis"*, Biblioteca Instituti Historici Societatis Iesu 28 (Rome: Institutum Historicum S. I., 1968); Codina Mir, "The Modus Parisiensis," in Duminuco, *Jesuit Ratio studiorum of 1599*, 28–49; Howard Gray, S.J., "The Experience of Ignatius Loyola," in Duminuco, *Jesuit Ratio studiorum of 1599*, 1–21.

13 O'Malley, *First Jesuits*, 200; Paul F. Grendler, "Jesuit Schools in Europe: A Historiographical Essay," *Journal of Jesuit Studies* 1, no. 1 (2014): 7–25, here 12; Grendler, "Laínez and the Schools in Europe," in *Diego Laínez (1512–1565) and His Generalate: Jesuit with Jewish Roots, Close Confidant of Ignatius of Loyola, Preeminent Theologian of the Council of Trent*, ed. Paul Oberholzer, S.J., Bibliotheca Instituti Historici S. I. 76 (Rome: Institutum Historicum Societatis Iesu, 2015), 654–58; Grendler, "Culture of the Jesuit Teacher," 20–21.

14 Karl Hengst, *Jesuiten an Universitäten und Jesuitenuniversitäten: Zur Geschichte der Universitäten in der oberdeutschen und rheinischen Provinz der Gesellschaft Jesu im Zeitalter der konfessionellen Auseinandersetzung* (Paderborn: Schöningh, 1981), 70.

studia inferiora, consisting of five or six classes, each of which had to be mastered before a pupil could advance to a higher level, and a *studia superiora*.[15] Classes would typically count fifty to a hundred pupils.

The emphasis in the Jesuit colleges' classes was on Latin and Greek authors, albeit sometimes in editions adapted to the pupils' age. The Jesuit teaching system allowed the study of extracurricular subjects, in many cases the pupils' mother tongue and national history, but also drama and music.[16] When the pupil had met the requirements of the final year, called rhetoric, he could advance to the *studia superiora*. Here, the student would become acquainted with the great philosophers, particularly the works of Aristotle, together with *some* science and mathematics.[17]

3 Mathematics in the Jesuit Curriculum

In 1548, Jerónimo Nadal (1507–80), the rector of the Messina College, planned a course of mathematics.[18] His program included Euclid's (*fl. c.*300 BCE) first three books, Oronce Finé's (1494–1555) *Arithmetica practica* (Practical arithmetic [1541]) and *De mundi sphaera, sive cosmographia* (The globe, or cosmography [1542]), Johannes Stöffler's (1452–1531) *Elucidatio fabricae ususque astrolabii* (Elucidation on the construction and the use of the astrolabe [1513]), and Georg von Peurbach's (1423–61) *Theoricae novae planetarum* (New planetary theory [manuscript 1454; first printed 1472]).[19] In this first proposal, Nadal

---

15   J. B. [Jean Baptiste] Herman, *La pédagogie des jésuites au 16e siècle: Ses sources, ses caractéristiques* (Leuven: Bureaux de recueil, 1914), 73; Codina Mir, *Aux sources de la pédagogie des jésuites*, 291; Hengst, *Jesuiten an Universitäten und Jesuitenuniversitäten*, 60–72.

   On the different curricula, see László Lukács, S.J., ed., *Monumenta pædagogica Societatis Iesu*, 7 Vols. (Rome: Institutum Historicum Societatis Iesu, 1965–92); Albert Krayer, *Mathematik im Studienplan der Jesuiten: Die Vorlesung von Otto Cattenius an der Universität Mainz (1610/11)* (Stuttgart: Franz Steiner Verlag, 1991), 24–25; Romano, *La Contre-Réforme mathématique*, 50–51; Codina Mir, *Aux Sources de la pédagogie des jésuites*, 256–336.

16   Codina Mir, *Aux sources de la pédagogie des jésuites*, 290–91; John W. O'Malley, S.J., and Gauvin Alexander Bailey, eds., *The Jesuits and the Arts 1540–1773* (Philadelphia: Saint Joseph's University Press, 2005), 11–12.

17   Garstein, *Rome and the Counter-Reformation in Scandinavia*, 49–52; Paul F. Grendler, *The Jesuits and Italian Universities, 1548–1773* (Washington, DC: Catholic University of America Press, 2017), 11.

18   Cosentino, "Mathematics in the Jesuit *Ratio studiorum*," 48–51, for a detailed history; on the Messina college and the university, see Grendler, *Jesuits and Italian Universities*, 37ff.

19   Cosentino, "Mathematics in the Jesuit *Ratio studiorum*," 49; Romano Gatto, "Christoph Clavius's *Ordo servandus in addiscendis disciplinis mathematicis* and the Teaching of

remained faithful to Ignatius's ideas; however, a couple of years later, he became more ambitious, writing a plan he hoped would be inserted in the *Constitutions*, as a result of which it would appear on the colleges' curriculum.[20] Yet he was unsuccessful, and it was the Roman College's model, in which mathematics, for the time being, was subordinate to physics, that was ultimately adopted.[21]

This need not come as a surprise. Two reasons can immediately be put forward for the rejection of the Messina program. The first was a lack of qualified teachers, a problem that would remain until the turn of the century.[22] The second, which in part was also the cause of the first, was that mathematics was not deemed necessary for the formation of theologians and missionaries, the Jesuit schools' main objectives.[23] The philosophers of the Collegio Romano, with a conservative Aristotelian orientation, and often ignorant of mathematics, were reluctant to assign any intrinsic scientific value to its reasoning.[24]

Mathematics in Jesuit Colleges at the Beginning of the Modern Era," *Science and Education* 15 (2006): 235–58, here 236.

20   Cosentino, "Mathematics in the Jesuit *Ratio studiorum*," 51; Gatto, "Christoph Clavius's *Ordo servandus*," 237.
21   Gatto, "Christoph Clavius's *Ordo servandus*," 237.
22   Cosentino, "Mathematics in the Jesuit *Ratio studiorum*," 52, 59; Romano, *La Contre-Réforme mathématique*, 69.
23   Cosentino, "Mathematics in the Jesuit *Ratio studiorum*," 53–54; Gatto, "Christoph Clavius's *Ordo servandus*," 238.
24   Frederick A. Homann, S.J., "Introduction," in Lukács and Cosentino, *Church, Culture, and Curriculum*, 1–16, here 5; Gatto, "Christoph Clavius's *Ordo servandus*," 238; Peter Dear, *Discipline and Experience: The Mathematical Way in the Scientific Revolution* (Chicago: University of Chicago Press, 1995), 36–37; Audrey Price, "Pure and Applied: Christopher Clavius's Unifying Approach to Jesuit Mathematics Pedagogy" (PhD diss., University of California at San Diego, 2017), 81. This is an echo of a sentiment present in Italian universities, where a number of prominent Italian philosophers denied that mathematics should be regarded as scientific knowledge in the Aristotelian sense. Dennis Chester Smolarski, "Teaching Mathematics in the Seventeenth and Twenty-First Centuries," *Mathematics Magazine* 75, no. 4 (2002): 256–62, here 257; Smolarski, "Jesuits on the Moon: Seeking God in All Things; Even Mathematics!," Seminar on Jesuit Spirituality, St. Louis, MO, 2005, 14–16; on mathematics at the universities, see Grendler, *Universities of the Italian Renaissance*, 408–29. Smolarski, "Teaching Mathematics," also gives examples of the low esteem in which the mathematics teacher was held (258). The seemingly low esteem in which mathematics was held vis à vis philosophy was not uncommon in the sixteenth century. For instance, the Flemish philologer and humanist Justus Lipsius (Joost Lips [1547–1606]) wrote to the Hungarian humanist Andreas Dudith (András Dudith de Horahovicza [1533–89]) about his compatriot Simon Stevin: "After all he is only a mathematician, without any other skill, indeed with hardly any knowledge of languages, in short the kind one more easily calls a practitioner than a theoretician of science" (*Mathematicus enim merus est sine alia arte, imo paene lingua, denique ex genere, qui μηχανοποιοι [mèchanopoioi]; potius quam θεωρητικοι [theorètikoi]*). Rudolf de Smet, "Simon Stevin en de paradox van

However, from 1553 onward, due to the efforts of Baltasar de Torres (1518–61) and Christopher Clavius (1538–1612), the Collegio Romano began organizing courses of mathematics, much along the lines of Nadal's first proposal.[25]

In his edition of Euclid (1574), Clavius put up a defense of mathematics.[26] In the preface, he drew attention to the fact that, as far back as ancient times, mathematics had been considered a subject above all others.[27] He assigned mathematics a central role in the process of learning, like the Platonists had done, as opposed to the Aristotelians, who claimed their philosophy was superior. The Aristotelians defined science as knowledge deriving from syllogisms on the basis of undeniable principles.[28] For the mathematicians, mathematical entities did exist in nature, but they were disembodied within mathematics to become abstractions. This point of view gave Clavius the opportunity to allocate mathematics an intermediary position between the physical and the metaphysical.[29] To make his point further, as a *scholion* to Euclid 1.1, he proved the property using only syllogisms.[30] Clavius's aim was to prove that mathematics could no longer be rejected on the basis that it was not a true science, and that it was indispensable for all sciences and therefore helpful to the Society's purpose.[31]

In 1580, Clavius proposed three mathematics curricula, the shortest of which was a two-year course with precise instructions on how much time should be

---

het gefragmenteerde humanisme," in *Simon Stevin 1548–1620: De geboorte van een nieuwe wetenschap*, ed. Wouter Bracke (Turnhout: Brepols, 2004), 27–34, here 27.

25   Cosentino, "Mathematics in the Jesuit *Ratio studiorum*," 57, 60; Romano, *La Contre-Réforme mathématique*, 72–83, on Torres's tenure, and 85–178, on Clavius's tenure. See also Antonella Romano, "Réflexions sur la construction d'un champ disciplinaire: Les mathématiques dans l'institution jésuite à la Renaissance," *Paedagogica historica* 40, no. 3 (2004): 245–59, here 255–59. On Clavius's ideas for a mathematics pedagogy, see Price, "Pure and Applied." On the Collegio Romano, see Ricardo García Villoslada, *Storia del Collegio Romano dal suo inizio (1551) alla soppressione della Compagnia di Gesù (1773)* (Rome: Università Gregoriana, 1954).

26   Price, "Pure and Applied," 87–88.

27   Gatto, "Christoph Clavius's *Ordo servandus*," 238–39; Price, "Pure and Applied," 75–77.

28   Gatto, "Christoph Clavius's *Ordo servandus*," 239; Frederick A. Homann, "Christopher Clavius and the Renaissance of Euclidean Geometry," *Archivum historicum Societatis Iesu* 52 (1983): 233–46, here 238–39.

29   Gatto, "Christoph Clavius's *Ordo servandus*," 241; Homann, "Christopher Clavius," 239–40; Peter Dear, "Jesuit Mathematical Science and the Reconstitution of Experience in the Early Seventeenth Century," *Studies in the History and Philosophy of Science* 18 (1987): 133–75, here 139–40; Rivka Feldhay, "Knowledge and Salvation in Jesuit Culture," *Science in Context* 1 (1987): 195–213, here 207–8; Price, "Pure and Applied," 81–82.

30   Gatto, "Christoph Clavius's *Ordo servandus*," 241–42.

31   Cosentino, "Mathematics in the Jesuit *Ratio studiorum*," 61–62; Gatto, "Christoph Clavius's *Ordo servandus*," 242.

spent on each subject.³² In most of his books, he stressed that mathematics was an efficient tool for use in other sciences. It is notable, however, that his preoccupation with applications did not extend to such subjects as commercial arithmetic, despite its importance in mercantile cities such as Antwerp.

In 1586, a draft plan of studies, the *Ratio studiorum*, was drawn up. Clavius was not a member of the commission that formulated the *Ratio* nor would he be part of the other committees that refined it. He did, however, propose a curriculum³³ and wrote letters to the commission in which he expressed his concerns.³⁴ In the first document, he lamented the low esteem in which many held mathematics and argued that the lack of interest in the subject was harmful to the Society.³⁵ Mathematics, he claimed, was also very useful for historians and politicians and would bring navigation and medicine within the Jesuits' reach. Although Clavius's position was considerably reinforced by his involvement in the Gregorian calendar reform, his program was not completely accepted.³⁶

At Superior General Claudio Acquaviva's (1543–1615, in office 1581–1615) request, the provinces were consulted on the text for a new *Ratio*, including Clavius's proposals for a curriculum of mathematics. The provinces were invited to write comments on the first draft, which they did in extensive commentaries.³⁷ But the Belgian province proved an exception to this rule, given that half of the Seventeen Provinces had been reconquered by the Spanish and that half was ravaged by war. Little wonder, then, that these Jesuits, who, for most part, had just returned to their former houses, had only superficial comments to make.³⁸

---

32 Christopher Clavius, "Historical Documents Part II: Two Documents on Mathematics," trans. Dennis Chester Smolarski, S.J., *Science in Context* 15, no. 3 (2002): 465–70; Dear, "Jesuit Mathematical Science," 136; Krayer, *Mathematik im Studienplan der Jesuiten*, 23–24; Lukács, *Monumenta pædagogica Societatis Iesu*, 7:110–18; James M. Lattis, *Between Copernicus and Galileo: Christoph Clavius and the Collapse of Ptolemaic Cosmology* (Chicago: Chicago University Press, 1994), 32–33.

33 Casalini and Pavur, *Jesuit Pedagogy*, 283ff.

34 Smolarski, "Jesuit *Ratio studiorum*," 450; see also Clavius, "Historical Documents Part II," and Casalini and Pavur, *Jesuit Pedagogy*, 281–300.

35 Smolarski, "Jesuit *Ratio studiorum*," 450; Clavius, "Historical Documents Part II,"465ff.; Smolarski, "Jesuits on the Moon," 27–28.

36 Krayer, *Mathematik im Studienplan der Jesuiten*, 38–39; Romano, *La Contre-Réforme mathématique*, 206ff.

37 Lukács, "History of the Jesuit *Ratio studiorum*," 33; see also Krayer, *Mathematik im Studienplan der Jesuiten*, 38–39, and Romano, *La Contre-Réforme mathématique*, 206ff. on the different opinions of the provinces.

38 Lukács, *Monumenta pædagogica Societatis Iesu*, vol. 7, does not mention an answer from the Belgian province. Giard, "Les collèges jésuites des anciens Pays-Bas," 101–2. In 1572, the

When the *Ratio* of 1586 mentioned mathematics, it was also noted that there was a shortage of skilled Jesuit mathematicians. The Jesuits tried to remedy this situation by bringing together around Clavius a number of promising young Jesuits who had studied mathematics, San Vicente among them. It was hoped that enough mathematicians would graduate to teach in all the order's colleges.[39] The comments of the provinces resulted in another draft of the *Ratio* in 1591, which was again sent to the provinces for comments. This procedure finally resulted in the definitive *Ratio* of 1599.[40]

In the "Rules for the Provincial," the *Ratio* states that "in the second year of philosophy, all the philosophers should also attend a forty-five-minute class of mathematics. Moreover, if there are any who are suitable for these studies and inclined towards them, they should work on them *in private classes after the course*."[41] The mathematics professor should teach Euclid's *Elements* for about two months. Then he should add something about geography, the sphere, or anything that might be of interest. Every month, or every other month, a student should give a lecture about a famous mathematical problem.[42] When discussing Aristotle's *De caelo* (On the heavens), the philosophy professor should only treat questions about its substance and influences and leave the rest to the professor of mathematics.[43]

It is easy to understand why the 1586 *Ratio* would state that "the less one knows about such things [i.e., mathematics], the more he disparages them" (*fit enim saepe, ut qui minus ista novit, his magis detrahat*).[44] Mathematics did prove to have one great advantage over the other subjects, however: surprisingly enough, it often held the key to establishing overseas missions, especially in China. In consequence, more Jesuit mathematicians and astronomers than might be expected were sent overseas, as it was the scientific knowledge of Jesuits like Matteo Ricci (1552–1610), Adam Schall von Bell (1591–1666), and Ferdinand Verbiest (1623–88) that allowed the Jesuits to operate in China.[45]

Flemish provincial Boudewijn Delange (1613–77) had written that "we do not have any mathematicians because we do not teach this science" (ARSI Germ.152, fol. 202ᵛ). This situation had changed by 1592. Poncelet, *Histoire de la Compagnie de Jésus*, 2:18.
39 De Dainville, *L'éducation des jésuites*, 325; Smolarski, "Jesuit *Ratio studiorum*," 451.
40 Edited and translated in Pavur, Ratio studiorum.
41 Pavur, Ratio studiorum, 19–20, my emphasis.
42 Pavur, Ratio studiorum, 109–10.
43 Pavur, Ratio studiorum, 103.
44 Smolarski, "Jesuit *Ratio studiorum*," 449.
45 Noël Golvers, "F. Verbiest's Mathematical Formation: Some Observations on Post-Clavian Jesuit Mathematics in Mid-17th-Century Europe," *Archives internationales d'histoire des sciences* 54 (2004): 29–47; Georg Schuppener, *Jesuitische Mathematik in Prag im 16. und 17. Jahrhundert (1556–1654)* (Leipzig: Leipziger Universitätsverlag, 1999), 47.

THE COLLEGE AND ITS SCHOOL OF MATHEMATICS 19

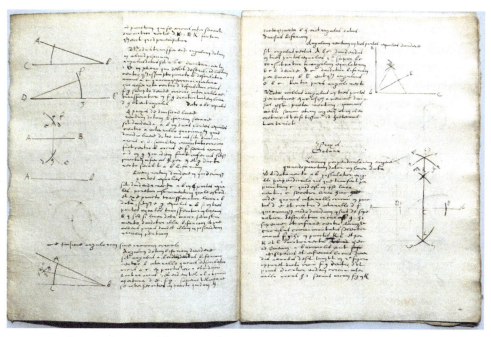

FIGURE 1.1　Course notes on Euclid written in the Antwerp College
　　　　　© ARAA T14/034 1879

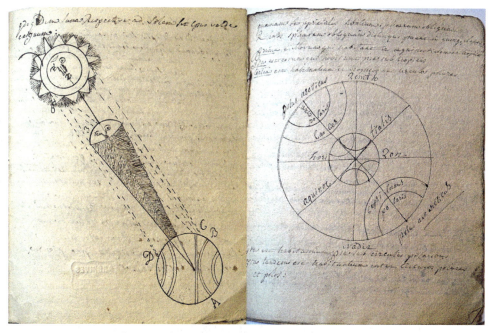

FIGURE 1.2　Two pages of notes on the astronomy course in the Antwerp College
　　　　　© ARAA T14/034 1879

4   The Academy of Mathematics at the Collegio Romano

No formal date for the erection of the Academy of Mathematics is known. It seems to have existed in the period from 1553 to 1560 when Torres was professor of mathematics at the Collegio Romano.[46] In 1563, Clavius was appointed as professor of mathematics. In a report of 1576, Ludovico (Ludovicus) Maselli (1574–83), rector of the college, wrote to Superior General Everard Mercurian (1514–80, in office 1573–80) that Clavius's academy had begun to bear fruits.[47]

Clavius's efforts to formalize the academy were successful in 1593–94, when it became a distinct pedagogical unit within the college. Admission required being nominated by the professor of mathematics in a college and by the superiors of the applicant's province.[48] It is not possible to identify all students of the academy, but the number must have been small. San Vicente was associated with the academy from 1606 to 1612 but was never a formal student. During the period Clavius ran the academy, Christoph Grienberger (1561–1636), Odo van Maelcote (1572–1615), Orazio Grassi (1583–1654), and Paul Guldin (1577–1643) were among his students.[49]

Clavius was succeeded by Grienberger. Yet, after Grienberger's demise, the academy went into decline, a trend that was not reversed until 1662, when Gilles-François (Aegidius) de Gottignies (1630–89), a student of San Vicente, took over the reins.[50]

---

46   Ugo Baldini, "The Academy of Mathematics of the Collegio Romano from 1553 to 1612," in *Jesuit Science and the Republic of Letters*, ed. Mordechai Feingold (Cambridge, MA: MIT Press, 2003), 47–98, here 51; Romano, *La Contre-Réforme mathématique*, 103.
47   Romano, *La Contre-Réforme mathématique*, 115.
48   Baldini, "Academy of Mathematics," 51–52.
49   Baldini, "Academy of Mathematics," 53, 79. On Grienberger, see Michael J. Gorman, "Mathematics and Modesty in the Society of Jesus: The Problems of Christoph Grienberger," in *The New Science and Jesuit Science: Seventeenth-Century Perspectives*, ed. Mordechai Feingold (Dordrecht: Kluwer Academic, 2003), 1–120.
50   Baldini, "Academy of Mathematics," 53. For a complete list of occupants of the chair of mathematics, see García Villoslada, *Storia del Collegio Romano*, 335, for the chair of physics, see 329–31. Baldini's view is not shared by all scholars. However, other scholars do not make the distinction between the occupant of the chair and the mathematical and astronomical research done in cooperation with students in the academy. They point out, for instance, that Athanasius Kircher was appointed to the chair of mathematics in 1634. Kircher was a veritable polymath, but he was not what we would call a theoretical mathematician. His interests lay in applied mathematics, in the broadest sense of the word. Kircher had a wide network of correspondents across Europe and beyond, making him one of the best-known Jesuit scientists of his age. These correspondents included San Vicente, della Faille, Moretus, and Ciermans. One of Kircher's biographers, however, states that "it is surprising that his mathematical studies were not more distinguished" (John Edward Fletcher, *A Study of the Life and Works of Athanasius Kircher, "Germanus*

During Clavius's tenure as professor, one of the main areas of research was geometry. The philosophy the Jesuit mathematicians seem to have adhered to was Archimedean rather than Euclidean or Apollonian, in accordance with the work of Federico Commandino (1509–75) and other Italian geometers.[51]

After 1590, when Tycho Brahe's (1546–1601) observations had become available, attention was also given to astronomy. This was especially so when a number of phenomena, such as the supernova of 1604 and the comet of 1607, appeared to be irreconcilable with Aristotelian physics.[52] At the college, observations were made with a telescope that had been given to Van Maelcote.[53] On the occasion of Galileo's visit to the college, the Jesuits compared their telescope with Galileo's and concluded that theirs was not inferior. Van Maelcote was full of praise for Galileo and his observations. He was backed up by the leading mathematicians Clavius and Grienberger, but also by San Vicente.[54]

Incredibilis": With a Selection of His Unpublished Correspondence and an Annotated Translation of His Autobiography [Leiden: Brill, 2011], 162). In the field of mathematics proper, his contributions remain limited to cryptology. Among the ten other occupants of the chair of mathematics before de Gottignies took over, only three merit attention: Giovanni Battista Giattini (1601–72) (as a linguist), Paolo Casati (1617–1707), and Gabriele Beati (1607–73). Beati seems to have been an able, but unimaginative, lecturer. It is possible that, at least at a technical level, the reversal began with Casati, who published books on mechanics and astronomy and did research in hydraulics. On Kircher, see John Edward Fletcher, *Athanasius Kircher und seine Beziehungen zum Gelehrten Europa seiner Zeit* (Wiesbaden: Harrassowitz, 1988). On Beati, see, e.g., Kerry V. Magruder, "Jesuit Science after Galileo: The Cosmology of Gabriele Beati," *Centaurus* 51 (2009): 189–212; Renée Raphael, "Teaching Sunspots: Disciplinary Identity and Scholarly Practice in the Collegio Romano," *History of Science* 52, no. 2 (2014): 130–52. On Casati, see, e.g., Rivka Feldhay, "On Wonderful Machines: The Transmission of Mechanical Knowledge by Jesuits," *Science and Education* 15, no. 2 (2006): 151–72; Feldhay and Ayelet Even-Ezra, "Heaviness, Lightness, and Impetus in the Seventeenth Century: A Jesuit Perspective," in *Emergence and Expansion of Pre-classical Mechanics*, ed. Rivka Feldhay, Jürgen Renn, and Matthias Schemmel, Boston Studies in the Philosophy and History of Science 333 (Cham: Springer International Publishing AG, 2018), 255–84; Michael Elazar and Rivka Feldhay, "Jesuit Conceptions of Impetus after Galileo: Honoré Fabri, Paolo Casati, and Francesco Eschinardi," in Feldhay, Renn, and Schemmel, *Emergence and Expansion of Pre-classical Mechanics*, 285–324; and also, in this monograph, p. 186.

51 Baldini, "Academy of Mathematics," 65.
52 Baldini, "Academy of Mathematics," 57.
53 Christiaan Huygens, *Oeuvres complètes: Tome II; Correspondance 1657–1659*, ed. David Bierens de Haan (The Hague: Martinus Nijhoff, 1889), 489–90. The telescope was supposedly invented in the Dutch Republic, with the first patent issued to Hans Lippershey (c.1570–1619) in 1608. The negotiators of the Twelve Years' Truce soon obtained a telescope and it found its way all over Europe.
54 "We [i.e., the Jesuit mathematicians] demonstrated with evidence, though to the scandal of the philosophers, that Venus circles around the Sun." San Vicente to Huygens,

Although the theologians of the Collegio Romano took a less lenient view, in mid-May 1610 Galileo was nevertheless welcomed to the college, where he received a most flattering oration from Van Maelcote.[55]

The college's mathematical culture would be presented to the outside world with the lectures on the 1604 supernova,[56] the lecture on Galileo's observations in 1611, and the reply to the Jesuit cardinal Robert Bellarmine (1542–1621), who had asked for the college's judgment on the observations.[57]

5   The College of Leuven

The Jesuits always aspired to integrate, or at least link, their higher education institutions to a university.[58] In the Southern Netherlands, they succeeded in organizing courses in cooperation with the University of Douai's College d'Anchin,[59] but they never got a foothold in the Netherlands' other university, Leuven,[60] which closely defended its monopoly. The Jesuits in the Flemish province never received royal permission to erect anything other than public secondary schools and were only allowed to provide higher education to their members, not the public at large.

The history of Antwerp's school of mathematics is, in part, linked to the fate of the Jesuits' college in Leuven. The college opened in 1565. From 1570 onward, alumni of the university's faculty of arts would complete their studies with courses in metaphysics, ethics, and mathematics at the Jesuit college. A full course in theology open to any student, for which one of the teachers was the future cardinal Bellarmine, was established as well.[61] At first, this did not pro-

---

October 4, 1659. Huygens, *Oeuvres complètes: Tome II*, 489–90; Lattis, *Between Copernicus and Galileo*, 193–94.

55   Mordechai Feingold, "The Grounds for Conflict: Grienberger, Grassi, Galileo, and Posterity," in Feingold, *New Science and Jesuit Science*, 121–57, here 124.

56   Which were written by Grienberger, although they or their recitation are often attributed to Odo van Maelcote. Feingold, "Grounds for Conflict," 124, and esp. Gorman, "Mathematics and Modesty," 82–91.

57   Baldini, "Academy of Mathematics," 55. The senior mathematicians of the Collegio Romano verified nearly all of Galileo's observations. Feingold, "Grounds for Conflict," 124.

58   De Dainville, *L'éducation des jésuites*; Hengst, *Jesuiten an Universitäten und Jesuitenuniversitäten*; Schuppener, *Jesuitische Mathematik in Prag*; Grendler, *Jesuits and Italian Universities*.

59   Hilde de Ridder-Symoens, "Het hoger onderwijs," in *België in de 17de Eeuw: De Spaanse Nederlanden en het Prinsbisdom Luik; Band I; Politiek*, ed. Paul Janssens (Brussels: Dexia Bank-Snoeck, 2006), 86–87.

60   Donnely, "Padua, Louvain, and Paris," 42–46.

61   Fernand Claeys Boúúaert, "Une visite canonique des maisons de la Compagnie de Jésus en Belgique (1603–1604): Rapports des visiteurs Olivier Manare et Léonard Lessius, envoyés

voke any conflicts with the university and, if there were any, these were juridical in nature.[62] One of the outcomes was that the Jesuits could confer degrees solely on their own members while other students had to sit exams—and pay for them—at the university.

The Jesuits overplayed their hand when they tried to revoke these restrictions after Farnese's re-conquest. The university invoked its monopoly over higher education, recognized by Charles V in 1530. The university proposed incorporating the college into the university, but this was rejected by Superior General Acquaviva.[63] The college's theological department and the theological faculty would remain at odds with each other for the ensuing years.

In 1594, Bishop Torrentius bequeathed a large amount of his fortune to the Leuven College in order to organize a full course in philosophy, including metaphysics and mathematics.[64] Having a rich competitor in the city provoked the university and its faculties, which obtained a papal brief decreeing that the university would have to approve the subjects the Jesuits could teach.[65] The course of mathematics did not materialize, whereas the metaphysics course did.[66] It was not until 1620 that the Jesuits began teaching mathematics at Leuven.

In 1624, the Jesuits tried to obtain permission to start a college for the humanities at Leuven, including a full philosophy course, a request that met with success the following year.[67]

6   The Antwerp College in the Sixteenth Century

When Nadal visited the Low Countries in 1567 as part of an inspection tour through Europe, he left detailed instructions for the organization of the colleges of Tournai and Dinant.[68] In 1574, the Belgian provincial wrote that it would be preferable to have a college in Antwerp, even if it was underfunded,

---

au P. Général Claude Acquaviva," *Bulletin de l'Institut historique belge à Rome* 7 (1927): 5–114, here 23; Roegiers, "Awkward Neighbours," 157. See also ARAA T14/034 2514, on the history of the Leuven college. Brouwers, "Aperçu historique sur la province belge," 2, ARAA T14/034 80.

62   Roegiers, "Awkward Neighbours," 157–58.
63   Roegiers, "Awkward Neighbours," 159–60.
64   ARAA T14/034 2529. Roegiers, "Awkward Neighbours," 161–62; Marinus, *Laevinus Torrentius*, 190–93.
65   Roegiers, "Awkward Neighbours," 162–63; Donnelly, "Padua, Louvain, and Paris," 42–46. On these issues in the period 1575–1627, see also ARAA T14/034 1785 and 1786 and ARAL T14 12.
66   Claeys-Boúúaert, "Une visite canonique," 23–24.
67   Roegiers, "Awkward Neighbours," 167–68. See also ARAL T14 10–11.
68   Giard, "Les collèges jésuites des anciens Pays-Bas," 94.

rather than a house or a novitiate. He claimed that Catholic families were suspicious of the orthodoxy of the schoolmasters.[69]

The Antwerp College opened its doors on March 12, 1575 and could boast three hundred pupils a couple of months later,[70] leading the rector Jan van Schoonhove (Schoonhovius [1540–1624]) to complain to Superior General Mercurian about the lack of qualified teachers.[71] The pupils came from all social strata, although unsurprisingly we find a disproportionate number of sons of the civic and commercial elite.[72] Ignatius had envisaged that each college would have a foundation, the proceeds of which would cover the costs of the buildings and staff.[73] In the Netherlands, however, it was extremely hard to capitalize on such a foundation,[74] as during the 1570s there was some hostility toward Jesuits, while after 1585 the Southern Netherlands was recovering from the ravages of war and the economy was in decline. The Antwerp Jesuits never succeeded in capitalizing on the foundation envisaged for the college.

In 1574, having obtained the support of the new governor, Luis de Requesens (1528–76), the Jesuits were able to buy a large house in Antwerp, Huys van Aecken (Aachen house),[75] for no less than thirty-four thousand guilders.[76] Within six months, the Jesuits had built a church—a single-nave chapel—in

---

69 Giard, "Les collèges jésuites des anciens Pays-Bas," 98; see also Ad Meskens, *Practical Mathematics in a Commercial Metropolis: Mathematical Life in Late 16th-Century Antwerp* (Dordrecht: Springer Science & Business Media B. V., 2013), 52, on the religious position of the teachers in the period 1560–85.

70 ARAA T14/034 10, fols. 1–8. Poncelet, *Histoire de la Compagnie de Jésus*, 1:230; Guido Marnef, *Antwerpen in de tijd van de Reformatie: Ondergronds protestantisme in een handelsmetropool 1550–1577* (Antwerp: Kritak, 1996), 170. A very detailed, but unpublished and slightly biased history of the college, can be found in Charles Droeshout, "Histoire de la Compagnie de Jésus à Anvers (6 vols.)" (unpublished MS, Kadoc ABML 3284–3289, n.d.). A list of pupils can be found in Delée, "Liste d'élèves."

71 Droeshout, "Histoire de la Compagnie de Jésus à Anvers," 2:339, referring to a letter dated August 20, 1575.

72 Marnef, *Antwerpen in de Tijd van de Reformatie*, 170.

73 On the funding of the Jesuit colleges, see Olwen Hufton, "Every Tub on Its Own Bottom: Funding a Jesuit College in Early Modern Europe," in O'Malley et al., *Jesuits II*, 5–23.

74 Brouwers, *Carolus Scribani 1561–1629*, 95.

75 See also ARAA T14/034 1916; T14/15.001 133. The name of the house referred to the fact that the family of its previous owner, Gaspar Schetz (1513–80), was from Aachen.

76 Droeshout, "Histoire de la Compagnie de Jésus à Anvers," 2:260–61, mentions two letters of Father Delange to Superior General Mercurian (dated March 9, 1574, and March 30, 1574), in which he informs the latter that in the first instance he refused his permission for the acquisition, but seeing that the Spanish colony had raised half of the required amount and not wanting to offend them, he finally gave his permission. See also ARAA T14/015.01 133 about the purchase of Huys van Aecken and ARAA T14/034 1916 about the amortization of the debts.

the garden of this mansion,[77] which was crowned with a tower the next year. The expenses for this chapel, five thousand ducats (twenty-five thousand guilders), were paid for by the Spanish merchant Ferdinand de Frias.[78]

The rector had to see to it that, through gifts, tax exemptions, and other means, the college kept afloat,[79] and the Jesuits also had to rely on small donations and legacies.[80] The college was closed in 1578, when the Jesuits were expelled from the city.[81] In 1585, after the sack of Antwerp, it was reopened, including a preparatory year.[82] Almost immediately, the Jesuits asked for (and obtained) exemptions from the city excises.[83] The college also obtained the support of the duke of Parma,[84] who instructed the city government to help the Jesuits. In 1593, they obtained a yearly grant of three thousand guilders for the cost of the college's upkeep,[85] although, by the end of the century, payment was not always guaranteed.[86] In 1592, the college asked the city council for permission for an extension and also asked it to buy a couple of houses on the college's behalf.[87] By 1600, the city was paying no less than 1,231 guilders in rents that were taken on the Society's property.[88] The Society also came into

---

77   Piet Lombaerde, "The Façade and the Towers of the Jesuit Church in the Urban Landscape of Antwerp during the Seventeenth Century," in *Innovation and Experience in the Early Baroque in the Southern Netherlands: The Case of the Jesuit Church in Antwerp*, ed. Piet Lombaerde (Turnhout: Brepols, 2008), 77–96, here 79; Joseph Braun, *Die Belgischen Jesuitenkirchen: Ein Beitrag Zur Geschichte des Kampfes* (Freiburg im Breisgau: Herder, 1907), 151.
78   Braun, *Die Belgischen Jesuitenkirchen*, 151. ARAA T14/15.001 133, in which the acquisition of premises adjacent to Huys van Aecken at a future date by Frias is settled (no specific amount is mentioned).
79   Brouwers, *Carolus Scribani 1561–1629*, 95. In *Afbeeldinghe van d'eerste Eeuwe der Societeyt Jesu* (Antwerp: Officina Plantiniana, 1640), 138–39, it is suggested that the duke of Alba had a visceral aversion for the Society of Jesus and that this had a negative influence on the Antwerp college.
80   Brouwers, *Carolus Scribani 1561–1629*, 95.
81   ARAA T14/034 10, fols. 12^{r}ff.
82   ARAA T14/034 10, fol. 18^r. Poncelet, *Histoire de la Compagnie de Jésus*, 2:10.
83   FA KK590, 1ste collegie, dated January 9, 1586; ARAA T14/034 126; 130; 1918; 3702; 3703; 3720. Marinus, "De financiering van de contrareformatie," 242, 246.
84   FA KK590, 1ste collegie, dated September 3, 1591.
85   FA KK590, 1ste collegie, nineteenth-century copy of Valckenisse V177; see also ARAA T14/034 1917.
86   Brouwers, *Carolus Scribani 1561–1629*, 96.
87   FA KK590, dated September 3, 1591.
88   To understand real estate holding in the Antwerp of the time, it is important to distinguish three kinds of property ownership. One could rent a house or own it, just as today, or else one could acquire a house in exchange for an interest, a so-called rent charge—roughly comparable to real estate certificates. See Hugo Soly, *Urbanisme en kapitalisme te Antwerpen in de 16de eeuw: De stedebouwkundige en industriële ondernemingen van Gilbert*

the possession of houses, some of which were demolished to build an exercise space for the gymnasium.[89]

The school averaged some five hundred pupils, peaking at 680 in the mid-seventeenth century. Unlike most Jesuit colleges, the Antwerp College accepted boarders, who were lodged in a house near the school, called a convict (*bursa*).[90] As early as 1575, the Spanish colony had agreed to accept boarders, but to no avail. In 1588, however, the superior general gave permission to accept boarders, lodged with a layman. Undoubtedly, he would have had pupils from the rebellious northern provinces in mind. The boarding school finally materialized in 1593.[91]

7 The Antwerp College in the Seventeenth Century

According to Olivier Manare's (1523–1614) visitation report, the Antwerp College suffered the burden of debts.[92] He concluded that the Jesuits had to rely on Providence and the generosity of the faithful to alleviate this burden. Little did he know that the vaingloriousness of the Antwerp Jesuits would burden the whole Flemish province with debts within two decades (see chapter 14).

In 1598, which was to prove an important year for the Antwerp College, a new rector was appointed: the energetic, yet choleric Carlo (Carolus) Scribani (1561–1629).[93] Meanwhile, the ailing, but highly competent mathematician Aguilón was appointed as confessor for the Spaniards residing at Antwerp

---

van Schoonbeke (Brussels: Gemeentekrediet van België, 1977), 54–59; Soly, "De schepenregisters als bron voor de conjunctuurgeschiedenis van Zuid- en Noordnederlandse steden in het Ancien Régime: Een concreet voorbeeld; De Antwerpse immobiliënmarkt in de zestiende eeuw," *Tijdschrift Voor Geschiedenis* 87 (1974): 521–44, here 523–27. The city had taken over rents in 1592 (on three houses), 1595, 1596, and 1597 (FA KK590, 1ste collegie). Interestingly, Engels Huys is also mentioned in this list, and while no date is mentioned, next to the list of beneficiaries of rents on this house a marginal note reads "is gequeten 5 april 93" (has been redeemed April 5, 1593). See also ARAA T14/015.01 175–77, for rents at the expense of the States of Brabant and the city of Antwerp.

89 ARAA T14/034 2028. Droeshout, "Histoire de la Compagnie de Jésus à Anvers," 2:298–99.
90 Poncelet, *Histoire de la Compagnie de Jésus*, 2:34. See also ARAA T14/034 1984 and 1981–91, for the convict in the seventeenth century.
91 Marianne Mochlig, "Het Jezuïetencollege te Antwerpen in de 17de en 18de eeuw" (Master's thesis, Katholieke Universiteit Leuven, 1988), 47.
92 Claeys-Bouuaert, "Une visite canonique," 16.
93 Brouwers, *Carolus Scribani 1561–1629*, 43, 93ff.

(see chapter 3).[94] Both were to play an important role in the foundation of the school of mathematics.

Scribani was born in Brussels on November 21, 1561 to a devout Catholic family.[95] During the troubles of 1576, the family left Brussels. It seems that in 1579 he was studying at the Jesuit college in Cologne. On March 17, 1582, he was admitted to the Jesuit order. After training as a novice for a year, he became a teacher in various Jesuit colleges, among them Molsheim, Trier, Liège, and Douai. In 1593, Scribani was sent to the Antwerp College to serve as prefect of studies.

By the turn of the century, the Huys van Aecken in which the school was housed had become too small for the number of students, and it was no longer possible to meet all the requirements of the *Ratio*—declamation exercises were out of the question, for example, because no room was large enough to house all the students; nor could sermons be delivered during Lent, because the college's chapel was too small.[96] The rector had already asked for the Jesuits to be given disposal of larger premises in 1599.[97] All of these remarks can also be found in Manare's visitation report,[98] giving improvements a sense of urgency.

Scribani wanted to move the college to the Hof van Liere, also called Engels Huys (English house), for which he needed the city council's approval. The premises had been in use by the English merchants, but when they left it became property of the city.[99] Engels Huys was nearly as large as a complete street block: it measured no less than 5,600 square meters, and its façade in Prinsstraat was seventy meters long with three carriage entrances.[100]

In a request to the town council, Scribani wrote that in a mercantile town such as Antwerp there should be a school where mathematics was taught. He promised that his school would, in due time, be teaching theology and mathematics. His plea on behalf of mathematics was only one of his arguments in persuading the city council. His motives may not have been so generous or selfless as they seem, as in part it may have been a move by the Jesuits to erect a school at university level, which did not compromise the monopoly of Leuven

---

94   Droeshout, "Histoire de la Compagnie de Jésus à Anvers," 2:295, 335, 391; August Ziggelaar, S.J., *François de Aguilón, S.J. (1567–1617): Scientist and Architect* (Rome: Institutum Historicum S. I., 1983), 37.
95   Biographical details from Droeshout, "Histoire de la Compagnie de Jésus à Anvers," 2:301–7; Brouwers, *Carolus Scribani 1561–1629*, see also *Afbeeldinghe*, 672–74.
96   Brouwers, *Carolus Scribani 1561–1629*, 98–99.
97   FA KK591, 2de collegie, not foliated, 1606 and 1607.
98   Claeys-Bouuaert, "Une visite canonique," 32–38.
99   FA KK591, April 14, 1607.
100 Brouwers, *Carolus Scribani 1561–1629*, 100.

FIGURE 1.3 The square in front of Huys van Aecken and Saint Charles Borromeo Church (now Hendrik Conscienceplein)
© KIK-IRPA B054921

but at the same time would certainly have hurt it.[101] Scribani also had other reasons for funding a school of mathematics: the members of the Academy of Church History had asked to be instructed in astronomy. Astronomy was

---

101 "Item dat meer is dat de vβ patres oock metter tijdt souden schicken te leeren mathematicam die noch te Louen noch elders in geene universiteiten en wordt geleert ende nochtans het fundament is van vele consten die in een coopstadt als dese meest van doene zijn niet alleen tot cieraet ende eere der seluer maer oock tot voorderinghe van degene die hen totte coopmanschap oft andere neiringe soude willen begeuen." FA KK591, 1606 or 1607. "Enseigner la mathematique (laquelle ne s'enseigne ny a Leuven ny ailleurs, estant toutefois le fondament et base de plusieurs arts liberaux le plus requit a une ville marchande comme ceste non seulement a l'embellissement et honneur d'Icelle mais aussy a l'aduancement de ceulx qui se vauldraoient adonner au (?) et marchandise" (Also, that the aforementioned fathers will in due time teach mathematics, which is taught neither at Leuven nor at any other university, but which nevertheless is the foundation of many of the liberal arts necessary in a merchant city, not only as a jewel in the city's crown but also to the advantage of those who would like to enter into commerce or into another craft). FA KK591, 1606 or 1607.

important for chronology, a subject that the academy itself studied.[102] At the behest of the archdukes, Mayor Hendrik van Etten (1587–1640) supported the claim.[103] Scribani was successful, and the college consequently moved to Engels Huys in 1608.[104] Unfortunately, however, the academy would cease to exist the year before the school was opened.

Although the college attracted mainly upper-class pupils, the Jesuits also directed their attention to the children of the working classes. With the support of the Marian congregations, they organized catechism classes parallel to the Sunday schools. In 1609, they taught no fewer than 3,200 children in thirteen chapels.[105]

In 1616, the Flemish Jesuits founded a *domus professa* (professed house) in the Huys van Aecken.[106] A professed house was a center for pastoral work without any fixed income, where Jesuits could devote their attention to ministry and to science, without being obliged to teach. As the main seat of the province, it became the nexus of the Jesuits' network in the Flemish province. Over the years, the house would welcome artists, writers, and scientists.[107] Heribert Rosweyde (1569–1629) and Jean Bolland (Johannes Bollandus [1596–1665]) turned the professed house into a veritable center of learning for hagiographers, with no fewer than fifty *Acta sanctorum* (Acts of the saints) on the lives of the saints being published.[108]

---

102 Omer van de Vyver, S.J., "L'école de mathématiques des jésuites de la province Flandro-Belge au XVIIe siècle," *Archivum historicum Societatis Iesu* 49, no. 97 (1980): 265–71, here 266; see also Droeshout, "Histoire de la Compagnie de Jésus à Anvers," 3:168. The "academy" for church history, presided over by Scribani, consisted of four members who had to find material that could be used in theological discussions. They quickly stumbled upon problems in chronology, relating to astronomy. A very direct application of astronomy in the ecclesiastical year is the determination of the date of Easter, i.e., the first Sunday after the first full moon of spring. The academy is also referred to in a letter of Vicar General Ferdinand Albert to Aguilón, February 24, 1615, copy by Bosmans in Kadoc ABML Bosmans V 1–1.

103 FA KK 591, April 14, 1607.

104 FA KK 591, Engels Huys; ARAA T14/034 1965; T14/015.01 136. Also *Afbeeldinghe van d'eerste Eeuwe der Societeyt Jesu* (Antwerp: Officina Plantiniana, 1640), 541; Brouwers, *Carolus Scribani 1561–1629*, 102; Lodewijk Brouwers, *Het Hof van Liere* (Antwerp: Loyola-vereniging, 1976), 27.

105 Mannaerts, *Sint-Carolus Borromeus*, 33.

106 ARAA T14/034 1919; 1992 and T14/015.01, 7.

107 See Brouwers, *Carolus Scribani 1561–1629*, 107–8, for a survey of scholars and artists, among them Daniel Seghers, the famous floral painter, and architect Peter Huyssens.

108 Mannaerts, *Sint-Carolus Borromeus*, 26; Brouwers, "Aperçu historique sur la province belge," 12–13.

8   The School of Mathematics

In 1615, the plan for creating a school of mathematics was finally being put into practice. Aguilón was granted permission by Superior General Acquaviva to start a special school of mathematics "in the interest of the Academy of Church History."[109] This was reaffirmed, shortly after Acquaviva's death, by Vicar General Ferdinand Albert.[110] By the end of 1615, San Vicente had arrived at Antwerp to help Aguilón write the curriculum.[111] By the end of 1617, the school of mathematics had opened.[112] Aguilón unfortunately did not live to see it open its doors.

During the period that San Vicente was in Antwerp, he wrote his first treatises, mainly on the parabola and the hyperbola (fig. 1.4).[113] A treatise about ellipses was written by Willem Boelmans (1603–38), one of his students at Leuven. At first, San Vicente based his work on that of Aguilón, but as he encountered more and more mathematical problems, he resolutely chose to resolve them. In his manuscripts written at Antwerp and Leuven, he treats the trisection of an angle, mean proportionals, conic sections and their properties, sequences, the logarithmic properties of hyperbolic segments, ductus figures, the determination of the volume of ductus figures and a special figure: the

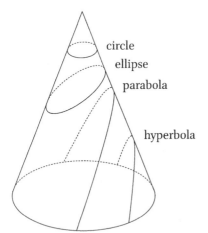

FIGURE 1.4
The intersection of the surface of a cone with a plane is called a conic section. Depending on the orientation of the plane, this can be a circle, an ellipse, a parabola, or a hyperbola

---

109   Van de Vyver, "L'école de mathématiques des jésuites," 266.
110   Kadoc ABML Fonds Bosmans 1. Transcription of a letter by Ferdinand Albert to Aguilón, February 21, 1615.
111   Ziggelaar, *François de Aguilón, S.J.*, 49.
112   Kadoc ABSE 104, copy of ARSI Fl. Belg. 420.
113   Conic sections are the intersections of a plane and a (double) cone. Depending on the orientation of the plane the resulting curve is an ellipse, parabola, or hyperbola.

ungula cilindrica (see chapters 7 and 8). Next to these purely mathematical subjects, he was also interested in statics and astronomy.

We do not have a curriculum for these years in Antwerp, but treatises by authors such as Ciermans, from the Leuven period, show that a wide range of subjects should have been covered, ranging from pure mathematics to physics and the art of warfare. It does seem that in the first Antwerp period the emphasis was on pure mathematics.

Laymen as well as students of the Jesuits could attend lessons in the school of mathematics. Unfortunately, we do not have a register of attendance, which makes it impossible to identify any laymen.[114] Della Faille and Nuyts were the first two Jesuits to take these classes in 1617.[115] The other Jesuits instructed by San Vicente at the Antwerp College were Ignatius (Abraham) Derkennis (1598–1656), Antonius Alegambe (1600–68), Jacob Durand (1598–1644), and Joannes Cox (1597–1622) (see chapters 10 and 11 for biographies of the most important students).[116]

Undoubtedly the most important pupil at the school of mathematics was della Faille. In 1616, he is mentioned as one of the twenty-three students studying "physics," meaning the second year of his philosophy studies.[117] He defended fifty theses on August 22, 1617.[118] Della Faille seems to have been the only one to follow the course for three consecutive years.[119]

From 1617 onward, San Vicente organized public disputations about mathematical subjects, which drew a considerable audience.[120] In 1618, the appearance of three comets triggered a stream of treatises. Because the last of these comets was the most impressive, it is often called *the* comet of 1618. It was first seen in mid-November emerging from the solar glare as a tail projected

---

114 ARAA T14/034 2/2, fol. 6ʳ mentions four private students for 1618. This document is a draft for the *historia domus*. The sentence was struck through and does not appear in the published *historia*.

115 The *Elogium* of della Faille (ARAA T14/034 486) mentions that he first attended classes with San Vicente in 1616. KB MS 20194; ARSI Fl.Belg. 44, fol. 32ʳ. Droeshout, "Histoire de la Compagnie de Jésus à Anvers," 3:199, 248; Herman van Looy, "Chronologie en analyse van de mathematische handschriften van G. a Sancto Vincentio (1584–1667)" (PhD diss., Katholieke Universiteit Leuven, 1979), 12–13.

116 KBR MS 20194. Van Looy, "Chronologie en analyse," 12–15; Delée, "Liste d'élèves," passim. Droeshout, "Histoire de la Compagnie de Jésus à Anvers," also mentions Jacob de Succa (1597–1634) as a mathematics student, but this is not attested by the archival material.

117 ARSI Fl. Belg. 44, fols. 8ʳ, 21ʳ; Kadoc ABSE1. Droeshout, "Histoire de la Compagnie de Jésus à Anvers," 3:151–52.

118 AFL 28.4.8.

119 ARSI Fl. Belg. 44, fols. 32ʳ, 46ᵛ; KB MS 20194. Droeshout, "Histoire de la Compagnie de Jésus à Anvers," 3:235, 259.

120 ARAA T14/034 2/1, 354 and 10, fol. 43ʳ.

above the horizon. The tail reached impressive proportions in mid-December, and the comet was last seen on January 22, 1619.[121] San Vicente and his students made observations and reported the results to Rome.[122] The Antwerp Jesuits observed the comet's vicissitudes and its entire course by means of telescopes.[123] San Vicente had his students defend theses on the nature of the comet.[124] These lectures seem to have been very successful, drawing large audiences.[125] Meanwhile, in Rome, Grassi anonymously published *De tribus cometis annus MDXVIII* (On the year of the three comets [1618]), in which he argued for the celestial nature of comets by citing parallax measurements made in Rome and Antwerp "scarcely ever exceeding one degree."[126] Other foreign expert observers seem to have preferred the Antwerp observations as well.[127] Already during the Christmas holidays of 1618, the Jesuits at the Collegio Romano organized discussions on this subject. Grassi proved, by comparing observations at different locations, that the comet was far more distant than the moon, thus shattering the Aristotelian doctrine that comets were atmospheric phenomena.[128]

---

121 Donald K. Yeomans, *Comets: A Chronological History of Observation, Science, Myth, and Folklore* (New York: Wiley, 1991), 51. According to Andreas Tacquet, *Opera mathematica* (Antwerp: Apud Henricum & Cornelium Verdussen, 1707), 231, the tail measured 90° at its maximum.

122 Erycius Puteanus [Erik de Put], *De cometa anni M.DC.XVIII: Novo mundi spectaculo, libri dvo.; Paradoxologia* (Leuven: Apud Bernardinum Masium, 1619), 168; Droeshout, "Histoire de la Compagnie de Jésus à Anvers," 3:246.

123 Kadoc ABSE 102, fol. 44, ARAA T14/034 2/1; 10, fol. 44^r. Giovanni Battista Riccioli, *Almagestum novum astronomiam veterem novamque complectens Observationibus aliorum, et propiis novisque theorematibus, problematibus, ac tabulis promotam, in tres tomos distributam quorum argumentum sequens pagina explicabit* (Bologna: Ex typographia haeredis Victorii Benatii, 1651), 101–2, 111; Tabitta van Nouhuys, *The Ages of Two-Faced Janus: The Comets of 1577 and 1618 and the Decline of the Aristotelian World View in the Netherlands* (Leiden: Brill, 1998), 240.

124 Despite being mentioned in Puteanus, it is uncertain whether these theses were printed. They are not mentioned in Pedro de Ribadeneyra and Philippe Alegambe, *Bibliotheca scriptorum Societatis Jesu: Post excusum anno M.DC.VIII catalogum R. P. Petri Ribadeneirae [...] Nunc [...] Ad annum M.DC.XLII. [...] Concinnata* (Antwerp: Joannes Meursius, 1643), unlike Ciermans's 1624 thesis.

125 Kadoc ABSE 102, fols. 43–44. Tabitta van Nouhuys, "Copernicus als randverschijnsel: De kometen van 1577 en 1618 en het wereldbeeld in de Nederlanden," *Scientiarum historia* 24 (1998): 17–38.

126 Yeomans, *Comets*, 57.

127 Van Nouhuys, *Ages of Two-Faced Janus*, 240.

128 Ziggelaar, *François de Aguilón, S.J.*, 10. Riccioli, *Almagestum novum*, after carefully comparing the observations of the 1618 comet, concluded that it was "not proven" that comets were superlunary, but that it was highly likely.

In 1620, the school of mathematics was transferred to Leuven.[129] Here, San Vicente had Willem Hesius (1601–90), Boelmans, Jacob van Rasseghem (1603–1662), Moretus, Jan Ciermans (1602–1648), Walter van Aelst (1603–38), Valentin le Vray (1604–?), and Reynier Pynappel (1600–36) as students (see chapter 12).[130]

As he had done in Antwerp, and as many of his students would do when they had become teachers themselves, San Vicente had his students defend theses. The first student to do so was della Faille, who in August 1620 defended his *Theses mechanicae* (Mechanics theses).[131] On July 29, 1624, Ciermans and Van Aelst defended statics theses.[132]

When in Leuven, San Vicente arrived at the idea that the ductus method might be a method with which the quadrature of the circle, understood as calculating the exact area of a circle with a given radius, could be solved. He wanted to publish on this subject and duly sent his manuscripts to Rome. In 1625, he was called to Rome to confer with Grienberger about the method.

From 1626, della Faille acted as a substitute professor for San Vicente. He taught physics to Jesuit students and mathematics to lay students.[133] Among the latter, he is reputed to have had the sons of the Pfalzgraf (count palatine) and several sons of Polish noblemen who had come to study at Leuven.[134] In 1628, della Faille was called upon to do his tertianship in Lier. At first, it was thought that Moretus would succeed him, but that very same year Moretus was sent to Bohemia and was never to return to the Low Countries.

---

129 ARAA T14/034 2/2 *Collegium Antwerpiense* large pages fol. 3 and small pages 1620. In the catalog of 1620, San Vicente is mentioned as a professor in Antwerp (KBR MS 20194), in that of 1621 as one in Leuven (KBR 20195).

130 Van Looy, "Chronologie en analyse," 15–20; Delée, "Liste d'élèves," passim. See also appendix 2.

131 Christiaan Huygens, *Oeuvres complètes: Tome I; Correspondance 1638–1656*, ed. David Bierens de Haan (The Hague: Martinus Nijhoff, 1888), 158, letter 105.

132 Gregorio a San Vicente and Walter van Aelst, *Theoremata mathematica [...] Defenda ac demonstranda in Collegio Societatis Iesv Louanij [...] Die 29; Iulij Ante Mtridiem, [!] Anno 1624* (Leuven: Henrici Hastenii, 1624). See also Jean Dhombres and Patricia Radelet-de-Grave, *Une mécanique donnée à voir: Les thèses illustrées défendues à Louvain en juillet 1624 par Grégoire de Saint-Vincent, S.J.*, De Diversis Artibus 82, n.s. 4 (Turnhout: Brepols, 2008).

133 ARSI Fl.Belg.44, fol. 152r; KBR 20197–98. Van de Vyver, "L'école de mathématiques des jésuites," 267; Ad Meskens, *Joannes della Faille, S.J.: Mathematics, Modesty, and Missed Opportunities* (Brussels: Belgisch Historisch Instituut te Rome, 2005), 43. No Jesuits are known to have taken mathematics classes with della Faille.

134 ARAA T14/034 486. Meskens, *Joannes della Faille, S.J.*, 44; Yves Schmitz, *Les della Faille* (Brussels: F. Van Buggenhoudt, n.d.), 3:76.

It is not clear who succeeded della Faille. It was probably Derkennis, as the 1633 catalog mentions that he taught mathematics for two years.[135] He was probably succeeded by Van Hees, who taught mathematics for one year.[136] Since he was transferred to Antwerp in 1634, he must have been the predecessor of Boelmans.[137] On August 8 and 9, 1634, Boelmans had theses defended on applied mathematics. One of the defendants was Andreas Tacquet (1612–60); guest of honor at the defenses was San Vicente.[138]

We have no indication of the identity of the mathematics teacher in the years 1635–36. In 1637, Ciermans succeeded Boelmans as professor at the school of mathematics.[139] In the theses Ciermans had defended, it becomes clear that his interest lay in practical mathematics and its applications. In 1640, the centenary of the Society, Ciermans published his course material under the title *Disciplinae mathematicae* (Mathematical sciences). The subjects the work dealt with again show his interest in practical mathematics.

When Ciermans left for the China mission (where he would never arrive), Alexander Barvoets (1619–54) was assigned to the post.[140] In 1644, Tacquet took over the reins and would remain the school's mathematics teacher for the rest of his life.[141] Tacquet taught mathematics and simultaneously finished his education by taking four years of theology.[142] He would follow all movements of the school of mathematics between Leuven and Antwerp.[143] In 1645, the school was transferred to Antwerp, to return to Leuven three years later, and to move definitively to Antwerp in 1656. After Tacquet's death in 1660, the school went into a state of lethargy, with it taking no less than five years simply to appoint Tacquet's successor.

After San Vicente had left, few of the students of the school of mathematics attained any fame in mathematics or science, Tacquet being the most notable exception. Verbiest's fame rests on his accomplishments in the Chinese mission.[144] Of the other students, only Ignatius de Jonghe (1632–92) published a mathematical treatise, *Geometrica inquisitio* (Geometric investigation [1688]). This treatise is about generalized parabola ($y^n = cx^p$) and hyperbolae

---

135 ARAA T14/034 34, 48/14; 35, 48/5; 36, 52/7.
136 ARAA T14/034 34, 48/9.
137 ARAA T14/034 34, 48/9; KBR MS 20203.
138 KBR MS 19337–38, fol. 69.
139 ARAA T14/034 37, 52/11; KBR MS 20204–6.
140 KBR MS 20207. Van de Vyver, "L'école de mathématiques des jésuites," 273.
141 KBR MS 20210–20.
142 ARAA T14/034 39, 47/5.
143 ARAA T14/034 40, 12/3.
144 See Golvers, "F. Verbiest's Mathematical Formation."

($y^n x^p = c$) and their quadratures, using the techniques of San Vicente and Tacquet and strongly influenced by Pierre de Fermat (1607–65).[145]

9   Michiel Coignet and the Jesuits

As late as 1638, the Englishman Ignace Stafford (1599–1642), who taught at the Royal Academy in Lisbon, wrote a treatise entitled *Arithmetica practica geometrica logarithmica* (Practical arithmetic with geometric progressions) in which he discusses the use of a *pantometer*, as he calls the sector.[146] This indicates that Iberian Jesuit mathematicians came into contact with the sector as a calculating instrument through Michiel Coignet's (1549–1623) work. This may have involved any Flemish Jesuit who came to Spain or to Portugal such as della Faille or Ciermans.

Although Stafford credits the invention of the sector to Clavius and claims that Coignet adapted the instrument, the direct influence of Clavius on Coignet seems unlikely.[147]

Yet Clavius and Coignet were not complete strangers to each other. There is no evidence of a direct correspondence, but we find Coignet's name mentioned a couple of times in Clavius's correspondence. Moreover, Marino Ghetaldi (1568–1626), who studied with Coignet around 1599—and later with

---

145   See Paul Bockstaele, "Ignatius de Jonghe et sa *Geometrica inquisitio in parabolas numero et specie infinitas*," *Janus* 54 (1967): 228–35; Bockstaele, "Een vergeten werk over de kwadratuur van parabolen en hyperbolen: Ignatius de Jonghe's *Geometrica inquisitio* (1688)," *Scientiarum historia* 4 (1967): 175–81.

146   See Samuel Gessner, "The Conception of a Mathematical Instrument and Its Distance from the Material World: The 'Pantometra' in Lisbon, 1638," *Studium* 4 (2011): 210–27.

147   On the history of the sector, see Ivo Schneider, *Der Propotionalzirkel: Ein universelles Analoginstrument der Vergangenheit* (Munich: Deutsches Museum, 1970); Stillman Drake, *Galileo Galilei: Operations of the Geometric and Military Compass 1606* (Washington, DC: Dibner Library Smithsonian Institution Press, 1978); Ad Meskens, "Michiel Coignet's Contribution to the Development of the Sector," *Annals of Science* 54 (1997): 143–60; Filippo Camerota, *Il compasso di Fabrizio Mordente: Per la storia del compasso di proporzione*, Biblioteca di Nuncius 37 (Florence: Leo S. Olschki, 2000).

All authors point to a shared parentage of the invention, including Jost Bürgi (1552–1632), Paolo Galucci (1538–c.1621), Guidobaldo del Monte (1545–1607), Jacques Besson (c.1540–73), Fabrizio Mordente (1532–c.1608), Christopher Clavius, and Galileo Galilei, among others. The development of Coignet's sector is well documented; for a detailed description, see Meskens, "Michiel Coignet's Contribution," and Meskens, *Practical Mathematics*, 118–37. In the development of Coignet's sector, there are three periods, a first with only rulers carrying scales, a second with an adapted reduction compass of the Mordente type, and a last with a proper sector, which bears a superficial resemblance to Galileo's sector.

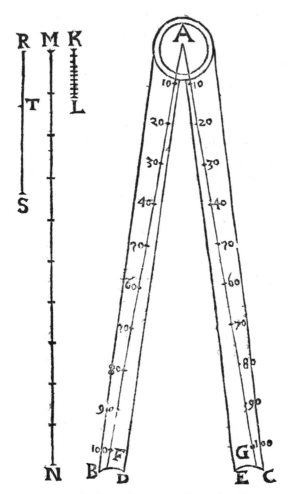

FIGURE 1.5　The figure of a sector in Christopher Clavius, *Geometria practica* (Rome: Aloisius Zannetti, 1604), 5

François Viète (1540–1603)—was a frequent visitor to the Collegio Romano during his stay in Rome between 1601 and 1605.[148]

As early as 1595, Adriaan van Roomen (1561–1615) informed Clavius that Coignet had not yet finished his *Theorica [Planetarum]* (Planetary theory).[149]

148　Baldini, "Academy of Mathematics," 56.
149　Paul Bockstaele, "The Correspondence of Adriaan van Roomen," *Lias* 3 (1976): 85–299, here 119.

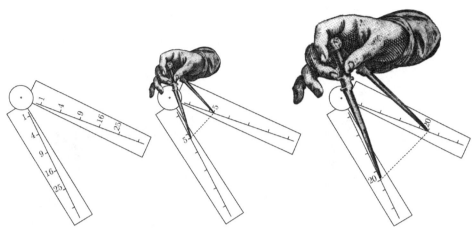

FIGURE 1.6 A sector consists of two pivoting arms with engraved scales, which can be used for computations. Using the sector, the equation $\dfrac{f(x_1)}{f(x_2)} = \dfrac{a}{b}$ can be solved for any of the letters x_1, x_2, a, b provided the other three numbers are known. If f represents the function square root, the numbers $1, 4, 9, \ldots, n^2$ are engraved at lengths $1, 2, 3, \ldots, n$ from the pivot. To solve the equation $\dfrac{\sqrt{5}}{\sqrt{x}} = \dfrac{1}{2}$, the scales for the square root on the sector are used. A line with length a is drawn, and the sector is opened such that the line fits between markings 5. With a set of dividers opened at $2a$, it can be determined that a line of length $2a$ fits between the markings 20, and therefore $x = 20$

Van Roomen's comments suggest that Coignet studied the Ptolemaic system.[150] For about twenty years, Coignet maintained that the publication of his astronomical treatise *Theorica planetarum* was imminent, but it is unclear whether it was ever published or even whether manuscript copies of it existed. Clavius must have taken an interest in the work and probably inquired about it. In February 1603, Jesuit John Hay (1546–1607) informed him that Coignet was still working on it, but that he was confident that it would be published before the Frankfurter Herbstmesse (Frankfurt autumn book fair).[151] His astronomical treatise was apparently completed by 1606, because he mentions it in a

---

150 This seems to be confirmed by his introduction to Cornelis de Jode's (1568–1600) atlas *Speculum orbis terrae* (1593), which described the Ptolemaic system with nine spheres. On the other hand, his entry in Antonio A. Anselmo, "Album Amicorum" (MS, Koninklijke Bibliotheek Den Haag 71 J 57, 1594–1602), fols. 30ᵛ–31ʳ, shows a Tychonic system. Meskens, *Practical Mathematics*, 192–94.

151 Ugo Baldini and Pier Daniele Napolitani, eds., *Christoph Clavius corrispondenza: Edizione critica; V (1602–1605)* (Pisa: Dipartimento di Matematica, 1992), 5:no. 204.

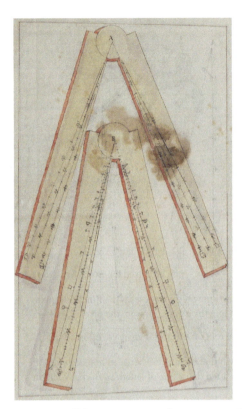 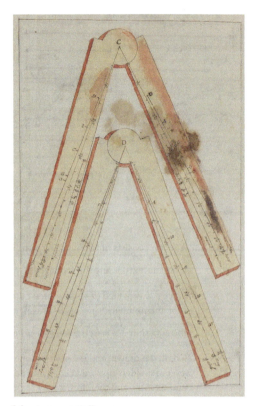

FIGURE 1.7   Coignet-type sectors, from a manuscript of the Antwerp College
© ARAA T14/034 1881

letter to Kepler.[152] In 1605, Hay again informed Clavius about Coignet, this time with personal news: Coignet had been forced out of his house on Grote Markt (Market square) by his neighbors, because three of his children had succumbed to the plague.[153]

It is little wonder that the Jesuits are found among the users of Coignet's instruments, as they literally lived next door to him. The file "geometrical figures" of the Jesuit archive in the Rijksarchief is actually a part of a manual for the use of Coignet-type sectors (fig. 1.7).[154]

---

152   Johannes Kepler, *Briefe 1604–1607*, Gesammelte Werke/Kepler, Johannes 15 (Munich: Beck, 1951), 544.
153   Baldini and Napolitani, *Christoph Clavius corrispondenza*; Ad Meskens, *Familia universalis: Coignet* (Antwerp: Koninklijk Museum voor Schone Kunsten, 1998), 183; Meskens, "Some New Biographical Data about Michiel Coignet," *Nuncius* 17 (2002): 447–54, here 449–50; Meskens, *Practical Mathematics*, 19–20.
154   ARAA T14/034 1881.

FIGURE 1.8 A putto carrying a sector. Detail from the frontispiece of Jan Ciermans et al., *Annus positionum mathematicarum* [...] (Leuven: Everardum de Witte, n.d.)
© EHC G 4894

At least two students of the school of mathematics seem to have used Coignet-type compasses: Moretus and della Faille. Toward the end of 1628, Moretus left the Low Countries to assume several positions in Central Europe.[155] It is possible that the "neunspitziger passer" (nine-pointed compass) manuscript came to Breslau (Wrocław) via Moretus, who at one time was stationed there.[156] The Prague manuscript on the sector may have been brought by either Moretus or

---

155 Van Looy, "Chronologie en analyse," 17–18. On Moretus in Prague, see Henri Bosmans, S.J., "Théodore Moretus de la Compagnie de Jésus, mathématicien (1602–1667): D'après sa correspondance et ses manuscrits," *De Gulden Passer* 6 (1928): 57–163.

156 Michiel Coignet, "Neunspitziger Passer: Fabricii Mordenti von Salerno Mathematici des Herzogen Alexandrii Fernesi Herzogen zu Parma" (MS, University Library Wrocław MS R461, n.d.).

FIGURE 1.9 Portrait of della Faille. A coignet-type sector can clearly be seen next to his right hand. Antoon van Dyck, "Jan-Karel della Faille," oil on canvas, 1629. Koninklijke Musea voor Schone Kunsten van België (Brussels), della Faille bequest, inv. no. 6254
© KIK-IRPA B117502 (DYCK 6254)

San Vicente. There was a manuscript on the sector in the library of the Besançon family Chifflet, whose son Laurent was a Jesuit in the Flemish province.[157]

Della Faille was sent to Spain, where he became a tutor to Don Juan of Austria (1545–78), Philip IV/III's (1605–65, r.1621–65/r.1621–40) bastard son.

---

157   Audenaert, *Prosopographia iesuitica Belgica antiqua*, 1:209.

During his time in Spain, he would appear to have bought sectors from Jacob de Coster (*fl.* 1640), who had previously worked in Coignet's workshop.[158]

Among the papers of della Faille in the della Faille family archives is a manual on Coignet's sector. One of the manuscripts in della Faille's collection that remained in the library of the Jesuit order in Madrid was entitled *Fábrica y uso del pantómetro* (Manufacture and use of the pantometer),[159] undoubtedly a manual on the use of Coignet's sector. The manuscript *Tratado de la división de las doce líneas rectas diversas de las pantómetras* (Treatise on the twelve different lines of the pantometer) was at one time in the collection of the Colegio Imperial,[160] which indicates the use of these calculating devices at the college.

Sir Anthony van Dijck's (1599–1641) 1628/29 portrait of della Faille features a Coignet-type sector, making it the sector with the largest viewing audience.[161]

---

158 AFL 28.15.20, also AFL 28.15.31. Della Faille acted on several occasions as an intermediary for Spanish noblemen to obtain mathematical instruments in Antwerp.

159 AFL 28.3.

160 BRAHM 9/2779–667. 2. Agustín Udías, S.J., "Los libros y manuscritos de los profesores de matemáticas del Colegio Imperial de Madrid, 1627–1767," *Archivum historicum Societatis Iesu* 74, no. 148 (2005): 369–448, here 444.

161 See Ad Meskens, "The Portrait of Jan-Karel della Faille by Anthony van Dijck," *Koninklijk Museum voor Schone Kunsten Jaarboek* 39 (1999): 124–37.

CHAPTER 2

The Seventeenth Century: The Dawn of a New Era

1   Conic Sections

The history of conic sections begins in antiquity. Two figures are important in our story: Archimedes (287 BCE–212 BCE) and his near contemporary Apollonius of Perga (*fl.* late third century BCE–second century BCE).

Archimedes does not seem to have written a treatise on conic sections; rather, many theorems about them can be found throughout his work. He knew how to find a right circular cone of which a given ellipse is a section.[1]

Archimedes also knew the fundamental properties defining the type of conic section. He gives the property that if from a point the tangents to a conic section and two intersecting straight lines parallel to them are drawn, the rectangles contained by the segments (of the chords) will be to one another as the squares of the tangents.[2] He is also familiar with the condition for similarity for conic sections and can prove that all parabolae are similar. On the parabola in particular, he would write *Tetragonismos paraboleis* (On the quadrature of the parabola).[3]

The pinnacle of studies on conic sections is undoubtedly Apollonius's *Konika* (Conics).[4] We know very little about Apollonius, save for his having been born in Perga in Asia Minor during the reign of Ptolemy Euergetes (*c.*280–222 BCE, r.247 BCE–222 BCE) and that he went to Alexandria to study mathematics at a very young age. The first four books of *Konika* contain a basic

---

1   *On Conoids and Spheroids*, propositions 7, 8, Archimedes, *Les œuvres complètes d'Archimède suivies des commentaires d'Eutocius d'Ascalon: Trad. du Grec en Français avec une introd. et des notes par Paul Ver Eecke*, ed. Paul Ver Eecke (Liège: Vaillant-Carmanne, 1960), 157–63.

2   *On Conoids and Spheroids*, proposition 3, Archimedes, *Les œuvres complètes d'Archimède*, 149–52; Thomas L. Heath, *Apollonius of Perga, Treatise on Conic Sections* (Cambridge: W. Heffer, 1961) xxxv; Kuno Fladt, *Geschichte und Theorie der Kegelschnitte und der Flächen zweiten Grades* (Stuttgart: E. Klett Verlag, 1967), 13.

3   See Archimedes, *Les œuvres complètes d'Archimède*, 377–404; Ken Saito, "Archimedes and Double Contradiction Proof," *Lettera matematica* 1, no. 3 (2013): 97–104.

4   See Apollonius Pergae, *Les coniques d'Apollonius de Perge: Œuvres trad. pour la première fois du grec en français, avec une intr. et des notes, par Paul Ver Eecke*, ed. Paul Ver Eecke (Brugge: Desclée-De Brouwer, 1923); Heath, *Apollonius of Perga*; Julian L. Coolidge, *A History of the Conic Sections and Quadric Surfaces* (Oxford: Clarendon Press, 1945), 13–25; Fladt, *Geschichte und Theorie der Kegelschnitte*, 14–41; Michael N. Fried and Sabetai Unguru, *Apollonius of Perga's Conics: Text, Context, Subtext* (Leiden: Brill, 2001).

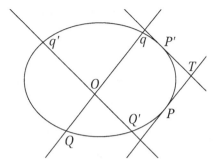

FIGURE 2.1
Archimedes's tangency theorem:
$|qO|.|OQ|.|TP'|^2 = |q'O|.|OQ'|.|TP|^2$

theory on the properties of conic sections, while the following four deal with specialized subjects such as maxima and minima, equal and similar sections of conics, and theorems involving the determination of limits.

Apollonius characterizes the three conic sections by a planimetric property, the length of the so-called *latus rectum* (the parameter, the chord parallel to the directrix and passing through the focus or one of the foci), which derives from the properties of plane sections of an oblique cone. He is quickly able to reduce these properties to expressions that we would recognize as the equations relative to the conjugate diameters of an ellipse or a hyperbola.[5] Conjugate diameters are diameters for which all chords parallel to one diameter are divided into two equal segments by the intersection point with the other diameter. These line segments are also called ordinates relative to the diameter. Moreover, tangents to the conic section at the extremities of a diameter are parallel to the ordinates relative to that diameter.

In the second book, Apollonius investigates the properties of asymptotes of hyperbolae, among other things to construct a hyperbola if its asymptotes and one of its points are known.[6] The third book deals with subjects we would categorize under poles and pole lines: if *TQ* and *Tq* are tangents to a conic section, then any line through *T* will intersect the line through the tangent points and the conic section in points that, when taken together with *T*, are a harmonic quadruple.

The fourth part of the third book deals with the segments that are cut off by a moving tangent on two fixed tangents. With these theorems, it becomes possible to draw a conic section given its tangents. Apollonius would expand on this theme in the seventh book. The rest of the third book deals with the foci of ellipses and hyperbolae to end with the locus of three or four straight lines. The

---

5   For a discussion of Apollonius's geometry in relation to algebra, see Fried and Unguru, *Apollonius of Perga's* Conics.
6   Heath, *Apollonius of Perga*, 56.

latter is the locus of a point moving in such a way that the ratio of the product of its distance to two given straight lines to the product of the distances to two other straight lines is constant.[7] The three-line locus is a special case of the four-line locus, in which two lines coincide. Book 4 deals with intersections of conic sections, in which, among other things, it is proven that two conic sections cannot intersect each other in more than four points. The fifth book deals with finding a minimum or maximum, often in relation to the normals to a conic section and often with the use of a *neusis*-like construction.[8] The sixth book deals with equal and similar conic sections, while in the seventh Apollonius investigates certain lines in relation to characteristic lines such as the axes.[9]

Apollonius's work would remain the standard for ages to come, the subject lying largely dormant until the sixteenth century. This does not mean that the subject was entirely forgotten, however, as it did have applications, for instance in gnomics.[10]

One of the first works to be printed containing a study of conic sections was Albrecht Dürer's (1471–1528) *Underweysung der Messung* (Instructions for measuring [1525]).[11] The work contains many drawings showing the orthographic projection of certain lines, such as the sine wave as the projection of the stairs of a spiral staircase. He also describes a practical way of obtaining conic sections using these projection techniques.

In the sixteenth and seventeenth centuries, it turned out that in describing physical phenomena, conic sections were everywhere, from the motion of celestial bodies (ellipses) to the motion of projectiles on Earth (parabolae) and the description of the refraction of light (hyperbolae).

---

7   Although, as Heath, *Apollonius of Perga*, cli, remarks, the four-line locus implies that the construction of a conic through five points is possible, Apollonius does not treat this problem.

8   Fried and Unguru, *Apollonius of Perga's* Conics, 146–204. On *neusis*, see Ad Meskens and Paul Tytgat, *Exploring Classical Greek Construction Problems with Interactive Geometry Software* (Basel: Birkhäuser Verlag, 2017), 23–25.

9   Fried and Unguru, *Apollonius of Perga's* Conics, 320–24.

10 In constructing a sundial, we assume that the sun revolves around the Earth once a day, following a path closely resembling an arc (there is a tiny departure due to the fact that the sun does not rise or set at the same point each day, but that the points of sunrise and sunset move along the horizon a little). Suppose we draw an imaginary line connecting the tip of the pointer to the center of the sun, then this line will sweep out the surface of a cone, with the apex at the tip of the pointer. Consequently, this cone will cut the Earth's surface (imagined to be a plane) in a conic section. Because the plane of motion of the sun slowly changes through the year, the conic section also changes, and in principle a line will be needed for any particular date.

11 Albrecht Dürer, *Underweysung der Messung* (Nuremberg: n.p., 1525).

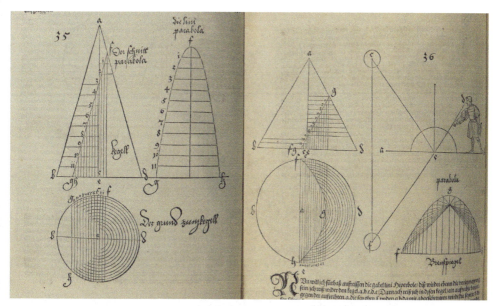

FIGURE 2.2 Albrecht Dürer's construction of a parabola as a conic section. To the right, he calls the parabola a "Brennspiegel," a burning mirror. Albrecht Dürer, *Underweysung der Messung* (Nuremberg: n.p., 1525), figs. 35, 36
© EHC H 202415

The subject therefore drew the attention of mathematicians, Jesuits among them, in a way it had not done for over fifteen hundred years. Two ancient treatises appeared in print just before 1550. Archimedes's work *Perí konoeidéon kai sphairoeidéon* (On conoids and spheroids) appeared nearly simultaneously in an edition by Niccolò Tartaglia (1499/1500–57) (following the Willem van Moerbeke [c.1215–c.1286] manuscript) in 1543 and by Thomas Gechauff (1488–1551) (following Johannes Müller von Königsberg's [known as Regiomontanus (1436–76)] reading of the text) in 1544.[12] The first edition of Apollonius's *Konika* was published in 1537 by Francesco Maurolico (Franciscus Maurolyci [1494–1575]), and a second in 1566 by Commandino. Unfortunately, these only contained the first four books—the study on normals and on maxima and minima was not known until 1661. As late as 1655, Claude Richard (1589–1664) would publish a version of Apollonius's *Konika* containing only the first four books.[13] In 1588, another important Greek work was published by

---

12 Jean Dhombres and Jacques Sakarovitch, *Desargues en son temps* (Paris: A. Blanchard, 1994), 59.

13 Claude Richard, S.J., *Conicorvm libri IV: Cvm commentariis* (Antwerp: Apud Hieronymum & Joannem Bapt. Verdussen, 1655–61).

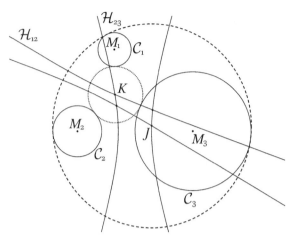

FIGURE 2.3 Adriaan van Roomen's solution to Apollonius's problem. Suppose the circles C_1, C_2, and C_3, in solid lines, centered at M_1, M_2, and M_3 and with radii r_1, r_2, and r_3 respectively are given. The hyperbolae that are the loci of the centers of circles tangent to C_1 and C_2 on the one hand and C_2 and C_3 on the other are drawn. The intersection point K is the center of a circle externally tangent to the three circles (dotted line). The intersection point J is the center of a circle tangent internally to the given circles (dashed line). The other intersection points do not yield solutions

Commandino, Pappus of Alexandria's (c.290–c.350) *Collection*, which would set the direction for geometrical problem-solving. The work would have a profound influence on early seventeenth-century mathematicians, including the Flemish Jesuits.[14] Interest in conic sections was revived, among others, by the Flemish mathematician Van Roomen, who, in response to a challenge by Viète, solved Apollonius's tangency problem with hyperbolae (fig. 2.3).[15]

Although the subject attracted the attention of mathematicians in the first half of the seventeenth century, 1630 marks a watershed in terms of publications on the subject. Thereafter, works by Claude Mydorge (1585–1647),

---

14   On Pappus's *Collection*, see Henk J. M. Bos, *Redefining Geometrical Exactness* (New York: Springer-Verlag, 2001), 37–57.

15   In Viète's challenge, a compass and straight edge construction was called for. Évelyne Barbin and Anne Boyé, *François Viète: Un mathématicien sous la Renaissance* (Paris: Vuibert, 2005), 10, 24–27; Bos, *Redefining Geometrical Exactness*, 110–12; Meskens and Tytgat, *Exploring Classical Greek Construction Problems*, 14–19.

Bonaventura Cavalieri (1598–1647), Frans van Schooten (1615–60), and John Wallis (1616–1703) saw the light of day. Among the Jesuits studying conic sections were Aguilón and San Vicente.

In 1639, Girard Desargues (1591–1661) published his *Brouillon-project d'une atteinte aux événements des rencontres du cone avec un plan* (Rough draft for an essay on the results of taking plane sections of a cone) in a mere fifty copies.[16] Desargues's style was very dense, and he introduced a terminology that was all but incomprehensible to his contemporaries. Yet his insights laid the basis for projective geometry. His work would only obtain fame after Noël Germinal Poudra's (1794–1894) edition of it was published in 1861. Blaise Pascal (1623–62) understood the possibilities of the new methods. In 1640, he published the theorem that we know by his name in his *Essai pour les coniques* (Treatise on conics): Pascal's mystic hexagram theorem. In the following century, Colin MacLaurin (1698–1746) would state the theorem as follows: "If three sides of a variable triangle rotate about three fixed points, and two vertices move along two fixed lines, then the third vertex will describe a conic section."

William Braikenridge (c.1700–62) would extend his theorem to a set of n instead of three points: "If n straight lines rotate about n fixed points and if $(n-1)$ of the $\dfrac{n(n-1)}{2}$ intersections move along a straight line, then the other $\dfrac{(n-1)(n-2)}{2}$ intersections describe a conic section." These theorems are equivalent to Pascal's mystic hexagram theorem.[17]

2 Squaring the Circle the Archimedean Way

Although San Vicente is usually associated with the squaring of the circle, this is not the main subject of this study. It is nevertheless important to sketch the history of the subject to be able to put San Vicente's efforts into context.

Archimedes was able to show that the area of a circle is equal to the area of a right-angled triangle with perpendiculars equal to the radius and to the circumference of the circle respectively.[18]

---

16   Judith V. Field and Jeremy J. Gray, *The Geometrical Work of Girard Desargues* (New York: Springer Verlag, 1987), 32.

17   See Fladt, *Geschichte und Theorie der Kegelschnitte*, 74–75; Coolidge, *History of the Conic Sections*, 90.

18   On Archimedes's method, see, e.g., Jean Christianidis, "Archimedes's Quadratures," in *The Genius of Archimedes: 23 Centuries of Influence on Mathematics, Science, and Engineering*,

If k is the length of the perimeter of a circle with diameter 1, then its area is $\frac{k}{4}$. Now the areas of circles are to each other as the squares of their diameters, so for a circle with radius r we find that $\frac{S_C}{k/4} = \frac{(2r)^2}{1^2}$ and $S_C = kr^2$.

Archimedes was unable to calculate the exact value of k, but he did propose good approximations for what we now call π. To approximate this number k, Archimedes used the fact that $S_{p_n} < S_C < S_{P_n}$, in which p_n is the regular n-gon inscribed in the circle and P_n the regular n-gon circumscribed about the circle. Archimedes was able to calculate the areas of the inscribed and circumscribed 96-gon and arrived at $3\frac{10}{71} < k < 3\frac{1}{7}$ or $3{,}1408 < k < 3{,}1429$.

It is clear that the larger n becomes, the closer the areas of the inscribed and circumscribed polygons will approximate the area of the circle.

Archimedes's method was revived in the late sixteenth century by Van Roomen in *Ideæ mathematica* (Mathematical ideas [1593]) and later in *In Archimedis circuli dimensionem expositio et analysis* (Exposition and analysis of the dimensions of the circle by Archimedes [1597]), in which he published a value for π to sixteen decimal places $\pi = 3{,}1415926535897931$. He used Archimedes's method for regular polygons having 2^{30} edges by consecutively doubling the number of vertices of an initial regular polygon. His result would soon be superseded by the work of his Dutch friend Ludolph van Ceulen (1540–1610).

In his 1596 work *Vanden Circkel* (About the circle), Van Ceulen published his ten-year research on the area of the circle and the areas of in- and circumscribed polygons. Van Ceulen used the same method as Van Roomen to calculate π. However, whereas Van Roomen's 1593 work puts forward "recipes" without formulae to calculate the edge, Van Ceulen's 1596 work explains the method in a symbolism that is still quite readable today.[19]

---

ed. Stephanos A. Paipetis and Marco Ceccarelli (Dordrecht: Springer Science & Business Media B. V., 2010), 58–67; Pier Daniele Napolitani, "Between Myth and Mathematics: The Vicissitudes of Archimedes and His Work," *Lettera matematica* 1, no. 3 (2013): 105–12; Lucio Russo, "Archimedes between Legend and Fact," *Lettera matematica* 1, no. 3 (2013): 91–95; Saito, "Archimedes and Double Contradiction Proof"; Vincent Jullien, "Archimedes and Indivisibles," in *Seventeenth-Century Indivisibles Revisited*, ed. Vincent Jullien (Basel: Birkhäuser, 2015), 451–57.

19 Steven Wepster, "In de Ban van de Cirkel," *Euclides* 85, no. 3 (2009): 98–100; Steven Wepster and Marjanne de Nijs, *Meester Ludolphs Koordenvierhoek* (Utrecht: Epsilon Uitgaven, 2010); Steven Wepster, "Hoe van Ceulen π insloot," *Pythagoras* 49, no. 3 (2010): 26–28; Wepster, "Ludolph van Ceulen (1540–1610): Meester der Rekenmeesters," *Pythagoras* 49, no. 3 (2010): 12–15; Meskens and Tytgat, *Exploring Classical Greek Construction Problems*, 95ff.

THE SEVENTEENTH CENTURY: THE DAWN OF A NEW ERA 49

FIGURE 2.4 Left: portrait of Ludolph van Ceulen; right: Van Ceulen's approximation of π, a lower limit in the upper half of the circle, an upper limit in the lower half of the circle. Ludolph van Ceulen, *De circulo et adscriptis liber* (Leiden: Colster, 1619), frontispiece
© EHC G 4867

It is easy to prove that $(2\sin\alpha)^2 = 2 - \sqrt{4-(2\sin 2\alpha)^2}$ and $(2\sin\alpha)^2 = 2 - \sqrt{2+\sqrt{4-(2\sin 4\alpha)^2}}$.

It was this formula that allowed Van Roomen and Van Ceulen to "quickly" calculate the edges of a 2^i-gon, in a circle with radius 1, knowing the edge of the n-gon. Putting the edge of a regular hexagon equal to x, then the edge of a regular dodecagon equals $s_{12} = \sqrt{2-\sqrt{4-x^2}}$.

Using the formula iteratively and beginning with a hexagon, the formula for a 192-gon (=2^5.6-gon) reads as $\sqrt{2-\sqrt{2+\sqrt{2+\sqrt{2+\sqrt{2+\sqrt{4-x^2}}}}}}$.

3 The Humble Beginnings of Infinitesimal Calculus

Infinitesimal methods had a humble beginning in the sixteenth century. In simple arithmetic books, one finds problems about the equilibrium of levers to which different shapes are attached.

For instance, in *Practique pour brievement apprendre a ciffrer* (Practical manual for learning ciphering easily [1565]), a work by Valentin Mennher (1521–70), a German arithmetic teacher working in Antwerp, problem 168 reads: "Given

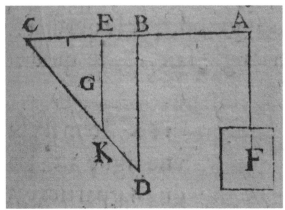

FIGURE 2.5 The use of a lever to determine the center of the mass of a triangle. Valentin Mennher, *Practique pour brievement apprendre a ciffrer, & tenir livre de comptes* [...] (Antwerp: Gillis I Coppens van Diest, 1565), problem 168
© MPM A3589

a lever *AC* with the pivot in the midpoint *B*. If a square weight is suspended in *A* and a right-angled triangle is suspended on *BC*, then the triangle and the square will have a ratio of 3:1."[20]

Mennher correctly goes on to point out that if the triangle is suspended in the center of mass, it can be rotated about this point at will without breaking the equilibrium of the lever. The problem refers to results that were already known to Archimedes, who proved that the center of mass of a triangle is on the median. To prove this, Archimedes inscribed parallelograms in the triangle. By construction, the centers of mass of these parallelograms are on the median. By augmenting the number of inscribed parallelograms, Archimedes can, using a *reductio ad absurdum*, determine that the center of mass of the triangle is indeed on the median.[21]

Similar reasoning can be found in Stevin's *Beghinselen der Weeghconst* (1586), although there is one fundamentally important difference[22] in that

---

20   Valentin Mennher, *Practique pour brievement apprendre a ciffrer, & tenir livre de comptes* (Anvers: Aegidius Diest [Coppens van Diest, Gillis I], 1565), fol. 9ʳ.

21   Archimedes, *De quadratura parabolae*, proposition 6, Archimedes, *Les œuvres complètes d'Archimède*, 382–83; see also Napolitani, "Between Myth and Mathematics," 109.

22   Simon Stevin, *De Beghinselen der Weeghconst* (Leiden: Christoffel Plantijn, by Françoys van Raphelinghen, 1586), 67–68; repeated in Simon Stevin, *De Beghinselen des Waterwichts* (Leiden: Christoffel Plantijn, by Françoys van Raphelinghen, 1586), part 4, 61–62, and Stevin, *Wisconstige Gedachtenissen: Inhoudende t'ghene Daer Hem in Gheoeffent Heeft* (Leiden: Jan Bouwensz., 1608), part 4, 57–58. For a detailed study, see Henri Bosmans, S.J.,

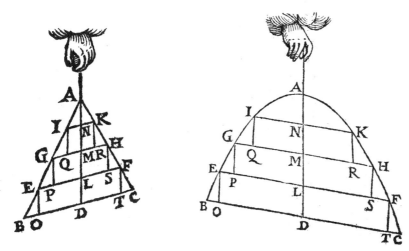

FIGURE 2.6 Simon Stevin's proofs for the determination of the centers of the mass of a triangle and a parabola. Simon Stevin, *De Beghinselen der Weeghconst* (Leiden: Christoffel Plantijn, by Françoys van Raphelinghen, 1586), 68, 78
© EHC G 76771

Stevin uses a direct method rather than a *reductio ad absurdum*. He states that the small triangles that remain at the sides of the inscribed parallelograms can become smaller than any given area as the number of parallelograms is allowed to increase. Therefore, if one takes *AD* to be the median, the weight of *ADC* would differ less from the weight of *ADB* than any given surface (fig. 2.6).

In the same fashion, he goes on to determine that the center of mass of a parabola is on the axis.[23] Again he argues that the difference between the areas of the parallelograms and the parabola can be made smaller than any given area. He also determined the centers of mass of prisms, cylinders, pyramids, and paraboloids. He applied the same methods to problems in hydrostatics.[24]

San Vicente was inspired by Stevin's argumentation in the *Theses mechanicae* he had defended at Leuven in 1624. However, with the exception of the Flemish Jesuits, the influence of Stevin's insights on European mathematics was minimal. This is most probably because his books were written in his vernacular Dutch. By the time a Latin edition appeared in 1608, and certainly when Albert Girard's (1595–1632) partial translation was published in 1634, his methods were known to mathematicians at large.

"Sur quelques exemples de la méthode des limites chez Simon Stevin," *Annales de La Société Scientifique de Bruxelles* 37 (1913): 171–99; Bosmans, "Le calcul infinitésimal chez Simon Stevin," *Mathesis* 37 (1923): 12–18, 55–62, 105–9.
23   Stevin, *De Beghinselen der Weeghconst*, 78–79.
24   Stevin, *De Beghinselen des Waterwichts*, passim.

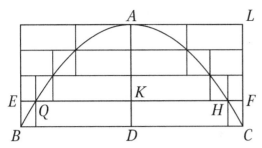

FIGURE 2.7 The determination of the center of mass of a parabolic segment using inscribed and circumscribed rectangles. After Luca Valerio, *De centro gravitatis solidorum libri tres* (Rome: B[artolomeo] Bonfadino, 1604)

Luca Valerio (1553–1618) published two books on the subject, *De centro gravitatis solidorum* (On the center of mass of solids [1604]) and *Quadraturae parabolae* (On the quadrature of the parabola [1606]). Although he wanted to determine the center of mass of a paraboloid, his reasoning was made with plane figures. He inscribes and circumscribes parallelograms about the parabola and notes that the sum of all the areas of the differences of areas of circumscribed and inscribed parallelograms equals the area of the base parallelogram. But the area of this base parallelogram can be made smaller than any given area as the number of parallelograms increases.

The revolution of these parallelograms about the axis of the figure results in cylinders. Archimedes, in comparing the parallelograms, concluded that adjacent inscribed and circumscribed parallelograms are equal, with the exception of the parallelogram at the base, which has no counterpart. Therefore, this parallelogram is equal to the sum of the excesses of the circumscribed parallelograms over the inscribed parallelograms. For Valerio, it is the top parallelogram that has no counterpart. For him, the sum of the rings of the excesses augmented with the top cylinder equals the base cylinder. Because this base cylinder can be made arbitrarily small, so can the difference between circumscribed and inscribed cylinders.

In his second book, Valerio states theorems that can be rendered as if E and F are variables and if $A - E$ and $B - F$ are simultaneously smaller than an arbitrarily small quantity, then if $\frac{E}{F} = \frac{C}{D}$ then also $\frac{A}{B} = \frac{C}{D}$. In modern terms, it could be rendered as a limit, but Valerio, like many of his contemporaries, expressed these limiting arguments as proportions.

Valerio used this theorem to determine the volume of a paraboloid as half the cylinder and one and a half times the cone, both having the same base and the same height as the paraboloid. Being the first to use infinitesimal arguments, the description of his procedure is in long and tedious sentences.

As Stevin had done twenty years earlier, Valerio uses primitive infinitesimal methods to prove that the center of gravity of a parabolic segment is on the diameter through the vertex. It is important to stress, however, that he only uses known geometrical methods.

4 Infinitesimals: The Keplerian Revolution

Two years after the death of his wife Barbara (c.1572–1611), Kepler married Susanna Reutinger (c.1589–1636) on October 30, 1613.[25] After the wedding, the wine cellars were stocked and, to determine the exact price, a gauger came in to determine their exact contents.[26]

Kepler wondered whether this gauging method, in which a cubic gauge was used, could result in accurate measurements. Kepler's amazement drove him to calculations that would ultimately lead to the first books in which infinitesimal methods were used. They are also the first books in which an attempt was made to found the wine-gauging methods on a more solid geometrical basis. His book *Nova stereometria doliorum vinariorum* (New stereometry of wine barrels) was available at the Frankfurt Herbstmesse (autumn fair) of 1615, while its German counterpart *Messekunst Archimedis* (The Archimedean art of measurement) was available at the following Frühlungsmesse (spring book fair) in 1616.

The first part of *Stereometria* deals with the Archimedean theorems on the surface area and the volume of solids. Kepler at first proves, in the classical way, that the area of a circle equals the area of a perpendicular triangle with the radius as base and the circumference as height. In the second theorem, however, he gives an alternative proof. As was also the case with many of his contemporaries, Kepler seems to have found the classical proof cumbersome

---

25 Johannes Kepler, *Briefe 1612–1620*, Gesammelte Werke/Kepler, Johannes 17 (Munich: Beck, 1955), letter 669.

26 See Johannes Kepler, *Außzug Außder Uralten MesseKunst Archimedis* [...] (Linz: Hansen Blancken, 1616), on wine-gauging, see also Menso Folkerts, "Die Entwicklung und Bedeutung der Visierkunst als Beispiel der praktischen Mathematik der frühen Neuzeit," *Humanismus und Technik* 18, no. 1 (1974): 1–41, and Meskens, *Practical Mathematics*, 97–112.

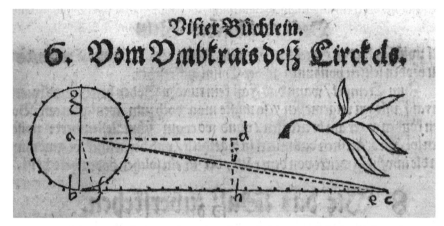

FIGURE 2.8 Johannes Kepler's transformation of a circle into a parallelogram. Johannes Kepler, *Außzug Außder Uralten MesseKunst Archimedis* [...] (Linz: Hansen Blancken, 1616), 5
© EHC G 86948

and troublesome. His second proof rests on him dividing the circumference in infinitely many arcs, each forming the arc of a circle sector, as if he were dividing a circular cake into infinitely many equal parts. Each of these sectors is considered to be equal to an equilateral triangle with the radius as height. By unwinding the circumference onto a straight line infinitely, many triangles stand on this line.[27]

Consider a circle, with radius r and perimeter P, which is cut up, like a cake, into n sectors. Take the sectors from the circle and arrange them as shown in figure 2.9.

Now consider a very small interior angle for each sector (i.e., let n become very large). Consider one such circular sector OAB and also $\triangle O'A'B'$. Kepler states that the height h of $\triangle O'A'B'$ is equal to the radius r of the circular sector OAB. Similarly, the base of $\triangle O'A'B'$ is nearly equal to $\frac{P}{n}$ or $|A'B'| = \Delta P_i \approx \frac{P}{n}$. Therefore, the area of $\triangle O'C'B'$, which is made up of the triangles, is nearly equal to the sum of the areas of the sectors and is

$$S = \sum_{i=1}^{n} \frac{1}{2} h \Delta P_i = \sum_{i=1}^{n} \frac{1}{2} r \Delta P_i = \frac{1}{2} r \sum_{i=1}^{n} \Delta P_i = \frac{1}{2} r P \cdot \left(= \frac{2\pi r}{2} \cdot r = \pi r^2 \right).$$

While Archimedes's proof is cumbersome but based on a solid geometric and logical basis, Kepler's proof is quite easy but not rigorous even by the

---

27 Kepler had already used similar methods in *Astronomia nova* (1609). Patricia Radelet-de Grave, "Kepler, Cavalieri, Guldin: Polemics with the Departed," in Jullien, *Seventeenth-Century Indivisibles Revisited*, 57–86, here 58–59.

THE SEVENTEENTH CENTURY: THE DAWN OF A NEW ERA

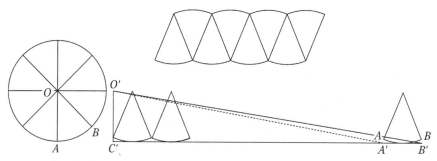

FIGURE 2.9 Johannes Kepler's transformation of a circle into a parallelogram. After Johannes Kepler, *Außzug Außder Uralten MesseKunst Archimedis* [...] (Linz: Hansen Blancken, 1616)

standards of his time. Kepler does not build his proof on a finite number of sectors but boldly states that there are an infinite number. In that case, he argues, the difference between a sector and a triangle is negligible. He considers this kind of reasoning to be legitimate.

In the second part of the book, Kepler treats solids of revolution generated by conic sections: circles, ellipses, parabolae, and hyperbolae. For circles, he distinguishes five possible solids (fig. 2.10). He further identifies another eighty-seven different kinds of solids when conic sections are rotated around axes in different positions.

In theorem 18, Kepler proves that the volume of a ring generated by a circle or an ellipse has a volume equal to the volume of a cylinder with the same circle or ellipse as base and the perimeter of the circle the center has traced out as height.

Kepler's argument, which resembles his reasoning on the area of a circle, runs as follows. Suppose the ring is cut by planes through the axis of revolution. These planes cut the ring perpendicularly and divide it into many small sections. Each section is smaller toward the center of revolution and wider away from the center. Its thickness in the middle is the average of the thicknesses at both extremities. Because all these sections are congruent to one another, they can be stacked into a kind of cylinder, putting the wider part of the ring on the smaller part of the underlying ring. Because these sections are infinitely small, the curvature can be neglected. The base of the cylinder therefore is the generating circle or ellipse. The height of the cylinder is the sum of the thicknesses at the middle, which obviously equals the circumference of the circle traced out by the center of the generating circle or ellipse.

The result also holds in the case of a closed ring (*annulus strictus*). Here, the thickness is zero at the inner end. On the other hand, the thickness at the outer

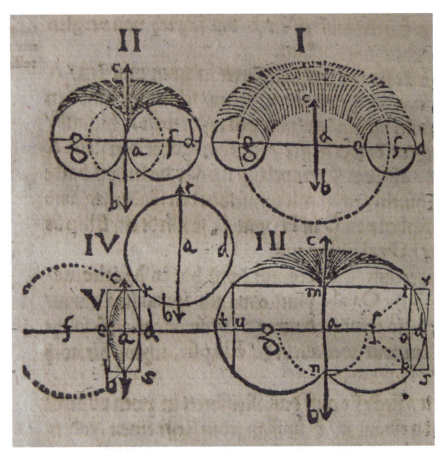

FIGURE 2.10   Solids of revolution with a circle as generating curve. 1 (I): if the axis of revolution does not intersect the circle, the resulting solid is an anchor ring or torus (*annulus*); 2 (II): if the axis is a tangent to the circle, one finds a closed ring (*annulus strictus*); 3 (III): if the axis intersects the circle, but does not contain the center, the largest segment is revolved around the axis and the resulting figure is an apple (*malum*); 4 (IV): if the axis of revolution passes through the center of the circle, the resulting solid is a sphere (*globus*); 5 (V): if the axis intersects the circle, but does not contain the center, the smallest segment is revolved around the axis and the resulting figure is a lemon (*malum citrium*). Johannes Kepler, *Außzug Außder Uralten MesseKunst Archimedis* […] (Linz: Hansen Blancken, 1616), 27
© EHC G 86948

end is twice that of the thickness in the middle. Therefore, the previous procedure still stands. These results are used to calculate the volume of an apple, one of the solids generated by revolving a circle segment around an axis (fig. 2.10, 3).

5   Cavalieri's Indivisibles

In 1635, Bonaventura Cavalieri published *Geometria indivisibilibus continuorum nova quadam ratione promota* (Geometry, developed by a new method using the indivisibles of the continua). Cavalieri was a priest of the Apostolic Clerics of Saint Jerome, also called Jesuats, who in 1629 obtained a professorship at the University of Bologna, which he held until his death. *Geometria indivisibilibus* would give Cavalieri fame as one of the fathers of infinitesimal calculus, although it has also been dubbed the most unreadable of all mathematical books, a fame the book shares with San Vicente's *Problema Austriacum* (see chapter 4).

In *Geometria indivisibilibus*, Cavalieri elaborates on what he calls *indivisibles*,[28] a concept he first entertained in the early 1620s. By November 1627, a first draft of *Geometria* had been written, but it would take another eight years before it was finally published.[29]

The fundamental idea behind Cavalieri's method is that if a line is moved parallel to itself, the indivisibles of a plane figure can be characterized by the intersection of the moving line with the figure. In a similar way, a plane moving parallel to itself will characterize the indivisibles in a solid. Cavalieri's method has drawn a great deal of criticism, not least because he encounters all the conceptual problems related to the infinite. One of his most articulate adversaries was Guldin.[30]

Cavalieri's method rests on some unproven implicit assumptions. Nevertheless, he would apply the method with great success in many problems, without being blind to its shortcomings.

---

28   On Cavalieri's method, see especially Kirsti Andersen, "Cavalieri's Method of Indivisibles," *Archive for History of Exact Sciences* 31, no. 4 (December 1985): 291–367; Andersen, Enrico Giusti, and Vincent Jullien, "Cavalieri's Indivisibles," in Jullien, *Seventeenth-Century Indivisibles Revisited*, 31–55.

29   Andersen, Giusti, and Jullien, "Cavalieri's Indivisibles," 33.

30   See especially Radelet-de Grave, "Kepler, Cavalieri, Guldin," on Guldin's main points of criticism.

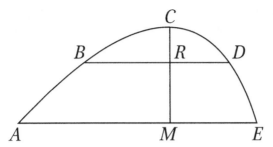

FIGURE 2.11   Using indivisibles, the areas under the curve to the left and to the right of *CM* are to one another as all the lines *BR* are to all the lines *RD*

One of these assumptions is the additive property: if a figure F can be decomposed into two figures F_1 and F_2, then "all the lines of F equal the sum of all the lines of F_1 and F_2" (fig. 2.11). Furthermore, if the two figures, F_1 and F_2, have the same altitude and if they have their bases on the same line *AME*, and if the equality $\frac{|BR|}{|RD|} = \frac{|AM|}{|ME|}$ holds for each line *BRD* parallel to *AE*, then $\frac{F_1}{F_2} = \frac{\text{all lines of } F_1}{\text{all lines of } F_2} = \frac{|AM|}{|ME|}$. A corollary to Cavalieri's principle is that if two figures are intersected by a line moving parallel to itself and every line intersects each figure in line segments of equal length, then the two figures have an equal area.

With this principle, Cavalieri is able to calculate the volumes of many of the solids introduced by Kepler in *Stereometria*: solids generated by rotating a conic section about an axis.

6   The Jesuits and Indivisibles

In Alexander's book *Infinitesimal*, the Jesuits take center stage as the culprits in trying to suppress the notion of indivisibles, at least in Italy.[31] David Sherry has proposed an alternative hypothesis explaining why the Jesuits were opposed to indivisibles. It was not because the very notion of indivisible would destroy the existing world order, but because it touches upon one of the central themes of

---

31   Amir Alexander, *Infinitesimal: How a Dangerous Mathematical Theory Shaped the Modern World* (London: Oneworld, 2015).

the Catholic faith: the Eucharist.[32] Both refer to a series of decrees issued by the Jesuit censors in which the following hypotheses were banned:[33]

> 1606: The continuum is composed from a finite number of indivisibles.[34]
> 1608: Christ exists in the Eucharist in a finitely multiplied manner, that is to say, as many times as there are indivisibles of the quantity of the sacramental species, out of which indivisibles that quantity is composed.
> 1615: The continuum is composed of a finite number of indivisibles [...]; nor should this proposition be allowed even if the number of indivisibles is asserted to be infinite.
> 1632: The permanent continuum consists entirely of physical indivisibles or atomic corpuscles, having mathematical parts, signifiable in themselves even if the said corpuscles can actually be distinguished from one another.

The 1606 and 1608 rulings were in response to questions sent by a house in the Belgian province. Unfortunately, we do not know which one, although Leuven and Douai spring to mind.

Two questions can be raised in connection with these Belgian questions: Were they of a natural philosophical or of a mathematical nature? Were these rulings on analytical concepts at all? After all, all these rulings predate Cavalieri's publication, and with the exception of the last one, they predate Cavalieri first mentioning mathematical indivisibles. No mathematical publication before Cavalieri's book uses indivisibles. Even if books, like Valerio's, can be termed a precursor to infinitesimal calculus, the methods used were variations of known geometrical methods.

With regard to these rulings, it seems best to use Ockham's razor[35] and (as Alexander did in part) to view the Jesuits, or at least their philosophers, as

---

32 See David Sherry, "The Jesuits and the Method of Indivisibles," *Foundations of Science* 23, no. 2 (2018): 367–92, and the response to it in Amir Alexander, "On Indivisibles and Infinitesimals: A Response to David Sherry, 'The Jesuits and the Method of Indivisibles,'" *Foundations of Science* 23, no. 2 (2018): 393–98.

33 See Sherry, "Jesuits and the Method of Indivisibles."

34 ARSI, Fondo Gesuitico 656 A–I, fol. 155ᵛ.

35 Ockham's razor (or Occam's razor) is a problem-solving principle attributed to the English Franciscan friar William of Ockham (c.1287–1347). According to Ockham, "it is futile to do with more things that which can be done with fewer," but it is more often cited as "entities are not to be multiplied without necessity." It is usually understood as "when there are competing hypotheses to solve a problem, one should select the solution with the fewest assumptions." As such, it is a perfect antidote to conspiracy theories. On Ockham, see, e.g., William J. Courtenay, *Ockham and Ockhamism*, Studien und Texte zur

staunch Aristotelians. In Alexander's story, the Jesuits try to *suppress* the very ideas at the basis of what was to become infinitesimal calculus. The simpler view is that, as Aristotelians, they *opposed* the notion of indivisible. In fact, in San Vicente the Jesuits found a mathematician who was to propose a theory that superseded Cavalieri's.[36]

In book 6 of *Physics*, Aristotle writes that no continuum can be composed of indivisibles, although he does not refute the notion of indivisible.[37] Philosophers in the Arab world, however, did entertain the idea of the continuum being composed of either a finite or infinite number of indivisibles (or atoms) that can be represented by geometric points. A refutation of the finitist viewpoint is al-Ghazâli's (Abū Ḥāmid Muḥammad ibn Muḥammad aṭ-Ṭūsīy al-Ġazālī [c.1058–1111]) wheel, a variation on Aristotle's wheel. Aristotle's wheel paradox is about two concentric circular wheels. If the wheels roll without slipping for a full revolution, then the paths traced out by the bottoms of the wheels are straight lines. The lengths of these straight lines seem to equal the wheels' circumferences. This implies that both wheels must have the same

---

Geistesgeschichte des Mittelalters (Leiden: Brill, 2008); Heinz-Helmut Möllmann, *Über Beweise und Beweisarten bei Wilhelm Ockham* (Amsterdam: John Benjamins, 2013); Elliott Sober, *Ockham's Razors: A User's Manual* (Cambridge: Cambridge University Press, 2015).

36   With regard to astronomy and especially Galileo's points of view, Roberto Buonanno (*The Stars of Galileo Galilei and the Universal Knowledge of Athanasius Kircher* [Cham: Springer International, 2014], 174), comes to a similar conclusion: "Clavius, Grassi and Scheiner did not play a role as enemies of modern science, but rather as scholars proposing different explanations of the phenomena you observe, a role foreseen by modern science. This role was actually indispensable along the complicated path of knowledge. The Jesuits' Collegio Romano was a place where the cultural heritage was preserved, where Galileo had found attentive colleagues and harsh critics who, sometimes restrained by their faith, reached more advanced conclusions than Galileo himself."

37   "Nothing that is continuous can be composed of indivisibles: e.g. a line cannot be composed of points, the line being continuous and the point indivisible […]; for if there could be any such thing it is clear that it must be either indivisible or divisible, and if it is divisible, it must be divisible either into indivisibles or into divisibles that are infinitely divisible, in which case it is continuous. Moreover, it is plain that everything continuous is divisible into divisibles that are infinitely divisible: for if it were divisible into indivisibles, we should have an indivisible in contact with an indivisible, since the extremities of things that are continuous with one another are one and are in contact. […] And if length and motion are thus indivisible, it is neither more nor less necessary that time also be similarly indivisible, that is to say be composed of indivisible moments: for if the whole distance is divisible and an equal velocity will cause a thing to pass through less of it in less time, the time must also be divisible, and conversely, if the time in which a thing is carried over the section A is divisible, this section A must also be divisible." See Aristotle, *Physics, Book VI*, ed. Robert P. Hardie, trans. Russell K. Gaye (Cambridge, MA: Massachusetts Institute of Technology, n.d.); http://classics.mit.edu/Aristotle/physics.6.vi.html (accessed November 12, 2019).

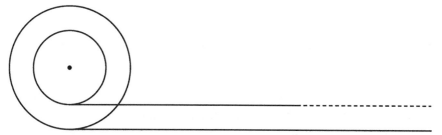

FIGURE 2.12 Aristotle's wheel, also called al-Ghazâli's wheel

circumference, which is a contradiction. The solution to the problem is that it is impossible for both wheels to roll without slipping. That there is a one-to-one correspondence between the points on the circles does not imply that their circumferences are equal. Treatises on these atomist philosophies and on their refutations circulated and were known throughout the High Middle Ages (1000–1300) and the early Renaissance (1300–1500).[38] We can therefore safely conclude that the rulings of the censors were mere reiterations of Aristotelian philosophy.

The analytical interpretation of the rulings of the censors would be made by Guldin, who, in discussing Kepler's and Cavalieri's work, completely refutes the idea that the continuum is composed of indivisibles.[39] Nevertheless, mathematicians and Jesuits alike would struggle with these issues for decades to come. Mathematicians on the one hand did not come to terms with the continuum and indivisibles, some even denying their existence, yet on the other hand they implicitly used the fact that the continuum does consist of indivisibles in a number of constructions. The paradoxes of the infinite would haunt mathematicians for the coming centuries.

It is equally true that there was a tension between philosophers and mathematicians within the Society of Jesus. Whereas the philosophers adhered strictly to Aristotelian natural philosophy, the mathematicians openly discussed the mechanistic worldview, even if they did not embrace it.

As we shall see, San Vicente did not encounter any problems when he proposed that a geometric series reaches its sum. Nor did Tacquet reject indivisibles outright; on the contrary, he used them but was distrustful of them and demanded that they be used as a tool. The results obtained by using them should then be rigorously proven using classical geometrical methods.

---

38   On this subject, see Jean Celeyrette, "From Aristotle to the Classical Age: The Debates around Indivisibilism," in Jullien, *Seventeenth-Century Indivisibles Revisited*, 19–30.

39   On Guldin's criticism, see Radelet-de Grave, "Kepler, Cavalieri, Guldin," especially 80–81.

CHAPTER 3

Francisco de Aguilón and Mathematical Optics

Francisco de Aguilón was born in Brussels in 1566,[1] son of Pedro (*fl.* 1570), a secretary to Philip II in the Low Countries, and Anna Pels.[2] In 1576, he was tonsured,[3] indicating that his father wanted him to join a religious order. In the autumn of 1576, the Aguilón family left the Low Countries. Aguilón studied at the Colleges of Paris and Douai. Before joining the Society, he had studied six years of humanities and one year of philosophy.[4] On September 15, 1586, he joined the Jesuit order in Tournai,[5] before being transferred to Kortrijk for a couple of months a year later.[6] In August 1587, he was back in Douai. From

---

1 Kadoc ABSE 34/1, 25; 31, 34; KBR MS 1016, 650. For a detailed biography of Aguilón, see Ziggelaar, *François de Aguilón, S.J.* New biographical information about Francisco's youth has been unearthed by Ruth Sargent Noyes, "'Per modum compendii a Leonardo Damerio Leodiensi in lucem editum': Odo van Maelcote and Leonard Damery's *Astrolabium aequinoctiale* and the Catholic Reformation Converting (Im)print," in *Winning Back with Books and Print: At the Heart of the Catholic Reformation in the Low Countries (16th–17th Centuries)*, ed. Renaud Adam, Rosa De Marco, and Malcolm Walsby (Leiden: Brill, forthcoming 2021).
   According to Ziggelaar, *François de Aguilón, S.J.*, 29, who refers to ARSI, Fl. Belg. 9, fols. 48ᵛ and 87ʳ, he was born in January 1567. In his autograph *curriculum vitae*, Aguilón himself mentions 1566 (KBR MS 1016, 650). The *Catalogi personarum* indicate January 1566 as his date of birth (e.g. in ARAA T14/034, 29). However, it is important to note that up until 1576 the so-called Easter style was in use in the Low Countries. In the Easter style, the New Year begins at Easter, and therefore January 1566 Easter style is January 1567 new style. Yet from the correspondence between his father Pedro and Cardinal Antoine Perrenot de Granvelle (1517–86), in particular a letter dated December 15, 1566, Ruth Sargent Noyes, "Odo van Maelcote," deduces that he was born in early December 1566. In these letters, she also found that Francisco's mother Anna Pels died in April 1568. Francisco seems to have had at least one brother, called Pedro (ARAA T14/015.01 186).
2 Kadoc ABSE 34/1, 25; ARAA T14/015.01 186; testament Juan de Aguilón; KBR MS 1016, 650. Juan de Cuellar (*fl.* 1550), the host of Ignatius on his visits to Antwerp, was married to Clara Pels. There is no firm evidence to support the claim that the two were sisters or relatives, but the common name does suggest so. Moreover, Cuellar is mentioned in Juan de Aguilón's will.
3 KBR MS 1016, 650; Kadoc ABSE 34/1, 25; ARAA T14/034 349, 99. In his autograph *curriculum vitae*, Aguilón writes that he was tonsured by Cardinal Granvelle. Noyes, "Odo van Maelcote," contests this view, since Granvelle was in Italy at that time. She suggests the tonsure was conferred by proxy.
4 ARAA T14/034 29.
5 Kadoc ABSE 34/1, 25; 12427, 34.
6 His mother did not live to see his entry into the Society. KBR MS 1016, 650. Ziggelaar, *François de Aguilón, S.J.*, 33.

August 1589 onward, he taught syntaxis and philosophy at the college,[7] while at the same time taking courses in theology. During his years in Douai, he saw the erection of the new church (which was built between 1583 and 1591). He was then sent to Spain, where he studied Scholastic theology at the University of Salamanca[8] before receiving his minor orders.[9] Apparently, his health deteriorated thereafter.[10]

In 1596, Aguilón was appointed vice-deacon at Ghent and later that year was ordained priest and made dean at Ypres.[11] In 1597, he was appointed as confessor for the Spaniards and Italians residing at Antwerp.[12] He taught for two years at the college. After 1600, he was prefect of health. Philippe Alegambe (1592–1652) praised him for having assisted his brethren, at the risk of his own health, during a plague epidemic that raged in Antwerp.[13] From 1601 to 1608, he was procurator, with responsibility for the college's finances.[14] On February 10, 1602, he took his four solemn vows.[15]

In 1611, when Scribani participated in a congregation of procurators, Aguilón became vice-rector.[16] When in 1612 the Belgian province was divided, Scribani became provincial of the Flemish province, remaining rector until 1614. He resigned in 1614, putting the responsibility on the frail shoulders of Aguilón.[17] In the winter of 1615–16, Aguilón was assistant provincial when Scribani was called to Rome to elect a new superior general after Acquaviva's death.[18]

Aguilón was involved, at least indirectly, with the construction of the city's fortifications. In a report to the city council, Coignet refers to plans of the *vesten* (moats), which were inspected by Aguilón and were revised as a result.[19] Aguilón's advice seems to have been sought because Jesuit premises were involved.

---

7   KBR MS 1016, 650. According to Droeshout, "Histoire de la Compagnie de Jésus à Anvers," 3:172–73, this cannot have been for longer than three years. According to the *Catalogus personarum* of 1615, he had taught four years of philosophy and one year of grammar over the whole of his career.
8   KBR MS 1016, 650. Droeshout, "Histoire de la Compagnie de Jésus à Anvers," 172–73.
9   KBR MS 1016, 650.
10 Ziggelaar, *François de Aguilón, S.J.*, 36–37.
11 KBR MS 1016, 650. Droeshout, "Histoire de la Compagnie de Jésus à Anvers," 3:172–73.
12 Droeshout, "Histoire de la Compagnie de Jésus à Anvers," 2:295.
13 Ribadeneyra and Alegambe, *Bibliotheca scriptorum Societatis Jesu*, 112.
14 ARAA T14/034 29. Droeshout, "Histoire de la Compagnie de Jésus à Anvers," 3:172–73; Ziggelaar, *François de Aguilón, S.J.*, 39.
15 Kadoc ABSE 47. Droeshout, "Histoire de la Compagnie de Jésus à Anvers," 3:172–73; Ziggelaar, *François de Aguilón, S.J.*, 39.
16 Ziggelaar, *François de Aguilón, S.J.*
17 ARAA T14/034 29.
18 ARAA T14/034 29. Ziggelaar, *François de Aguilón, S.J.*, 41.
19 FA Pk712, fols. 48^v–49^v.

1 *Opticorum libri sex*

In 1613, the widow and the sons of Moretus published Aguilón's magnum opus: *Opticorum libri sex* (The six books of optics).[20] The 684-page book is richly illustrated and has a frontispiece and illustrations designed by Rubens himself.[21] The etchings were done in the workshop of Theodoor Galle (1571–1633).[22] Shortly after the plates were delivered, Balthasar (I) Moretus (1574–1641) had 1,263 prints of it made[23] and 1,100 copies of the book printed.[24] Aguilón received two hundred guilders from the city council after the publication of his book.[25] Despite its length, the book does not treat the whole of optics, but just those phenomena that are related to direct rays of light: no reflection, no refraction.

Optics, which was part of the course of mathematics for the Jesuits, is described with mathematical precision in this book, which accepts outright the results of experiments done with specially designed apparatus. These

FIGURE 3.1
Frontispiece of Francisco de Aguilón, *Opticorum libri sex philosophis iuxtá ac mathematicis utiles* (Antwerp: Ex officina Plantiniana, apud viduam et filios Jo. Moreti, 1613)
© EHC G 5050

---

20 For a detailed description, see Ziggelaar, *François de Aguilón, S.J.* That the book was published by the house of Moretus comes as no surprise; in 1593, the provincial elders had given it a monopoly on publishing Jesuit books.
21 MPM Arch 128, 178, right.
22 MPM Arch 123, fol. 21^r–22^v. The original plate of the etching is kept in the museum under the number KP81D. See also Max Rooses, *Le Musée Plantin Moretus* (Antwerp: Zazzarini, 1913), 281.
23 MPM Arch 123, fol. 23^r.
24 MPM M39, fol. 23^v; M324, fol. 55^v.
25 Floris Prims, "Letterkundigen, Geleerden en Kunstenaars," *Verslagen en Mededeelingen van de Koninklijke Vlaamsche Academie voor Taal- En Letterkunde* (March 1931): 171–221, here 189, no. 118, referring to city accounts of 1613, fol. 282.

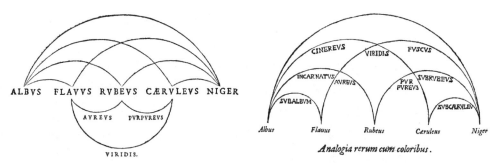

FIGURE 3.2 Left: color scheme in Francisco de Aguilón, *Opticorum libri sex philosophis iuxtá ac mathematicis utiles* (Antwerp: Ex officina Plantiniana, apud viduam et filios Jo. Moreti, 1613), 40
© EHC G 5050
Right: Aguilon's color scheme in Athanasius Kircher, *Ars magna lucis et umbrae in decem libros digesta* (Rome: sumptibus Hermanni Scheus, 1646), 49
© EHC G 5052

experiments had to be repeated several times to be sure that no mistakes were made during the observations.[26] Experience and experiment are necessary to formulate the postulates on which the theory is built.

During the Renaissance, optics was not only the science of light rays but more generally the science of vision. The first book of *Opticorum* duly treats this topic by discussing the anatomy of the eye. Aguilón also proposes a theory of colors, which he bases on primary colors (yellow, blue, and red) to which are added "whiteness" and "blackness." This suggests that he was influenced by painters, Rubens among others,[27] which he himself admits: "Nobody knows these things [i.e., mixing colors] so precisely as painters."[28]

In the second book, Aguilón defines the horopter, the line that passes through the fixation point and parallel to the line that connects the centers of the eyes. The horopter plane is defined as the plane through the horopter and perpendicular to the plane of the optical axis. Aguilón states that the horopter plane is like a window that receives images of objects as if they were projected on it (again, notice the analogy with painting). It is this concept that Aguilón uses in books 3 and 4 to explain some of the phenomena that trouble our

---

26 On Aguilón's opinion of experiments, see Dear, "Jesuit Mathematical Science," 147–48, 161ff.

27 On this subject, see, among others, Charles Parkhurst, "Aguilonius's *Optics* and Rubens's Colour," *Nederlands Kunsthistorisch Jaarboek* 12 (1961): 35–49; Michael Jaffe, "Rubens and Optics: Some Fresh Evidence," *Journal of the Warburg and Courtauld Institutes* 34 (1971): 362–66.

28 Francisco de Aguilón, *Opticorum libri sex philosophis iuxtá ac mathematicis utiles* (Antwerp: Ex officina Plantiniana, apud viduam et filios Jo. Moreti, 1613), 41.

FIGURE 3.3 Frontispiece of book 5 in Francisco de Aguilón, *Opticorum libri sex philosophis iuxtá ac mathematicis utiles* (Antwerp: Ex officina Plantiniana, apud viduam et filios Jo. Moreti, 1613). The frontispieces were designed by Peter Paul Rubens and executed by Theodore Galle. The drawing depicts an experiment on the attenuation of light. Although the figure seems to suggest an inverse linear relation between the light intensity and the distance, Aguilón knew that this was not the case. The novel arrangement of the experiment with a primitive photometer was designed to establish the law of attenuation of light.
© EHC G 5050

eyesight and mislead our senses. An important, but physically wrong, result was the theorem that states that to see two lines as two parallel straight lines, one of the lines has to be a hyperbola.[29] Aguilón also describes, using a circular horopter, and using Euclid 4.5, that we can see things at different distances under the same angle.[30]

In book 5, Aguilón shows that the attenuation of light is not proportional to the distance. The last part of the book is devoted to the *Camera obscura*.

---

29 Aguilón's mistake lies in his suppositions, not in his reasoning. He assumes that we in fact see a horizontal plane as a horizontal plane. In reality, we see a slightly curved surface (which may be approximated by a slightly inclined plane).
   Ad Meskens, *Wiskunde tussen Renaissance en Barok, aspecten van wiskunde-beoefening te Antwerpen 1550–1620* (Antwerp: Stadsbibliotheek Antwerpen, 1994), 101–2.

30 Thorne Shipley, Kenneth Neil Ogle, and Samuel C. Rawlings, "The Nonius Horopter: I. History and Theory," *Vision Research* 10, no. 11 (1970): 1225–62, here 1228.

FIGURE 3.4   Frontispiece of book 2 in Francisco de Aguilón, *Opticorum libri sex philosophis iuxtá ac mathematicis utiles* (Antwerp: Ex officina Plantiniana, apud viduam et filios Jo. Moreti, 1613). In this book, Aguilón defines the horopter plane to explain stereoscopic vision and uses this concept to investigate how we estimate distances.
© EHC G 5050

Aguilón reaches some important results and shows that the image of the object will be inverted and smaller than the original object if the distance of the object is larger than the distance from the hole to the screen.

In the last part of the book, Aguilón deals with projections and perspective. He treats various projection methods (parallel, central, and shadow projections), but he also treats their applications to sundials and astrolabes. Again, the way he treats this subject shows he was in close contact with painters.

It is difficult to assess the influence of Aguilón's book. It was used in various Jesuit colleges as a handbook and influenced many Jesuits. His most important contributions to science seem to have been his color theory and his theory of vision. The first part of the book is especially important in the history of psychology, because it introduces new concepts such as the circular formulation of the horopter and the non-linear relationship between physical and perceptual space.[31] His ideas survived up to the nineteenth century, mainly in the works of English physio-philosophers who influenced later psychologists.

---

31   Shipley, Ogle, and Rawlings, "Nonius Horopter: I," 1228–29.

2 Aguilón's *Catoptrica* Manuscript[32]

Among the manuscripts of San Vicente kept at the Royal Library in Brussels is a hundred-page mathematical treatise by Aguilón.[33] Although the manuscript does not contain any reference to optics, it becomes clear that the lemmas Aguilón proposes are in fact the basis for theorems in geometrical optics.

In the first part of this manuscript, Aguilón deals with trisections of an angle. Whether or not they were to be part of one of his next books on optics is not clear. Here, we find the solutions of Hippias of Elis (*fl.* late fifth century BCE), Archimedes, Nicomedes (c.280 BCE–c.210 BCE), Pappus (c.290–c.350 CE), and Alhazen (Abū 'Alī al-Ḥasan ibn al-Ḥasan ibn al-Haytham [c.965–c.1040]) for the trisection of an angle. These solutions all have in common that they make use of verging methods or *neusis*.[34]

Aguilón then ventures into uncharted territory. This time, two line segments need to be of an equal length, which is unknown at the start (MS 1, fol. 236^r).

Let O be the center of a circle C (fig. 3.5). Let F be the point from which a straight line a is drawn. This line intersects the circle at B and the line AO at E. Then it is possible to rotate the line a into such a position that $|BE| = |EO|$.

Aguilón uses this result to trisect an angle (fig. 3.6). Suppose an angle $\angle KAC$ has to be trisected. Draw a circle with diameter $[AC]$ that intersects AK at F. Now determine the intersection E of AC and a line through F for which $|BE| = |EO|$. Draw BO, which intersects the circle at G, then $\angle CAG = \frac{1}{3} \angle CAK$. $|BE| = |EO|$ therefore $\triangle BEO$ is isosceles. $\angle EOB = \angle EBO$, now $\angle EOB$ is a central angle

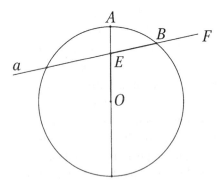

FIGURE 3.5
One of Aguilón's neusis-like constructions: given a line through F, it is possible to find a point E on a radius such that $|BE| = |EO|$

---

32 Co-authored by Herman van Looy.
33 KBR MS 5780, MS 1.
34 On *neusis*, see Meskens and Tytgat, *Exploring Classical Greek Construction Problems*, 23–25, and on the contribution of the Flemish Jesuits, 65–71.

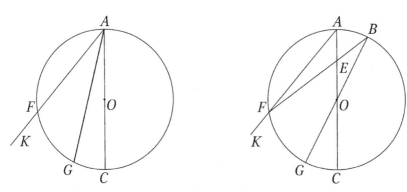

FIGURE 3.6 The trisection of an angle using neusis-like constructions. Using the construction of figure 3.5, it is possible to trisect the angle ∠ CAF

and ∠FBG is an inscribed angle. ∠FOG = 2∠FBG = 2∠EOB ⇒ $\widehat{FG} = 2\widehat{AB}$. Now $\widehat{AB} = \widehat{GC}$ and $\widehat{FG} = 2\widehat{GC}$, therefore ∠CAF = 3∠CAG.

The *neusis*-like lemmas become more and more complicated.

This complexity culminates in lemma 13 (MS 1, fol. 239; fig. 3.7): given a circle with a chord *CF*, which is not a diameter, and a second circle centered at *F* and with radius *r* =|*CF*|. Let *L* be the intersection of *CF* with the second circle, let a line through *F* intersect the first circle in *G* and the second in *Q*, let the line *CG* intersect the second circle in *N*. Determine the point *G* such that $\widehat{NL} = \widehat{CQ}$. From his corrections, strikethroughs, and rewordings, it seems clear that Aguilón struggled with this problem.

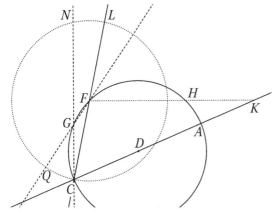

FIGURE 3.7 The trisection of an angle for which the properties of inscribed and central angles in two circles are used

FIGURE 3.8 Top: Hans Vredeman de Vries, "Interior of a Gothic Church," 1594, private collection, detail
© KIK-IRPA KM004054
Bottom: Pieter Jansz. Saenredam, "The Nave and Choir of Mary's Church in Utrecht," oil on canvas, 1641, Rijksmuseum Amsterdam, detail
© RIJKSMUSEUM, AMSTERDAM, SK-A-851

From $\overset{\frown}{FC} = 3\overset{\frown}{GF}$, it follows that this construction can be used for the trisection of an acute angle.[35]

In the second part of the manuscript (MS 1, fols. 256–82), Aguilón, using cross-ratios, explores the properties of four points A, B, D, and C, in which D is the harmonic conjugate of C with respect to A and B. In Aguilón's formalism, this means $\dfrac{|AB|}{|CB|} = \dfrac{|AD|}{|DB|}$ with A, D, B, and C in this order. Obviously, the orientation of the line, which in our formalism determines the sign of the cross-ratio, was a concept that in Aguilón's time still had to be developed.

Aguilón begins his exposé with the well-known results of Pappus and Apollonius but finds results that can best be called elementary projective geometry. This would remain an object of study for the other Flemish Jesuit mathematicians for the next half century.

One may wonder whether Aguilón's involvement with artists such as Rubens influenced him in this field as well. This was the age of a nascent theory of perspective, as can clearly be witnessed from the paintings of gardens by Hans Vredeman de Vries (1527–c.1607) and of church interiors by Pieter Jansz.

---

35   The proof is remarkably simple using inscribed and central angles. $\angle QFC = \angle GFC = 2\angle GCF$ in the circle with center F. In the circle with center D, these are both inscribed angles, from which it follows that $3\angle FDG = \angle FDC$.

Saenredam (1597–1665). The underlying geometrical structure of tilings, for instance, naturally leads to mathematical questions that open the door to the theory of conic sections on the one hand and to projective geometry on the other (see also chapter 10).

Aguilón begins this part of his treatise with a definition of harmonic conjugate points (MS 1, fol. 256ʳ), or a harmonic quadruple for short. When a line segment is divided into three segments such that the whole segment is to one of the outermost segments as the other outermost segment is to the middle one, the line segment is divided *in proportione totius et partium* (in extreme and mean proportion). This definition begs comparison to the golden ratio in which a line segment is divided into segments in *ratio totium et partium* (in extreme and mean ratio). Although Aguilón emphasizes the difference between both proportions, later on in the manuscript he will use both terminologies for the cross-ratio. Aguilón further remarks that these harmonic conjugates have been used by Pappus,[36] Apollonius,[37] and Alhazen.[38] It is especially the theorems by Alhazen that have influenced him when moving from a quadruple of points to a four-rayed pencil.

In some marginal notes, San Vicente remarks on the terminology and suggests *ratio extremorum et mediorum* (extreme and mean ratio) and *extrema et media ratio proportionali* (extreme and mean proportional ratio), terminology that he would later use in his book *Problema Austriacum* (1647).

Notice that, for Aguilón, a line segment [AC] is divided *in ratione totius et partium* if $\frac{|AC|}{|CB|} = \frac{|AD|}{|DB|}$. In our terminology, we would say that the points A and B are harmonically divided by C and D. Disregarding the sign, this means $\frac{|CA|}{|CB|} \div \frac{|DA|}{|DB|} = 1$ or $\frac{|AC|}{|CB|} = \frac{|AD|}{|DB|}$, and with the sign notation we immediately see that either C or D is located between A and B.

---

36   *Collectiones Mathematicae* 6:52 and 7:passim. Pappus Alexandrinus, *La collection mathématique*, ed. Paul Ver Eecke (Brugge: Desclée De Brouwer, 1933).

37   *Konika* 1:34, 36. Apollonius Pergae, *Les coniques d'Apollonius de Perge*.

38   *Perspektiva*, 6:5, 6, 7, A. Mark Smith, "Alhacen on Image-Formation and Distortion in Mirrors: A Critical Edition, with English Translation and Commentary, of Book 6 of Alhacen's *De aspectibu*, the Medieval Latin Version of Ibn Al-Haytham's *Kitāb Al-Manāzir*; Volume One; Introduction A," *Transactions of the American Philosophical Society* 98, no. 1 (2008): i–153, here 26ff; Smith, "Alhacen on Image-Formation and Distortion in Mirrors: A Critical Edition, with English Translation and Commentary, of Book 6 of Alhacen's *De aspectibus*, the Medieval Latin Version of Ibn Al-Haytham's *Kitāb Al-Manāzir*; Volume Two; English Transl.," *Transactions of the American Philosophical Society* 98, no. 1 (2008): 155–393, here 176ff.

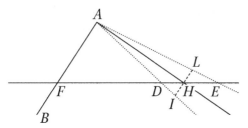

FIGURE 3.9 The construction of a harmonic quadruple if a right angle is given

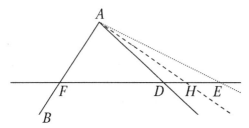

FIGURE 3.10 The construction of a harmonic quadruple if an acute angle is given

Aguilón then turns his attention to a harmonic pencil of four rays, a harmonic pencil for short. A pencil is called harmonic if the points of intersection with any line determine a harmonic quadruple.

Given two intersecting lines, Aguilón constructs two more lines such that all intersections with a third given line are in harmonic position (lemma 33, MS 1, fol. 258; fig. 3.9).

The first configuration has two lines AB and AC at right angles. The problem is solved if two lines AD and AE are drawn for which AC is the bisectrix of $\angle DAE$. Let H be the intersection of AC with the third given line and let F be the intersection of AB with this line.

Draw $IL \parallel AB$ through C, then $\dfrac{|AF|}{|HL|} = \dfrac{|AF|}{|HI|}$ because $|HL| = |HI|$.

Now from $\triangle AGF \sim \triangle LGH$ whence $\dfrac{|AF|}{|HL|} = \dfrac{|FG|}{|GH|}$ on the one hand, and $\triangle AFK \sim \triangle IHK$ whence $\dfrac{|AF|}{|IH|} = \dfrac{|FK|}{|KH|}$ on the other, we find $\dfrac{|FG|}{|GH|} = \dfrac{|FK|}{|KH|}$, which we had to prove.

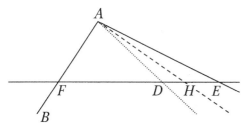

FIGURE 3.11   The construction of a harmonic quadruple if an obtuse angle is given

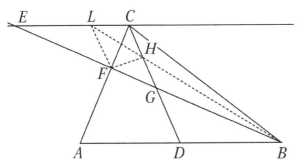

FIGURE 3.12   The construction of a harmonic quadruple using a triangle

When the angle between the rays [AB and [AD is an acute angle (fig. 3.10), Aguilón constructs [AC perpendicular to [AB. This reduces the problem to the former.

In the last case, the angle between the rays [AB and [AE is obtuse (fig. 3.11), and Aguilón again constructs [AC perpendicular to [AB, which reduces the problem to the former.

In the following lemmas, Aguilón lets CD be the median in a triangle △ABC and lets *l* be a parallel with AB through C (fig. 3.12). Several properties are now explored, such as: if a line intersects *l* in a point E, then (B, F, G, E) is a harmonic quadruple. If BF ⊥ AC, then FC is the bisectrix of ∠LFH, for any L on *l*. If, moreover, FH //AB, then L is the midpoint of [EC].

On the following pages (MS 1, 262^v–264^r), Aguilón makes use of Apollonian properties to construct a harmonic quadruple in a circle (fig. 3.13). Suppose AC is a tangent to the circle and CE is a chord perpendicular to the diameter through A. For any line through A, the intersection points of AC with the circle F and G and the intersection point AC with the chord CE and A will be a harmonic quadruple.

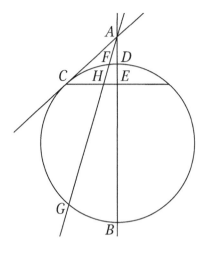

FIGURE 3.13
The construction of a harmonic quadruple in a circle

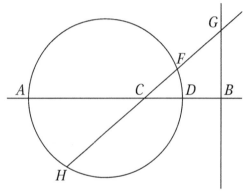

FIGURE 3.14
The construction of a harmonic quadruple in a circle

In a marginal note, San Vicente remarks that this property also holds if the circle is replaced by any conic section.[39] What we have here is nothing other than what would later be called pole and polar with respect to the conic section.

Now Aguilón wants to construct a harmonic quadruple if the given point is inside the circle (fig 3.14). Aguilón states that if (A, B, C, D) are in harmonic position and BG is perpendicular to AB, then for any line through C, the intersections of the line with the circle F, H and the intersection of the line with the perpendicular, G, and the point C will be a harmonic quadruplet.

---

39   These properties are theorems 27 to 31 in Gregorio a San Vicente, *Problema Avstriacvm plvs vltra qvadratvra circvli* (Antwerp: Joannes et Jacobus Meursius, 1647), 185–87. The properties are present in Apollonius, book 3, see Heath, *Apollonius of Perga*, passim.

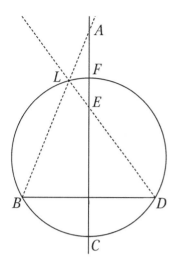

FIGURE 3.15
The construction of a harmonic quadruple in a circle

Aguilón also gives a construction for the harmonic cross-ratio on the diameter of a circle (fig. 3.15). Let BD be a chord perpendicular to the diameter AC, let E be the intersection of DL with AC, then (A, E, F, C) is a harmonic quadruplet.

Although Aguilón has dealt with cross-ratios in a purely mathematical way, the relation with optics is clear. The theory of harmonic quadruples and pencils is very well suited to study reflection in concave and convex spherical mirrors, as was already known by Alhazen.[40]

Let C be the center of a spherical mirror and O the position of an object (fig. 3.16). The light ray emanating from O and meeting the eye of the observer at E is reflected on the mirror at R. The line ER intersects OC at B, while the tangent t at R intersects OC at Q. Then (O, Q, B, C) is a harmonic quadruplet.[41]

---

40   Alhazen's problem (*Perspectiva*, 5:18) states: "Given a light source and a spherical mirror, find the point on the mirror where the light will be reflected to the eye of an observer." Mathematically, this means drawing lines from two points in the plane of a circle meeting at a point on the circumference that make equal angles with the normal at that point. See A. Mark Smith, "Alhacen on the Principles of Reflection: A Critical Edition, with English Translation and Commentary, of Books 4 and 5 of Alhacen's *De aspectibus*, the Medieval Latin Version of Ibn Al-Haytham's *Kitāb Al-Manāzir*; Volume One; Introduction and Latin," *Transactions of the American Philosophical Society* 96, no. 2 (2006): i–288; Smith, "Alhacen on the Principles of Reflection: A Critical Edition, with English Translation and Commentary, of Books 4 and 5 of Alhacen's *De aspectibus*, the Medieval Latin Version of Ibn Al-Haytham's *Kitāb Al-Manāzir*; Volume Two; English Translation," *Transactions of the American Philosophical Society* 96, no. 3 (2006): 289–697, here 415.

41   By definition, BR and ER make equal angles with the tangent t, which is at right angles with the radius. This is nothing else than the configuration described by Aguilón in lemma 33.

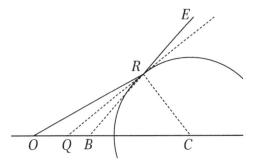

FIGURE 3.16
Alhazen's problem: given a light source O and a spherical mirror, find the point R on the mirror where the light will be reflected to the eye of an observer at E

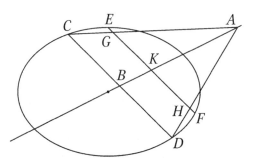

FIGURE 3.17
The equality of certain lines in an ellipse |EG|=|HF|

The property had already been proven by Alhazen, but his proof was cumbersome and difficult to follow.[42]

The following theorem would play an important role in Aguilón's further investigations (MS 1, fol. 284^r; fig. 3.17): let l be a diameter of an ellipse, let A be on l, let $[EF]$ and $[CD]$ be two ordinates on l, and let a triangle $\triangle ADC$ be drawn. Let G be the intersection of AC and EF, and H, the intersection of DA and EF, then $|EG| = |HF|$.[43]

In the following lemmas, Aguilón studies a new situation (MS 1, fol. 287^r; fig. 3.18). He first proves that if AC and BD are two non-conjugate diameters, $[AE]$ an ordinate on BD, and $[BF]$ an ordinate on AC, and I and L the midpoints of $[AB]$ and $[DC]$ respectively, then AE, BF, and IL are concurrent.

Indeed, draw the tangents at D and C respectively and let their intersection point be K. KH divides the tangent chord and $[AB]$ into two equal segments and $AB // CD$. Therefore, I and L are on KH. Moreover, $DK // AE$ and $CK // BF$

---

42 See, e.g., A. Mark Smith, "Alhacen's Approach to 'Alhacen's Problem,'" *Arabic Sciences and Philosophy* 18, no. 2 (2008): 143–63.

43 The proof rests on the similarity of triangles:

$$\triangle ACB \sim \triangle AGK : \frac{|GK|}{|CB|} = \frac{|AK|}{|AB|}, \triangle ADB \sim \triangle AHK : \frac{|HK|}{|DB|} = \frac{|AK|}{|AB|} \Rightarrow \frac{|GK|}{|CB|} = \frac{|HK|}{|DB|} \text{ or } \frac{|GK|}{|HK|} = \frac{|CB|}{|DB|}$$

Since $|CB| = |BD|$, we have that $|GK| = |HK|$ and $|EG| = |HF|$.

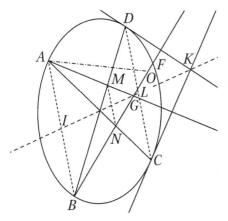

FIGURE 3.18
Properties of certain lines in an ellipse

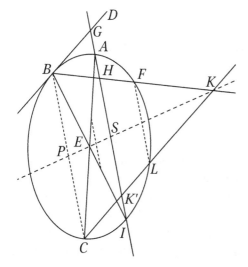

FIGURE 3.19
The construction of a harmonic quadruple on the diameter of an ellipse

because, if, from a point of tangency, the diameter is drawn, then all lines parallel to the tangent are ordinates on that diameter (proven by Aguilón on fol. 283ʳ). Now suppose that the intersection point, O, of AE and BF is not on IL and is the point O. Connect A to O, then AO will be coincident with AG or, if this is not the case, AO will be parallel with DK; but if AO // DK and AO // DK, then AO and AG are coincident.[44]

---

44 If the intersection point of AE and BF is not on HK, then BF has to intersect HK in G. Draw AG and, using similar triangles, it can be proven that AG // DK and AG is coincident with AO.

In the next lemma, using the same figure, Aguilón proves that *AN* and *AM* are divided by *H* into segments of the same proportion. Furthermore, *MN // AB* and *EF // AB* because *M* and *N* are the midpoints of [*AE*] and [*BF*] respectively.

In an analogous figure (MS 1, fol. 289ʳ; fig. 3.19), we have that *BD* is a tangent in *B*, [*CL*] is an ordinate on *BI*, *P* and *S* are the midpoints of *BC* and *AI* respectively. *CL* intersects *PS* at *K*. Then [*BF*] is an ordinate on *CA*.

Let *H* be a point on [*BF*] for which *GH // BC* then |*GA*| = |*AH*|.

Using this lemma, Aguilón is able to construct a harmonic quadruple on the diameter of the ellipse. He notes that this had already been done by Apollonius in *Konika* 1.36, using a proof by contradiction. He now proceeds differently.

If [*BH*] is an ordinate on *AC*, and *BD* is a tangent to the ellipse in *B*, then [*DC*] is divided by *A* and *E in ratione totius et partium*. If the line *FA // BC* is drawn, then according to the previous lemma |*FA*| = |*AG*|. And $\frac{|BC|}{|FA|} = \frac{|CE|}{|GA|}$, but according to lemma 12 of his *Optics* 4 (△*BEC* ∼ △*GEA*) this implies $\frac{|BC|}{|AG|} = \frac{|CE|}{|EA|}$ and so $\frac{|BC|}{|FA|} = \frac{|CE|}{|EA|}$. Now, because $\frac{|BC|}{|FA|} = \frac{|CD|}{|DA|}$ (△*DFA* ∼ △*DBC*), we have $\frac{|CE|}{|EA|} = \frac{|CD|}{|DA|}$, which needed to be proven.

Aguilón has now attained his first goal: constructing a harmonic quadruple on a chord of an ellipse, as he had previously done for a circle.

After a couple of lemmas concerning conjugate diameters and corresponding tangents, Aguilón proposes that if in a right-angled triangle △*ABC* the point *D* is given, to construct an ellipse in which [*AB*] and [*BC*] are the axes and to which *AC* is a tangent in *D*, the fundamental lemma follows: *Data quacunque coningatione diametrorum ellipsis axes inveniere* (To find the axes of an ellipse if a pair of conjugate diameters is given).

Before proceeding to the proof, Aguilón remarks that he has already solved the problem of describing an ellipse if two conjugate diameters are given and if the major axis of the ellipse is equal to the diameter of the circle in *Opticorum* 6, lemma 25.[45] This, according to Aguilón, amounts to: given two conjugate diameters and knowing the length of the major axis, construct this major axis. This theorem was important in the orthographic projection of circles.[46] He has to generalize this lemma because he deals with cylindrical, conical, and spherical mirrors. A solution to this problem had already been

---

45   Aguilón, *Opticorum libri sex*, 479.
46   Aguilón, *Opticorum libri sex*, 503. If a circle, not perpendicular to the direction of projection, is orthographically projected onto a plane, the image will be an ellipse, two

FRANCISCO DE AGUILÓN AND MATHEMATICAL OPTICS 79

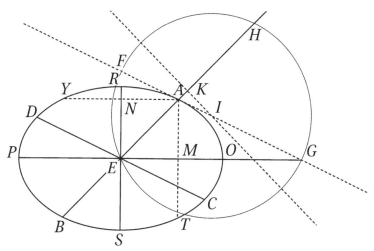

FIGURE 3.20   In *Collectiones mathematicae* 8, proposition 14, Pappus of Alexandria gives a construction for the axes if conjugate diameters are known. If [AB] and [CD] are two conjugate diameters, draw l_A //DC and construct the point H of AB for which $|DE|^2=|EA|.|AH|$. If K is the midpoint of [EH], then erect a perpendicular $KI \perp HE$ that intersects l_A in I. With I as center, describe a circle through H and E. This circle determines the positions of F and G on l_A. By drawing FE and GE, the directions of the axes are found. Construct $|EO|^2=|GE|.|EM|$ (= $|EP|^2$) and $|ER|^2=|FE|.|EN|$ (= $|ES|^2$), from which one can conclude that [OP] and [RS] are the axes of the ellipse

given by Pappus, he claims (MS 1, fol. 294ᵛ; fig. 3.20), but without a proof. This is a major indication that Aguilón had another publication on optics in mind and that these lemmas are the mathematical basis for his investigations.

After this fundamental lemma, which for Aguilón is the main lemma, we find some lemmas on diameters and their tangents.

On fol. 302ʳ (MS 1), Aguilón begins with a new subject: similar ellipses. Aguilón proves that the intersection of a plane parallel to the axis of rotation with the spheroid is an ellipse similar to the ellipse that generated the spheroid. He also proves that if the plane is not parallel to the axis, the ellipse will be "dissimilar."

The last theorem states that the tangent lines from a given point to the spheroid will have points of tangency that are in one plan. The result has some bearing on the scenographic projection, a projection that Aguilón introduced

---

perpendicular diameters of the circle will become two conjugate diameters of the ellipse, and the diameter parallel to the plane of projection will become the major axis of the ellipse.

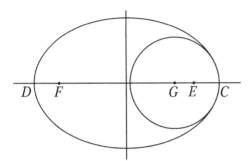

FIGURE 3.21 The construction of the largest internal circle tangent to the ellipse at an extremity of the major axis

in the sixth book of his *Opticorum*. It follows from these theorems that one sees the spheroid as an ellipse, similar or not, with the generatrix of the spheroid.

After a blank page, Aguilón begins a last and unfinished subject: to construct the largest internal circle tangent to the ellipse at an extremity of the major axis (fig. 3.21). He claims that if E and F are the foci of the ellipse, the center G of the circle has to satisfy the relation $\dfrac{|FC|}{|CE|} = \dfrac{|FG|}{|GE|}$. A proof is lacking, but Aguilón remarks that this is the desired circle if one can prove that $|GC|$ is the shortest distance from G to the ellipse.

The proof of this assertion was later inserted by Moretus. Aguilón also wrote down the counterpart of this theorem: to describe the smallest circle about an ellipse and touching it in an extremity of the minor axis.

In conclusion, it is clear that this manuscript was a preparatory text for a second book, the one on catoptrics. Clearly, the main themes are harmonic quadruples and rays and (perspectival) projections, subjects that are closely related to painting.

Aguilón's work on optics would not be continued by San Vicente, but it would remain a constant theme with his students, as can be ascertained from the manuscripts of Boelmans and della Faille.

CHAPTER 4

Gregorio a San Vicente: An Ignored Genius

1   A Tragic Life

Gregorio a San Vicente was born on September 8, 1584 in Bruges.[1] He was most probably the son of Gregorio a San Vicente (d. after 1616).[2]

San Vicente went to secondary school in Bruges, studied philosophy in Douai, and joined the Society of Jesus on October 21, 1605 in the Sant'Andrea novitiate at Rome.[3] Before entering the Society, he had studied seven years of humanities and two years of philosophy, for which he was called master of arts.[4] After joining the Society, he studied another year of philosophy and four years of theology.[5] The Jesuit superior general had a post in Sicily in mind for San Vicente once he had completed his studies, but Clavius managed to have him stay in Rome to study mathematics. Although Clavius no longer held the chair of mathematics at the Collegio Romano, it is very likely that San Vicente studied under him. However, he was never a formal student of the school of mathematics at the Collegio Romano, although in one of his letters to

---

1   ARAA T14/034 31, 46/10.
2   Germain Bonte and François Jongmans, "Sur les origines de Grégoire de Saint Vincent," *Mededelingen van de Koninklijke Academie van België: Klasse der Wetenschappen* (1998): 295–326, have traced San Vicente's lineage to Spanish *judeoconversos* residing in Bruges for some generations. In the *Liber novitiorum* of Sant'Andrea (no. 526; copied in ABME Bosmans Cahiers 124), written in Italian, he is called Gregorio de san Vincenzo. *Judeoconversos* were Jews who had converted to Catholicism and their descendants. The Spanish often called them *marranos*, suggesting they were false converts and thus often persecuted by the Inquisition. Many of them fled the country, including to the Low Countries. The position of Jews (or Muslims) and *marranos* was a difficult issue for the Society. The Society initially did not give in to pressure to follow suit. When accepted into the order, candidates from one of these groups were sent to Italy to receive their education in an Italian house and college. In 1593, the policy was reversed: from that moment on and up to 1946, despite mitigations in 1608 and 1923, descendants of Jews and Muslims were excluded from membership. Because San Vicente entered the Society in 1605—after the ban on candidates of Jewish origins had been made law in 1593—his *judeoconverso* origins must have been unknown or hidden. On this subject, see Robert A. Maryks, *Jesuit Order as a Synagogue of Jews: Jesuits of Jewish Ancestry and Purity-of-Blood Laws in the Early Society of Jesus* (Leiden: Brill, 2009).
3   *Liber novitiorum* of Sant'Andrea (no. 526), copied in ABME Bosmans Cahiers 124; ARSI Rom.54, fol. 205, as cited in Baldini, "Academy of Mathematics," 96n106. Bonte and Jongmans, "Sur les origines de Grégoire de Saint Vincent," 310.
4   ARAA T14/034 31, 46/10.
5   ARAA T14/034 31, 46/10.

Christiaan Huygens (1629–95) he writes: "Also Father Paulus Guldin who was, as was I, a student in the house of Clavius."[6] His relationship with Clavius therefore remains unclear.

During San Vicente's stay in Rome, the college organized an academic session with Galileo in attendance. Together with Clavius's other students, San Vicente became an adept of Galileo's new concepts. Shortly after Clavius's death, San Vicente returned to the Low Countries. On March 23, 1613, he was ordained priest in Leuven,[7] and he held several posts in Brussels, 's-Hertogenbosch,[8] and Kortrijk. He also seems to have been chaplain to the Spanish troops.[9] In 1615, he arrived in Antwerp.[10] Toward the end of 1615, on December 18, he asked for permission to go to China as a missionary.[11] On February 6, 1616, Superior General Muzio Vitelleschi (1563–1645, in office 1615–45) granted him permission, but San Vicente never embarked on the journey. He had renounced his request in favor of Gillis Carpentier (1582–1632),[12] possibly because of pressure put on him by Aguilón and other confrères.

From 1617 to 1621, San Vicente was a mathematics teacher at the Antwerp Jesuit college. Together with his school of mathematics, he was transferred to Leuven in 1620/21. He took his four solemn vows on May 3, 1623 in Antwerp.[13] Up to the moment he left for Rome, the history of the school of mathematics and the biography of San Vicente almost coincide. This period was his mathematically most creative one. All his later work, after returning from Rome, would be devoted to applying his new theories to the quadrature of the circle.

By 1624, San Vicente had become convinced that the method he had developed, which he called *ductus plani in planum* ([multiplication] of a plane in a plane), could lead to the quadrature of the circle. It also convinced him that he could write a book on the subject that he hoped could be published. He duly asked Vitelleschi for permission.[14] Vitelleschi, however, was cautious, as

---

6   "Etiam Pater Paulus Guldin, qui vna mecum domesticus auditor Clauij fuit." San Vicente to Huygens, November 1, 1651. Huygens, *Oeuvres complètes: Tome I*, 153, letter 101.
7   ARAA T14/O34 348, 59.
8   ARAA T14/O34 31, 46/10.
9   Audenaert, *Prosopographia iesuitica Belgica antiqua*, 2:281, and appendix 9.
10 Droeshout, "Histoire de la Compagnie de Jésus à Anvers," 3:180.
11 ARSI, *Index indepetae*, referring to FG21, fol. 6, December 18, 1615.
12 Edmond Lamalle, S.J., "La propagande du P. Nicolas Trigault en faveur des missions de Chine (1616)," *Archivum historicum Societatis Iesu* 9 (1940): 49–120, here 82. Carpentier, despite the insistence of Vitelleschi, did not embark on the journey.
13 Kadoc ABSE 47.
14 To guarantee a degree of uniformity, a system of censorship was imposed. This censorship should be viewed as an early kind of peer review, combined with a strict editorial policy. All manuscripts intended for publication had to be sent for Rome's approval. Revisers

FIGURE 4.1 Left: the well-known portrait of Gregory of Saint-Vincent
© MPM. PK.OP.11007
Right: a less well-known portrait
© RIJKSPRENTENKABINET, AMSTERDAM, RP-P-1906-689

up to then the quadrature of the circle had defeated even the most brilliant mathematical minds. He therefore advised San Vicente to send his work to Grienberger, Clavius's successor at the Collegio Romano, for peer review.[15] To convince Grienberger, San Vicente, together with his students and former students, wrote a number of treatises summarizing his results.[16] Grienberger must have been pleased with the results, but at the same time he also appears to have had some reservations. He was not entirely convinced or, perhaps

---

would often make comments on several aspects of the book, but if the manuscript was not rejected it sufficed to take the comments and suggested corrections into account in order to publish the book without further ado. When, due to the sheer volume of manuscripts sent to Rome, it became impossible to read all of them, permission was given based on the reports by the provincial revisers. On this censorship, see Marcus Hellyer, *Catholic Physics: Jesuit Natural Philosophy in Early Modern Germany* (Notre Dame, IN: University of Notre Dame Press, 2005), 35ff.

15 Kadoc ABML Bosmans 124 Sancto Vincentio.
16 Some of these are part of KBR MS 5770–72.

because of the absence of proof, had difficulties understanding the intricacies of San Vicente's new methods.

On April 19, 1625, Vitelleschi wrote to the Flemish provincial, Floris van Montmorency (1580–1659), to state that San Vicente was allowed, at his convenience, to travel to Rome to confer with Grienberger.[17] On September 9, 1625, San Vicente embarked on the journey.[18] In a letter of October 11, 1627, Grienberger wrote to Vitelleschi that San Vicente had indeed found a method that might lead to the circle quadrature[19] but added that his ideas were not yet sufficiently well developed: "Because nothing was in an orderly fashion, I could not delve deep into the content," Grienberger wrote to Vitelleschi.[20] A little further, we read:

> But because neither in the beginning, nor further along, I found a trace of the quadrature, which I expected impatiently, I read on sloppily and in the end, seeing there were still many difficulties, I have interrupted my work, not to never return to it, but to be able sometime to read it with an open mind. Nevertheless, as I see it now, this will not end right.[21]

San Vicente did not take up his teaching post again. At the request of Emperor Ferdinand II (1578–1637, r.1619–37), he was sent to Prague, where he became confessor at the imperial residence. Shortly after his arrival in Prague, he suffered a stroke.[22] Hardly had he settled in Prague when the order came that he was to go to Madrid, but because of his ill health he did not embark on the journey. Although San Vicente had no teaching duties in Prague, it seems likely that he gave private lessons to talented students and laymen.[23] While residing there, he drew up plans for the Klementinum, the Jesuit college at Prague. Because of his poor health, he asked for an assistant, preferably della Faille or Moretus, to help him with further research on the quadrature of the circle, a problem he was still hoping to solve. Because della Faille had been sent

---

17   Kadoc ABML Bosmans 124 Sancto Vincentio.
18   KBR MS 5770–72, fol. 380ᵛ. See also Elisabeth Sauvenier-Goffin, "Les manuscrits de Grégoire de Saint-Vincent," *Bulletin de la Société royale des sciences de Liège* 20 (1951): 413–738, here 422.
19   "Nam, ut jam, intelligo, scriptum illud continet succum quadraturae." Henri Bosmans, S.J., "Lettre inédite de Christophe Grienberger sur Grégoire de Saint-Vincent," *Annales de La Société d'émulation de Bruges* 63 (1913): 41–50, here 48.
20   Bosmans, "Lettre inédite de Christophe Grienberger," 49.
21   Van Looy, "Chronologie en analyse," 194, referring to a letter dated October 11, 1627, in ARSI 534, fols. 52–53; see also Bosmans, "Lettre inédite de Christophe Grienberger."
22   San Vicente, *Problema Avstriacvm, praefatio*, b3ᵛ.
23   Schuppener, *Jesuitische Mathematik in Prag*, 102.

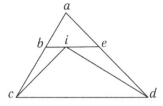

FIGURE 4.2
Bonaventura Cavalieri asked Saint-Vincent to prove that in a triangle $\triangle acd$: $|ca|+|ad|-|cd| \geq 2(|cb|+|be|+|ed|-|ci|-|id|)$

to Madrid, Vitelleschi assigned Moretus to the task,[24] and Moretus grudgingly obliged.[25]

In 1631, the Saxons pillaged Prague and set fire to many parts of the city. During this period, San Vicente lost several of his manuscripts. Part of his library was spared and was later brought to Vienna by Father Rodrigo (Rodericus) de Arriaga (1592–1667), where it would remain for another ten years before it was returned to San Vicente.[26] However, three volumes of print-ready manuscripts dealing with statics and geometry were destroyed.[27] After the sacking of Prague, San Vicente went to Vienna[28] and Graz.[29] On February 18, 1632, he was at Laibach (Ljubljana) and was directed to Rome by the Austrian provincial, Father Christoph Grenzing (1567–1636).[30] Vitelleschi, on the other hand, was of the opinion that "teaching at the Collegio Romano would be too great a burden" and ordered him back to the Netherlands. In 1632, San Vicente settled at Ghent.[31]

By that time, it had become known in Italy that San Vicente was working on the quadrature of the circle. This becomes clear from a letter of Stephanus Ghisoni dated November 23, 1636, in which he informs Giannantonio Rocca (1607–56) that he did not know whether San Vicente's book had already been published or not, nor did he know its title.[32] During the 1630s, Bonaventura

---

24   Kadoc ABME Fonds Bosmans V. 8, 261, referring to letters of Muzio Vitelleschi, dated April 7, 1629, and July 21, 1629, in Bohemia Epist. Gen. 1628–37 (277, 291–95).
25   APUG MS 534, fols. 71–72. Van Looy, "Chronologie en analyse," 7.
26   San Vicente, Problema Avstriacvm, praefatio, b3^v–b4^r.
27   San Vicente, Problema Avstriacvm, praefatio, b3^v; Schuppener, Jesuitische Mathematik in Prag, 102.
28   San Vicente, Problema Avstriacvm, praefatio, b3^v; Van Looy, "Chronologie en analyse," 7 and passim.
29   APUG 534, fol. 34^r, letter from Durandus to Grienberger. San Vicente, Problema Avstriacvm, praefatio, b4^r.
30   Henri Bosmans, S.J., "Saint-Vincent (Grégoire de)," Biographie Nationale (Brussels: Académie royale de Belgique, 1911), c.141–71, col. 149.
31   San Vicente, Problema Avstriacvm, praefatio, b4^r.
32   Giovanni Antonio Rocca and Cosimo Gaetano Rocca, Lettere d'uomini illustri del secolo XVII a Giannantonio Rocca: Filosofo, e matematico Reggiano (Modena: Presso la Societa' Tipografica, 1785), 62.

Cavalieri (1598–1647) wrote San Vicente a letter, but the precise dating is uncertain (fig. 4.2).[33]

San Vicente's first book *Problema Avstriacvm plvs vltra qvadratvra circvli* (The Austrian problem, extended with the quadrature of the circle) was published in 1647, but this was a very belated publication.[34] It is the final version of the ideas he worked on in the 1610s and that he systematized in the treatises for inspection in Rome in the 1620s.

After the publication of *Problema Austriacum*, Huygens entered into a correspondence with San Vicente.[35] Huygens asked permission to publish remarks on the book, which was granted by San Vicente.[36] As in his letters, in the last part of *Theoremata de quadratura* (Theorems on the quadrature [1651]), Huygens points out an error in the circle quadrature. He then proves the error by assuming that the erroneous theorem is true and, invoking other theorems of *Problema Austriacum*, arrives at absurdities.

Both mathematicians would meet each other on July 13, 14, and 15, 1652 during Huygens's visit to the Southern Netherlands. Huygens, however, was very disappointed with the results of their deliberations.[37]

For his second publication, *Opus geometricum* (Geometrical work [1668]), San Vicente easily obtained permission to publish from Superior General Giovanni Paolo Oliva (1600–81, in office 1664–81).[38] However, San Vicente never saw the finished book, as he suffered a stroke and died on January 27, 1667.

---

33   Remark in a letter of November 1, 1641 to Rocca. Rocca and Rocca, *Lettere d'uomini illustri del secolo XVII*, 266.

   Cavalieri also corresponded with San Vicente's student Moretus. Georg Schuppener, "Theodor Moretus (1602–1667): Ein Prager Jesuiten-Mathematiker," in *Bohemia Jesuitica 1556–2006*, ed. Petronilla Cemus (Prague: Nakladatelství Karolinum, 2010), 2:661–75, here 652.

34   Because the half title or bastard title of the book is *Opus geometricum quadraturae circuli et sectionum coni, decem libris comprehensum*, the book is often cited as *Opus geometricum*. Since the (real) title, *Opus geometricum posthumum ad mesolabium per rationum proportionalium novas proprietates* (Ghent: Balduinus Manilius, 1668), begins with the very same words, there is some confusion between the two books. This monograph will therefore consistently use *Problema Austriacum* for the 1647 book.

35   Huygens, *Oeuvres complètes: Tome I*, letters 111, 112, 117, 122, 125; Huygens, *Oeuvres complètes: Tome II*, letters 96, 99, 100, 101, 102, 105, 106.

36   Huygens, *Oeuvres complètes: Tome II*, letters 96, 99, 100, 101, 102. Later, the correspondence turns somewhat sour. Although San Vicente expresses his admiration for Huygens, the latter seems to be disappointed that he is not explicitly praised for his work.

37   Huygens, *Oeuvres complètes: Tome I*, letter 137, to Tacquet, dated November 4, 1652.

38   Henri Bosmans, S.J., "Compte rendu de R. Guimarâes: 'Les mathématiques en Portugal' (1909)," *Revue des questions scientifiques* 67 (1910): 636–46, here 642–43.

The Jesuits called San Vicente "our Apollonius." As we shall see, he can indeed be seen as the last of the ancient geometers, someone who could see the promised land of calculus and Cartesian geometry and began to venture into it.

2   Mathematical *Oeuvre*

Although San Vicente gives the impression of being a prolific author, his manuscripts deal solely with a few genial ideas, paving the way for the calculus. The Royal Library Albert I at Brussels holds no fewer than seventeen of his manuscript volumes,[39] roughly counting about ten thousand pages. His manuscripts have been bound regardless of theme or subject. Van Looy has organized these texts in chronological order (see appendix 1).[40]

All of the manuscripts, save one, can be seen as preparatory treatises for what were to be his two books: *Problema Austriacum* (1647) and *Opus geometricum* (1668). One treatise is in the hand of Aguilón and seems to have been a preparatory manuscript for his book on catoptrics. Two volumes, 5789 and 5793, must be considered as a whole. These volumes are often referred to by San Vicente as *Tomus primum* and *Tomus secundum*. One treatise is about trisections, the other about conic sections. Both were written in his Antwerp period (1617–20), with some additions by Boelmans on the ductus-method written in Leuven. Volume 5770–72 and a part of volume 5786–88 were written in Leuven, in part by San Vicente himself and in part by one of his students. Volume 5770–72 contains the manuscripts that were sent to Rome. Volume 5786–88 contains the treatise *De progressionibus* (On sequences), written by Boelmans. Volumes 5790 and 5791, known as *Chartae Romanae* (Roman files), were begun during San Vicente's stay in Rome (1625–27) and were finished in Prague (1628–29). Volume 5780 and parts of volume 5786–88 were written by Moretus in Prague (1630–31). Volumes 5784 and 5785 are known as the *Chartae Gandenses* (Ghent files) and are closely related to the preparatory manuscripts for *Problema Austriacum*. Parts of volumes 5770–72, 5773–75, 5780, and 5786–88 contain preliminary work on *Problema Austriacum*. After he had received the manuscripts salvaged after the Sack of Prague, San Vicente began drafting

---

39   KBR 5770–72 to 5793.

40   Van Looy, "Chronologie en analyse"; Herman van Looy, "Chronologie et analyse des manuscrits mathématiques de Grégoire de Saint Vincent (1584–1667)," *Archivum historicum Societatis Iesu* 49 (1980): 279–303; Van Looy, "A Chronology and Historical Analysis of the Mathematical Manuscripts of Gregorius a Sancto Vincentio (1584–1667)," *Historia mathematica* 11 (1984): 57–75.

 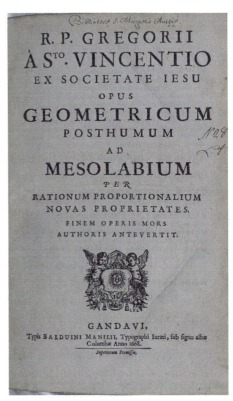

FIGURE 4.3 Left: the frontispiece of Gregory of Saint-Vincent, *Problema avstriacvm plvs vltra qvadratvra circvli* (Antwerp: Joannes et Jacobus Meursius, 1647)
© EHC G 4869

Right: the frontispiece of Gregory of Saint-Vincent, *Opus geometricum posthumum ad mesolabium per rationum proportionalium novas proprietates* (Ghent: Balduinus Manilius, 1668)
© EHC G 4868

The frontispiece of *Problema austriacum* displays a projection of sunlight, which, although passing through a square body, projects as a circle on the ground. Notice that on a ray of light the motto "Mutat quadrata rotundis" (The square is transformed into a roundel) can be read

the texts that would result in *Problema Austriacum*, which are volumes 5776–77, 5778, 5779, and 5783.

The other manuscripts were written after the publication of *Problema Austriacum*. Volume 5782 contains a defense against the arguments of the opponents of his circle quadrature. The other volumes relate to the duplication of the cube. These were all written in Ghent (1650–68).

The destruction of a number of San Vicente's manuscripts when the Saxons pillaged Prague makes it difficult to ascertain which of the ideas explored in the

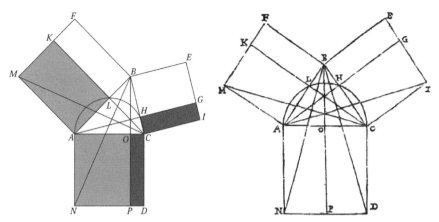

FIGURE 4.4 Gregory of Saint-Vincent's generalization of Pythagoras's theorem. Right: Gregory of Saint-Vincent, *Problema avstriacvm plvs vltra qvadratvra circvli* (Antwerp: Joannes et Jacobus Meursius, 1647), 1:32
© EHC G 4869

post-1631 manuscripts had come to full fruition before 1631. It is clear, however, that anything written after 1635 may have been influenced by Cavalieri's book.

It would not be until 1647 *Problema Austriacum* was published. It remains a book that is hard to read. San Vicente does not use formulae, making the text long and tedious. His methods would only be rendered legible in the works of one of his successors at the school of mathematics, Tacquet. Yet the book would influence Huygens and Gottfried Wilhelm Leibniz (1646–1716), if only by showing which road not to take in infinitesimal calculus. Unfortunately, San Vicente's circle quadrature contains what we would call an integration error, which Huygens noticed.[41] It blemished San Vicente's reputation ever after.

In studies of San Vicente, scholars have always concentrated on his work in infinitesimal calculus, but he has also given us simple theorems that are at the level of secondary-school pupils, though they nevertheless merit some attention. For instance, we owe a generalization of Pythagoras's theorem to San Vicente (fig. 4.4).[42] In an acute-angled triangle, erect the squares on the

---

41   Van Looy, "Chronologie en analyse," 286–94; Joseph E. Hofmann, *Das Opus geometricum der Gregorius à Sancto Vincentio und seine Einwirkung auf Leibniz*, Abhandlungen der Preußischer Akademie der Wissenschaften 13 (Berlin: Berlin Verlag der Akademie der Wissenschaften in Kommission bei Walter de Gruyter u. Co., 1941), 70–71.

42   San Vicente, *Problema Avstriacvm*, 1:33, proposition 55; Thomas L. Heath, *Euclid: The Thirteen Books of the Elements (3 Vols.)* (New York: Dover Publications, 1956), 1:404; Alexander Ostermann and Gerhard Wanner, *Geometry by Its History*, Undergraduate Texts in Mathematics (Dordrecht: Springer Verlag, 2012), 351.

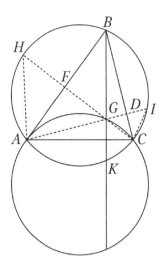

FIGURE 4.5
The proof that the perpendiculars of a triangle intersect one another in the orthocenter G

sides, draw the altitudes from each vertex and extend these into the squares. These altitudes divide the squares into two rectangles. If you shade these rectangles as in figure 4.4, then the regions in the same shade of gray have the same area. He also notes that $\square ACDN = \square ABFM + \square CBEI - 2\square BEGH$, which is the geometric equivalent of the law of cosines.[43]

In one of his manuscripts, written by Moretus after San Vicente had left for Rome, we find in theorem 460 (MS 2.1, fol. 202^v) the property: "In any triangle, the perpendiculars are concurrent." The proof is present in MS 6.9 (fol. 318^r), one of the preparatory texts for *Problema Austriacum*, and in *Problema Austriacum* itself.[44] This proof is supposed to be the oldest proof in the Western world (fig. 4.5).[45]

Draw the circumscribed circle C about triangle $\triangle ABC$ and draw a congruent circle C_{AC} through A and C. Draw the perpendicular GE to AC. Let G be the intersection point of GE with C_{AC}. Now draw AG and CG. Because the sum of two interior angles equals the exterior angle of the opposing vertex, we have in triangle $\triangle BCG$ that $\angle BCG + \angle KBC = \angle KGC$. Because $\triangle KGC$ is isosceles we also have that $\angle KGC = \angle BKC$, which expressed as arcs reads $\overarc{HB} + \overarc{KC} = \overarc{BI} + \overarc{IC}$. The congruence of the circles implies $\overarc{KC} = \overarc{GC} = \overarc{IC}$, therefore $\overarc{HB} = \overarc{BI}$ and AB is the bisectrix of $\angle HAG$.

43　Ad Meskens, "Veralgemening van de stelling van Pythagoras," *Wiskunde en Onderwijs* 45, no. 177 (2019): 26–29.
44　San Vicente, *Problema Avstriacvm*, 3:197, proposition 53.
45　Hofmann, *Das* Opus geometricum *der Gregorius à Sancto Vincentio*, 30.

Furthermore, because $\angle HCA = \angle HCA$, which is an angle common to both circles, we have $\widehat{AG} = \widehat{AH}$ and $|AG| = |AH|$. Therefore, $\triangle AGH$ is isosceles and $AB \perp CH$ (because median, bisectrix, and perpendicular in A coincide).

Analogously one can prove that $BC \perp AI$, from which we conclude that the perpendiculars are concurrent in G, the orthocenter.

3   The Mechanics Theses

The theses of Van Aelst and Ciermans (1624) stand out in San Vicente's *oeuvre* because they deal with mechanics. In 1630, San Vicente was about to publish a book on statics, but the manuscript perished during the Sack of Prague. The only trace of his ideas on mechanics is found in these theses, which, in the Jesuit tradition, were written by the president of the session, in this case San Vicente.[46] In this book, San Vicente treats the statics of inclined planes and free fall. In the most important proposition, San Vicente claims that the kinetic energy depends *only* on the height at which an object is dropped. As was usual in these publications, the proposition was not proven. In other propositions, a distinction is made between energy and inertia, and it is stated that the acceleration along an inclined plane is constant and independent of the initial velocity and that the energy that is gained is independent of the trajectory. Proposition 16 claims that the kinetic energy that is gained during a fall allows an object to again retain its initial position, a proposition that is a limited case of the conservation of energy theorem. This shows that San Vicente and his students were at the cutting edge of mechanics research.

---

46   San Vicente and Van Aelst, *Theoremata mathematica*; Gregorio a San Vicente and Ciermans, *Theoremata mathematica* [...]: *Defenda ac demonstranda in Collegio Societatis Iesv Louanij* [...] *Die 29. Iulij Ante Mtridiem,* [!] *Anno 1624* (Leuven: Henrici Hastenii, 1624). For an extensive description and analysis of these theses, see Dhombres and Radelet-de-Grave, *Une mécanique donnée à voir*. The authors claim that the emblems in the thesis were designed by Ciermans and note a resemblance with the work of Otto van Veen (Vaenius [c.1556–1629]). This theme is further explored in Ralph Dekoninck and Agnes Guiderdoni, "Knowledge in Transition: A Case of 'Scientific Emblematics' (Ciermans and Vaenius) at the Turn of the Seventeenth Century," in *Embattled Territory: The Circulation of Knowledge in the Spanish Netherlands*, ed. Sven Dupré et al. (Ghent: Academia Press, 2015), 279–96. For one drawing, they also note a peculiar resemblance of a celestial object with Saturn and its rings.

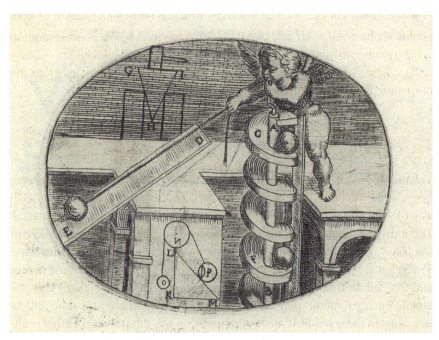

FIGURE 4.6 The proof that speed is independent of the path. Gregory of Saint-Vincent and Walter van Aelst, *Theoremata mathematica* [...]: *Defenda ac demonstranda in Collegio Societatis Iesv Louanij* [...] *Die 29. iulij ante mtridiem,* [*!*] *Anno 1624* (Leuven: Henrici Hastenii, 1624), 9 left
© UGENT BHSL.RES.1449

GREGORIO A SAN VICENTE: AN IGNORED GENIUS

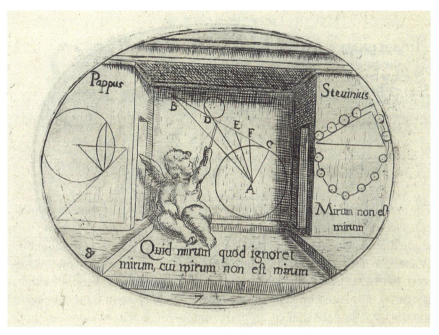

FIGURE 4.7 Comparison between Pappus of Alexandria's and Simon Stevin's theories on inclining planes
The Latin text reads: *"What wonder is it that he, for whom the wonder is no wonder, is ignorant of the wonder."* It is a reference to Stevin's motto—*"Wonder en is gheen wonder"* (*[A] wonder is no wonder*), which can be read in the Latin version (*Mirum non est mirum*) underneath the clootcrans (*string of balls*). Gregory of Saint-Vincent and Walter van Aelst, *Theoremata mathematica* [...]: *Defenda ac demonstranda in Collegio Societatis Iesv Louanij* [...] *Die 29. iulij ante mtridiem,* [*!*] *Anno 1624* (Leuven: Henrici Hastenii, 1624), 8 right
© UGENT BHSL.RES.1449

CHAPTER 5

The Creative Antwerp–Leuven Period

The following paragraphs give an outline of San Vicente's most important achievements (see appendix 1 for the scheme used to refer to the manuscripts).[1]

1 Trisection of an Angle

The main subject of the oldest manuscript (MS 2.1) is the trisection of an angle. San Vicente was clearly aware that the territory had to a great extent been charted already: "De hac materia agunt Vieta, Clavius, Pappus, Aguillonius"[2] (Viète, Clavius, Pappus [of Alexandria], and Aguilón have worked on this subject). To trisect an angle, San Vicente uses verging and pseudo-verging methods similar to those used by Aguilón (see chapter 3). Because of their complexity, the constructions are a far cry from the simple Archimedean *neusis* methods. San Vicente's propositions 5 and 6 (MS 2.1, fol. 3^{r-v}) are literal transcriptions of Aguilón's lemmas 13 and 14. In other propositions, Aguilón's influence is clear, but it is less clear to what extent both mathematicians cooperated before Aguilón's death.

As an example of a pseudo-*neusis*, we have (MS 2.1, fol. 2^v):

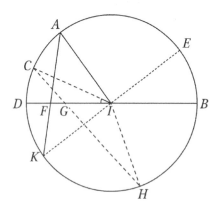

FIGURE 5.1
Saint-Vincent's trisection of an angle using a neusis-like construction

---

1 Co-authored with Herman van Looy.
2 These references are probably to Viète's *Supplementum geometriae* (Tours: Excedubat Iametius Mettayer, 1593), Clavius's edition of Euclid (*Euclidis Elementorum libri XV* [Rome: Apud Vincentium Accoltum, 1574]), and Commandino's edition of Pappus's *Collectiones mathematicae* (Pappus of Alexandria et al., *Pappi Alexandrini mathematicae collectiones* [Pesaro: Apud Hieronymum Concordiam, 1588]).

Suppose the arc $\overparen{AB}$ has to be trisected (fig. 5.1). Choose a point C on $\overparen{AD}$ in such a way that $\overparen{AC}$ is smaller than one-sixth of the circle perimeter. Now construct ∠HCI equal to ∠CID and draw AK such that $\frac{|CG|}{|FK|} = \frac{|AK|}{|CH|}$. Finally draw KI, which intersects the circle in the required point E.[3] The proof can be summarized as: △CIG ~ △CHI. $|CI|^2 = |CH| \cdot |CG| = |FK| \cdot |KA|$ (by construction) and $|KI|^2 = |CI|^2 = |FK| \cdot |KA|$ so $\frac{|FK|}{|KI|} = \frac{|KI|}{|KA|}$, from which △FKI ~ △AIK (isosceles). Hence ∠FKI = ∠FKI and ∠AKE = ∠AKE. Now $\frac{1}{2}\overparen{AE} = \overparen{EB}$ thus $\overparen{EB} = \frac{1}{3}\overparen{AB}$.

After this series of trisections, San Vicente proposes a number of elementary theorems, which give the impression that he is making a list of properties of the circle that can be used in trisections.

In his first series of propositions, San Vicente studies the equality of arcs subtended by straight lines. In proposition 30 (MS 2.1, fol. 16r), we find the first and easiest problem, which is elaborated in the following theorems (fig. 5.2): draw a line from a point A, outside a circle, for which the intersection points subtend a chord equal to a given chord [EF]. The construction is simple: with G as center, a circle tangent to EF is drawn, then a tangent from A to this circle is drawn. This tangent cuts off the desired arc.

In the next theorems (MS 2.1, propositions 31–36, fols. 16v–19r; fig. 5.3), two points A and B are chosen and from each point two secants to the circle are drawn. If $\overparen{IF} = \overparen{LM}$ and $\overparen{HD} = \overparen{KN}$, then $\overparen{KL} + \overparen{MN} = \overparen{IH} + \overparen{FD}$.

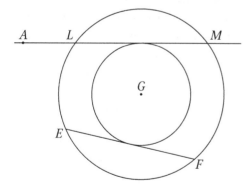

FIGURE 5.2
The construction of a chord [LM] on a line through A, equal to [EF]

---

3 In this pseudo-neusis, we need to find the intersection of the circle with the locus of the points K, such that for a given point A and a straight line DB we have $|AF| \cdot |FK| = k^2$, as F moves along DB. Choosing the x-axis through A and parallel to BD, and putting a equal to the distance of A to DB, we have that $|AF| = a/\sin\theta$. This leads to $|AK| = r = a/\sin\theta + k^2\sin\theta/a \Rightarrow r\sin\theta = a + (k^2r^2\sin^2\theta)/(r^2a) \Rightarrow r^2a(r\sin\theta - a) = k^2r^2\sin^2\theta \Rightarrow a(x^2 + y^2)(y - a) = k^2y^2$.

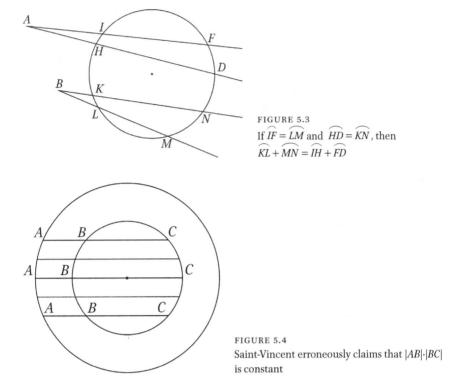

FIGURE 5.3
If $\overset{\frown}{IF} = \overset{\frown}{LM}$ and $\overset{\frown}{HD} = \overset{\frown}{KN}$, then $\overset{\frown}{KL} + \overset{\frown}{MN} = \overset{\frown}{IH} + \overset{\frown}{FD}$

FIGURE 5.4
Saint-Vincent erroneously claims that $|AB|\cdot|BC|$ is constant

San Vicente then cites some similar self-evident theorems and theorems on the equality of line segments in concentric circles (MS 2.1, propositions 36–46, fols. 19^v–24^r).

It is worth noting the following—erroneous—assertion (MS 2.1, proposition 47, fol. 24^v; fig. 5.4): if two concentric circles are given, then for parallel lines $|AB|\cdot|BC|$ is constant. The assertion bears some importance because this is the first time that San Vicente uses parallel line segments. This type of line segment will be essential in San Vicente's transformation theorems of conic sections and for his integration method *ductus plani in planum*.

In one theorem (MS 2.1 proposition 116, fol. 55^r; fig. 5.5),[4] San Vicente claims, and proves, that if on the sides of a triangle circle segments, similar to the segment of the circumscribed circle on the longest side, are described and a line is drawn through a vertex *B*, which intersects the circle segments on [*BA*] and [*BC*] in *G* and *F* respectively and the circumscribed circle in *H*, then $|GB| = |HF|$.

---

4 Referring to fig. 5.5, draw *FC*, which intersects the circumscribed circle again in *I*. Then ∠*AIC* = ∠*ABC*, but ∠*AIC* = ∠*ABC* (because they are on similar chords). Therefore ∠*AIC* = ∠*BFC* and *GF* // *AI*. Similarly, ∠*AGF* = ∠*IFG* and thus |*IH*| = |*AB*|, |*IG*| = |*AF*| and |*FH*| = |*GB*|.

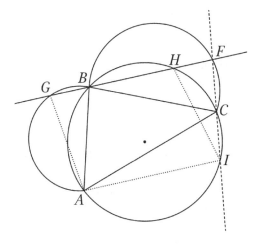

FIGURE 5.5
If the circle segments on the sides are similar to the segment of the circumscribed sector on the side [AC], then |GB|=|HF|

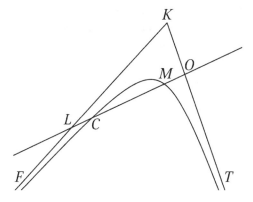

FIGURE 5.6
Witelo's theorem: If |LC| = |MO|, then |LM| = |OC|

On this theorem, San Vicente remarks (fig. 5.6): "This theorem can serve to solve proposition 129 in book 1 of Vitello." Witelo's (c.1230–c.1300) theorem reads: "In an angle ∠FKT a point M is given, through which a straight line has to be drawn making the segment [LM] equal to [CO]. Which will be the case once |LC| = |MO|."

Witelo was a friar, natural philosopher, and mathematician. He wrote *Perspectiva* ([The science of] perspective), which was largely based on the work of Alhazen and in turn influenced later scientists.[5] Witelo himself does

---

5   *Perspectiva* consisted of ten books, of which the first is a mathematical introduction to the other nine books that deal with *Optics*. For his age, Witelo displayed an uncommon mastery of geometry. The first book is based on Apollonius's *Konika*, Euclid's *Elements*, and Alhazen's *Optics*, but also on Eutocius's commentary on Archimedes's *On the Sphere and the Cylinder* and perhaps even some parts of Pappus's *Collection*. See Sabetai Unguru, "Witelo and

not provide a proof for the theorem but refers to Apollonius, *Konika* 2.4.[6] In this theorem, Apollonius constructs a hyperbola given a point on the curve and with two given rays *KL* and *KT* as asymptotes. It is obvious that San Vicente is still following in Aguilón's footsteps and has optics in mind.

2   Mean Proportionals

In MS 2.1, after his studies on the trisection, San Vicente turns his attention to the construction of mean proportionals of two given magnitudes, which again indicates that he is pursuing optical problems. On the other hand, the study of mean proportionals will lead him to one of his greatest discoveries, the logarithmic properties of a hyperbola.

To construct a mean proportional between [*AB*] and [*BC*] (MS 2.1, proposition 203, fol. 97r; fig. 5.7), construct the rectangle *ABCD* and draw the circumscribed circle. Draw a line through the vertex *C*, which intersects *AB* at *E*. Let *G* be the intersection of *EC* and *AD*. Draw a line through *D*, parallel to *EG*. This line intersects the circle in *F*. If |*GE*| = |*DF*|, then, according to San Vicente, [*BH*] and [*HF*] are two mean proportionals between the given line segments [*AB*] and [*BC*].[7]

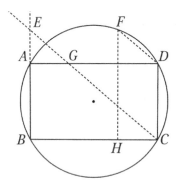

FIGURE 5.7
The construction of two mean proportionals between [*AB*] and [*BC*]

Thirteenth-Century Mathematics: An Assessment of His Contributions," *Isis* 63, no. 4 (1972): 496–508.
6   Heath, *Apollonius of Perga*, 56, and Apollonius Pergae, *Les coniques d'Apollonius de Perge*, 121.
7   The proof of the proposition is straightforward. It is obvious that △*EBC* ~ △*ABH* ⇒ |*EB*|/|*BC*| = |*AB*|/|*BH*| ⇒ |*HF*|/|*BC*| = |*AB*|/|*BH*| (1). Elongate both *FD* and *BC* and call their intersection *J*. $\widehat{BFD}$ is an angle on a diameter and therefore a right angle. So △*BFH* ~ △*FJH* ⇒ |*FH*|/|*BH*| = |*HJ*|/|*HF*| = |*HF*|/|*BH*| = |*BC*|/|*FH*| (2). From (1) and (2), it follows that |*AB*|/|*BH*| = |*BH*|/|*FH*| = |*FH*|/|*BC*|.

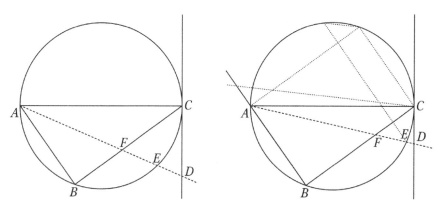

FIGURE 5.8 In proposition 266 (MS 2.1, fol. 127ᵛ), Saint-Vincent states: "If [AB] and [BC] are given, construct a circle on the diameter [AC]. Construct the tangent in C and a secant AF that determines the point D on the tangent such that $\frac{|AF|}{|FE|} = \frac{|FE|}{|ED|}$. Then two mean proportionals between [AB] and [BC] are determined." He fails to mention what these mean proportionals are, but if the drawing, in which the conditions are met, is superimposed on the drawing of proposition 203, this immediately becomes clear (right)

He adds: "Demonstratio pendet a Pappi 31 libri 4" (The demonstration is based on [theorem] 31 in the fourth book of Pappus).[8] In Pappus's proposition, [CE] and [DF] are constructed with the aid of a circle and a hyperbola.

San Vicente performs the construction of two mean proportionals another seventeen times, often without any proof.[9] In many cases, we find a reference to Pappus's *Collectio mathematica* (Mathematical collection), sometimes adding the reference to Commandino's edition.[10] A verging or pseudo-verging method lies at the basis of most of his proofs, which are variations of constructions proposed by Heron of Alexandria (c.10 CE–c.70 CE) in *Mechanics*[11] and

---

8   For this proposition, see Heike Sefrin-Weis, *Pappus of Alexandria: Book 4 of the* Collection, Sources and Studies in the History of Mathematics and Physical Sciences (London: Springer, 2010), 146, 284–85.
9   Propositions 221, 222, 231, 241, 243, 244, 245, 249, 250, 252, 259, 262, 266, 289, 291, 292.
10 KBR V5005 is San Vicente's own copy of Commandino's edition of Apollonius (1566), which he obtained in 1623.
11 Bos, *Redefining Geometrical Exactness*, 28–29. Essentially, Heron's method runs as follows: if the mean proportionals between two line segments of length *a* and *b* are required, construct a rectangle *ABGD* with |AB| = *a* and |BG| = *b* and elongate the sides [BA] and [BG]. Let *H* be the center of the rectangle. Draw a line through *D*. Let the intersections of this line with *BA* and *BG* be *Z* and *E* respectively. Rotate the line about *D* until |HE| = |HZ|, then [AZ] and [GE] are the required mean proportionals. Heron Alexandrinus, *The Mechanics of Heron of Alexandria*, ed. Jutta Miller (1999); http://www.faculty.umb.edu/gary_zabel/

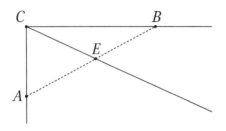

FIGURE 5.9
Nicomedes's construction of two mean proportionals

are often present in Pappus's *Collectio mathematica*. In looking for two mean proportionals, San Vicente sometimes has to construct three line segments in continued proportion (fig. 5.8).

The construction that is ascribed to Nicomedes (c.280–c.210 BCE), and was used by Viète, is also cited by San Vicente.[12]

San Vicente only records the most essential (MS 2.1, proposition 447, fol. 197^v; fig. 5.9):

> Nicomedes's construction method [of the trisection of an angle] is very general and can be used both for the solution of finding two mean proportionals and for performing the trisection [of an angle]. After all, the solutions of both problems are based on the lemma: "Given an angle *BCD* and a point *A* outside it, draw a line *AB* through *A* that cuts off [a line segment] *EB* equal to the known [line segment] *CA*."

San Vicente again asserts the equivalence of finding mean proportionals and the trisection of an angle (MS 2.1, fol. 213^v): "[He] who finds two mean proportionals also finds the division of the angle into three parts and also finds the reflection point." Finding the reflection point is still San Vicente's main aim (MS 2.1, proposition 512, fol. 218^v): "If two points *A* and *B*, outside the circle *CDE*, are given, the reflection point, for which the shortest path is obtained, is required" (fig. 5.10).

San Vicente does not give an explanation for his construction, just a brief note: "Vide 426." Proposition 426 (MS 2.1, fol. 191^v; fig. 5.11) mentions that, if *D* is the middle of [*AC*] in the triangle △*ABC* and if *BF* is parallel to *AC*, then $\frac{|FE|}{|ED|} = \frac{|FG|}{|GD|}$. In other words, in the figure (*F, D, E, G*) are a harmonic quadruple.

Courses/Bodies,%20Souls,%20and%20Robots/Texts/mechanics_of_heron_of_alexandria.htm (accessed March 28, 2020), *Mechanica* 11.

12 Archimedes, *Archimedis opera omnia cum commentariis Eutocii*, ed. J. L. [Johan Ludvig] Heiberg (Leipzig: Teubner, 1910–15), 3:106; Sefrin-Weis, *Pappus of Alexandria*, 148, 285–86.

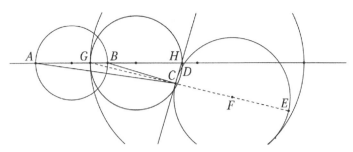

FIGURE 5.10   Saint-Vincent's attempt to solve Alhazen's problem

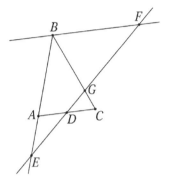

FIGURE 5.11
The construction of a harmonic quadruple using a triangle

The relation with proposition 512 becomes clear from San Vicente's remark: "Fundamentum huius speculationis consistet in hoc ut sit BG ad GA ita BH ad AH" (The basis of this hypothesis is the fact that BG is to GA such as BH is to AD). This implies that in figure 5.10 (A, B, G, H) are a harmonic quadruple and (CA, CB, CG, CH) a harmonic pencil of rays.

With this theorem, San Vicente's attempts to solve Alhazen's problem came to an end. He is gradually drifting away from Aguilón's program, and in a last effort he turns to trisections using a cissoid (MS 2.1, propositions 453–55, fols. 200^v–201^r) or a quadratrix (MS 2.1, proposition 483, fol. 208^v). From now on, his attention will be focused on conic sections.

To find the two mean proportionals between [OD] and [OF] (fig. 5.12), again draw the rectangle ODGF and its circumscribed circle. Now draw a straight line NM through G such that either $|LM| = |LN|$ (MS 2.1, proposition 609, fol. 209^v) or $|LM| = |OL|$ (MS 2.1, proposition 611, fol. 241^v).

In both cases, San Vicente states that E, the other intersection point of NM with the circle, determines the two mean proportionals.

Again, this is a variant on Heron's construction, who indicates that the mean proportionals are [MD] and [NF]. Everything therefore revolves around determining the point E.

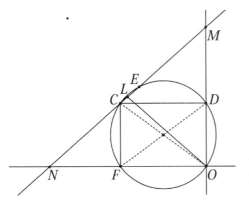

FIGURE 5.12
The construction of two mean proportionals using a rectangle and the circumscribed circle

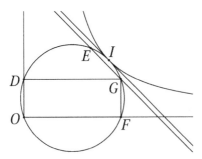

FIGURE 5.13
The construction of two mean proportionals using a circle and a hyperbola

This is solved in a new proposition (MS 2.1, proposition 612, fol. 242ʳ; fig. 5.13): if a hyperbola is constructed with *OD* and *OF* as asymptotes and it is tangent to the circle in *I*, then a line through *G* and parallel to the common tangent of the circle and the hyperbola will intersect the circle in the desired point *E*.

These are the first instances in San Vicente's manuscripts where conic sections are explicitly used to solve the problem of the two mean proportionals. This variation on Heron's solution again finds its inspiration in Pappus's *Collection*. In book 3, after proposition 4, Pappus states that Heron and Philon (*fl.* fourth century BCE) perform this construction in the same way as Apollonius, using conic sections.[13] He did not give this solution, which may explain why San Vicente tried his hand at it.

After some theorems about ratios of line segments and about the *latus rectum* (the parameter; the focal chord parallel to the directrix) and *latus transversum* (the line segment of the principal real diameter between the vertices) in a parabola and a hyperbola, San Vicente tries to determine two mean proportionals using a parabola (MS 2.1, proposition 628, fol. 256ʳ; fig. 5.14).

---

13   Pappus of Alexandria, *La collection mathématique*, 39.

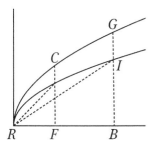

FIGURE 5.14
The construction of two mean proportionals between [IB] and [BR] using parabolae

Consider the parabolae RG and RI, with common vertex and common axis. If the parabola GRB (with |RB| = |BG|) is intersected by CF in such a way that the parabolic segments RIB and RCF are equal, then [RF] and [FC] are two mean proportionals between [IB] and [BR]. To San Vicente, this property is self-evident as it immediately derives from *Konika*.

From the similarity of the parabolic segments RIB and RCF and the corresponding triangles, it follows that $|RB|.|BI| = |RF|.|FC|$. Moreover, $\dfrac{|RF|}{|RB|} = \dfrac{|CF|^2}{|BG|^2}$ (property parabola) with $|BG| = |RB|$, so $|RF| = \dfrac{|CF|^2}{|RB|}$ or $|RF|.|RB| = |CF|^2$. Combining these two relations, we have $\dfrac{|LB|}{|RF|} = \dfrac{|RF|}{|CF|} = \dfrac{|CF|}{|RB|}$.

3   Properties of Conic Sections

It is often underscored that San Vicente was among the first mathematician to study conic sections in depth. Again, the belated publication of his book adds to the impression that he was not original. However, most of his work on conic sections dates from before 1625. Already in this period, some characteristic features are visible.[14]

By the time San Vicente's book was published, René Descartes (1596–1650) had published *La géométrie* (Geometry [1637]), the book in which he introduced coordinate geometry. In the 1610s and 1620s, however, geometry was still confined by the limits posed by the ancient Greeks, for conic sections in particular by Apollonius. Yet San Vicente had broken free of these bonds and

---

14   For a detailed analysis of these features, see Jean Dhombres, "L'innovation comme produit captif de la tradition: Entre Apollonius et Descartes, une théorie des courbes chez Grégoire de Saint-Vincent," in *Geometria, flussioni e differenziali*, ed. Marco Panza and Clara Silvia Roero (Naples: La Città del Sole, 1995), 17–102, 42ff.

was thus able to find new and interesting results. Very primitive steps toward coordinate geometry can be discerned in the early work of San Vicente.

Although San Vicente does not use a coordinate system as such, there are privileged straight lines serving as axes from which an abscissa and an ordinate are defined. The intersection of these axes is extremely important to San Vicente, because it allows him to compare proportions of abscissae and the related ordinates.

San Vicente tries to find, usually by seeking a constancy of a certain proportion between six elements, a relationship between abscissae and ordinates of a curve. This often allows the elementary study of the curve at hand. It is one of the first instances where a path toward the algebraization of geometry becomes visible.

In most cases, the curves are characterized by a relation between a geometric series of the abscissae and an arithmetic series of the ordinates.

The proportions furthermore allow San Vicente to avoid a parametrization of the curve. These procedures, however, force San Vicente to view each curve as a different species, whereas in Descartes's formalism all conic sections can be regarded as second-degree curves.

Another characteristic is that San Vicente uses line segments to transform one figure into another. For instance (fig. 5.15), suppose that $\triangle ABC$ is a right-angled triangle.[15] Draw line segments in the triangle parallel to AB. These lines intersect AC in E_i and BC in D_i. Perpendicular to AC, line segments $[E_i F_i]$ are erected, such that $|BD_i|.|D_i E_i| = |E_i F_i|^2$.

We can view this construction of an ellipse as an early example and a special case of the MacLaurin–Braikenridge theorem,[16] *if* we accept a point at

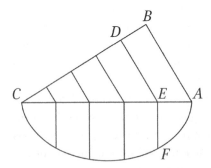

FIGURE 5.15
Transforming a triangle into an ellipse using the *per subtensas* method: $|BD_i|.|D_i E_i|=|E_i F_i|^2$

---

15 San Vicente, *Problema Avstriacvm*, 4:322, proposition 149.
16 If n straight lines rotate about n fixed points and if $(n-1)$ of the $n(n-1)/2$ intersections move along a straight line, then the other $(n-1)(n-2)/2$ intersections describe a conic section.

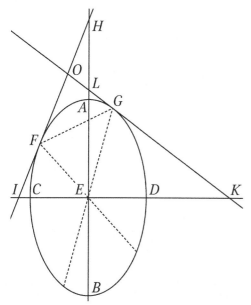

FIGURE 5.16
Properties of associated diameters and tangents in an ellipse as an aid for constructing mean proportionals

infinity. The lines D_iE_i are parallel lines, as are E_iF_i and D_iF_i. These pencils of lines all "rotate" about their point at infinity, D_i and E_i move along a line, thus F_i describes a conic section. It is an example of San Vicente's transformation method *per subtensas* (by subtension, a transformation using parts of chords).

The part of MS 2.1 dealing with conic sections (fols. 240–75) is a continuation of the part on the trisection of an angle and the determination of two mean proportionals. It is a new line of inquiry for San Vicente: Can these problems be solved using conic sections?

In the first instance, San Vicente's attention is drawn to the ellipse and the properties of associated diameters and tangents. For instance (fig. 5.16), in propositions 630 and 631 San Vicente considers the conjugate diameters [EF] and [EG]. The tangent lines in E and G intersect the axes in I, H and L, K respectively. The segments [HI] and [KL] are divided by E and G respectively in the same ratio. Moreover, $\dfrac{|LK|}{|IH|} = \dfrac{|EF|}{|EG|}$ (MS 2.1, fols. 257^v–259^r). The proofs make use of similar triangles.

Proposition 641 (MS 2.1, fol. 261^v; fig. 5.17) is a construction of the tangents to an ellipse.

For a given ellipse, construct the foci ("poli"). Elongate the minor axis [CD] and draw a circle centered on this elongation.

Let K and M be the intersections of CD and the circle. Let L be an intersection of the ellipse and the circle. The proof is then very short: "*KLM* is a right

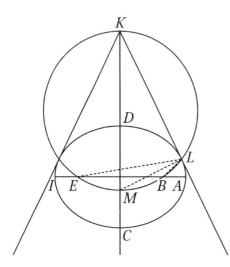

FIGURE 5.17
The construction of tangents from a point *K* to an ellipse

triangle and ∠*ELM* = ∠*MLB*." This explanation suffices, because the normal bisects the angle between the lines to the foci.

It is clear that San Vicente is repeating the properties of an ellipse taken from Apollonius's *Konika*. In doing so, he has lost sight of the goal he had set himself, determining the mean proportionals. He is now studying the properties of conic sections *per se*.

A part of MS 2.2 is an unfinished project about conic sections, but it is not clear whether it is a continuation of MS 2.1 or whether it was intended as a course. In MS 2.2, we find a new line of attack to determine mean proportionals.

In proposition 671 (MS 2.2, fol. 10^r; fig. 5.18), San Vicente tries to find mean proportionals in a parabola: "To construct two mean proportionals between lines that are to each other such as 3 is to 18, proceed like this: if a parabola [*BAC*] has diameter *AD*, divide *AB* and *AC* into equal segments *AEB*, *AFC* at *E* and *F* and draw *EF*. I say that the cube on *BC* is to the cube on *EF* as 18 is to 3," which leads to the conclusion that |*BC*| and |*EF*| are two mean proportionals between 18 and 3.

For an analogous division of *EIB* and *FKC*, San Vicente states that $\dfrac{|IK|^3}{|EF|^3} = \dfrac{73}{22}$ and that |*IK*| and |*EF*| are two mean proportionals between 73 and 22. This assertion is wrong. In the margin, he notes that "this calculation is wrong, but it will be corrected in the [final] proof for the book." This is most probably the reason why he investigates the following propositions involving ordinates in a parabola.

THE CREATIVE ANTWERP–LEUVEN PERIOD   107

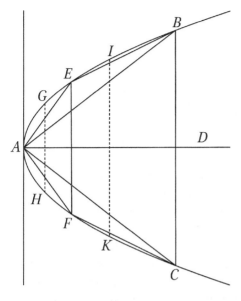

FIGURE 5.18
The construction of two mean proportionals in a parabola

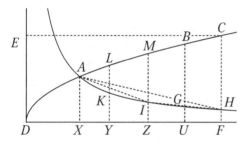

FIGURE 5.19
The construction of two mean proportional in a parabola using a hyperbola

Initially, San Vicente tries to construct segments in a parabola that are in continuous proportion. He tries to solve this problem using a hyperbola (MS 2.2, proposition 673, fol. 8^r; fig. 5.19):

To divide the parabolic segment *ABC* (i.e., the segment bounded by the chord *AC*) in four segments in a continuous proportion, San Vicente constructs a hyperbola through *A* with *DF* (the axis of the parabola) and *DE* (the tangent at the vertex) as asymptotes.

The hyperbolic segment *AH* is divided in the middle by *I*, and the segments *AI* and *IH* are divided into equal parts by *K* and *G*. San Vicente now draws *KL*, *IM*, and *GB* parallel to *DE* and concludes that the segments with chords *AL*, *LM*, *MB*, and *BC* are in continuous proportion.

In the next proposition, San Vicente asserts another relation between the hyperbola and parabolic segments (fig. 5.19). He claims that the previous

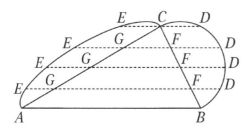

FIGURE 5.20
Construction of an ellipse segment *AEC* with an area equal to that of a semi-circle *BDC*

division can be achieved if *DF*, *DU*, *DZ*, *DY*, and *DX* are in continuous proportion. He then claims that the hyperbolic segments *AXYK*, *KYZI*, and so on are equal. All these assertions are correct.[17] He most probably reached the equality of hyperbolic segments using the equality of inscribed rectangles *AX.XY*, *LY.YZ*, and so on, but this cannot currently be ascertained from the manuscripts. It is nevertheless an important new result: to conclude from the proportionality of the abscissae that the corresponding hyperbolic segments are equal. In fact, San Vicente has unwittingly stumbled upon what is perhaps his greatest achievement, the discovery of the logarithmic properties of a hyperbola (see chapter 7).

The first theorems in MS 2.2 deal with the comparison of the ellipse with the circle.[18] An important remark in this context is what is written in the corollary to proposition 1 (MS 2.2, fol. 345^v): "Circles and ellipses do not differ more from each other than ellipses among themselves." More than anything else, San Vicente investigates the similarities and the relations of circle and ellipse segments.

An important theorem in this matter reads as follows:

> On the side [*BC*] of a triangle, right-angled or not, a circle is described. If the parallel lines [*DE*] are drawn parallel to the base [*AC*], and the lines of [*FD*] are transferred to [*GE*], then
> 1. an ellipse will be described by the points *E*,
> 2. the area of this ellipse segment will be equal to the area of the circle segment [MS 2.2, proposition 6, fol. 343^r; fig. 5.20].

This theorem is the first of many transformation theorems that will prove to be ideally suited for the exhaustion method (see chapter 6). For now, San Vicente hardly uses his transformation theorems, and he will later move lines of a conic

17   These propositions are repeated in propositions 694 and 695 on fol. 289.
18   While the propositions are numbered in ascending order, the pages are numbered in descending order from 345 to 1 in this manuscript.

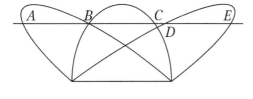

FIGURE 5.21
The subtended chords on a line parallel to the common base of semi-ellipses have an equal length

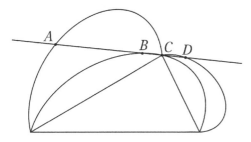

FIGURE 5.22
An attempt to generalize proposition 116 (fig. 5.5): Is the proposition still valid if circles are replaced by ellipses?

section (actually ordinates) to become the ordinates of another conic section. The displacement of these lines can be used when comparing areas, especially if these lines are considered in the sense of Cavalieri's *indivisibilia*. However, San Vicente is extremely careful when using these "lines" and only sees them as the bases of rectangles that lead to an exhaustion of the surface. Sometimes, San Vicente uses exhaustion without explicitly mentioning it.[19] This is the case in the proof of the second assertion of the previous theorem: the ellipse segment is equal to a semicircle.

Comparing the lines is again a subject in proposition 24 (MS 2.2, fol. 333^r, fig. 5.21): if one compares equal semi-ellipses on the same base, then their maximal distances to the base are equal. If parallel lines to the base are drawn, then the subtended chords are equal (i.e., $|AB| = |BC| = |DE|$).

San Vicente now puts to himself the question of whether, as is the case for similar circle segments on the sides of a triangle, $|AB|$ is equal to $|CD|$ in ellipse sectors as well (MS 2.2, fol. 327^r; fig. 5.22).

San Vicente now moves on to calculate the area of an ellipse. A first assertion in this direction is: if ellipse segments *ACB* and *ADB* have equal areas, then this is also the case for the ellipses themselves (MS 2.2, proposition 57, fol. 316^r; fig. 5.23).

Another assertion is given: if [*AC*] and [*BD*] are two conjugate diameters and similar ellipses are constructed with [*AB*] and [*BC*] as conjugate diameters,

---

[19] His prudence with regard to this issue becomes clear in MS 4.2, written in Prague, in which he seems to have entertained ideas similar to those of Cavalieri but rejected them because they sometimes led to false results (fols. 60^v, 79^v).

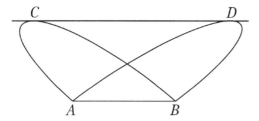

FIGURE 5.23
If the ellipse segments *ABC* and *ABD* have an equal area, then so will the ellipses themselves

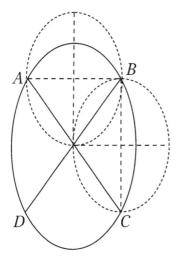

FIGURE 5.24
Properties of conjugate diameters

then the sum of the areas of these ellipses is equal to the area of the given ellipse (MS 2.2, proposition 93, fol. 300ʳ; fig. 5.24). A similar problem involves finding a circle that has an area equal to a given ellipse (MS 2.2, proposition 97, fol. 298ʳ).

Here, a square, which is equal to the parallelogram with two conjugate diameters as diagonals, is constructed, and then a circle through the four vertices of the square is described. A source of inspiration for these theorems can be found in Archimedes's *On Conoids and Spheroids*, in which he proves that two ellipses relate to one another as the products of their axes.[20] In all probability, San Vicente came to these conclusions by comparing the inscribed and circumscribed parallelograms that can be constructed with the use of the conjugate diameters. In this case, one needs another property to relate the product (i.e., the rectangle) of the axes to the parallelogram on the two conjugate diameters. This property is found later (MS 2.2, proposition 88, fol. 302ʳ): in any ellipse, the parallelograms on two conjugate diameters are equal. This theorem

---

20   Archimedes, *Les œuvres complètes d'Archimède*, 155.

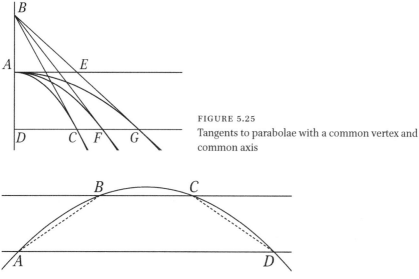

FIGURE 5.25 Tangents to parabolae with a common vertex and common axis

FIGURE 5.26 Parabolic segments between parallel lines are equal

is referred to as Apollonius's theorem. That the theorem is a proposition in *Konika 7*, which was first published in 1661,[21] suggests that manuscript copies of this book circulated.

Most of the theorems in the following part (MS 2.2, propositions 101–209, fols. 297–247)[22] deal with the parabola and the hyperbola. As in the previous part, attention goes to the calculation of the area.

At first (fig. 5.25), San Vicente considers parabolae that have a common vertex *A* (and, moreover, a common axis *AD*), and lets *BC* be tangent to one of the parabolae.

Let *DC* be the tangent chord. And let *F* and *G* be the intersection points of *DC* with the other parabolae, then *BG* and *BF* are tangent to these parabolae. Moreover, the parabola segments are to one another as the triangles △*BDC*, △*BDF*, and △*BDG* are to one another (MS 2.2, proposition 102, fol. 296^r).

In the next theorem (MS 2.2, proposition 104, fol. 294^v; fig. 5.26), we find: "If *AD* and *BC* are parallel, then the segment *AB* has the same area as the segment *CD*. This property holds for any conic section." This is an assertion that again

---

21   Giovanni Alfonso Borelli, *Apollonius Pergaeus Conicorum lib. V. VI. VII* (Florence: Giuseppe Cocchini, 1661).
22   The page numbers go backward in this manuscript.

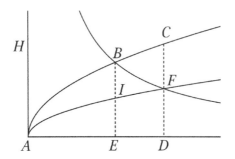

FIGURE 5.27
The proportion between the areas of the parabolic segments *ABE* and *AIFD* can be expressed in terms of the lengths of [*DF*] and [*FC*]

makes clear that San Vicente had a good command of the theory of conic sections (MS 2.2, proposition 123, fol. 288ᵛ; fig. 5.27).

San Vicente then goes on to consider two parabolae with the same axis and the same vertex. Let *B* be a point on the parabola. Draw the hyperbola through *B*, which has the tangent at the vertex and the axis of the parabola as asymptotes. Let the hyperbola intersect the second parabola in *F*. Then the parabolic segments *ABE* and *BCDE* are to one another as *DF* is to *FC*. The proof is very short: the segment *ABE* is equal to the segment *AIFD*, and the segments *AIFD* and *ACD* are to one another as [*DF*] and [*FC*]. This suffices as proof because the rest of the proof is easily reconstructed:

$$\frac{\text{segm } AIFD}{\text{segm } ACD} = \frac{|DF|}{|DC|} \Rightarrow \frac{\text{segm } ACD - \text{segm } AIFD}{\text{segm } AIFD} = \frac{|DC| - |DF|}{|DF|}$$

$$\Rightarrow \frac{\text{segm } ACD - \text{segm } ABE}{\text{segm } ABE} = \frac{|CF|}{|DF|} \Rightarrow \frac{\text{segm } BCDE}{\text{segm } ABE} = \frac{|CF|}{|DF|}$$

That San Vicente is certain that the segment *ABE* is equal to the segment *AIFD* has already been asserted a few pages earlier (MS 2.2, proposition 106, fol. 295ʳ) with the proof: it is valid for the triangles △*AEB* and △*AFD* in the hyperbola. To him, this last assertion is the most elementary because |*AE*|.|*EB*| = |*AD*|.|*DF*|, which is nothing else than the "equation" of the hyperbola.

A last theorem about areas in a hyperbola states: "Hyperbolic segments with the same vertex and axis are to one another as the bases that are on the same straight line" (MS 2.2, proposition 202, fol. 250ʳ).

Two transformation theorems are new. The first (MS 2.2, proposition 199, fol. 252ʳ) reads: "If *AB*, *BC*, and *CD* are in continuous proportion and if the points *D* are on a straight line, then the points *C* are on a hyperbola with *AF* as *latus rectum* [parameter]" (fig. 5.28).

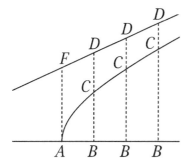

FIGURE 5.28
If *AB*, *BC*, and *CD* are in continuous proportion, and if the points *D* are on a straight line, then the points *C* are on a hyperbola

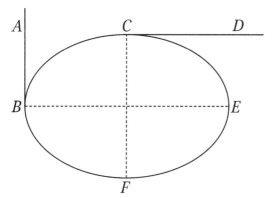

FIGURE 5.29
The construction of an ellipse with [*AB*] and [*CD*] as *latera recta*

San Vicente here transforms the straight line *DD* into a hyperbola *CC*. There is no proof, but the conclusion is correct.[23] He goes on to construct two mean proportionals between [*AB*] and [*CD*] with the aid of conic sections (MS 2.2, proposition 205, fol. 248^v; figs. 5.29 and 5.30). To do so, San Vicente constructs an ellipse of which the axes have [*AB*] and [*CD*] as *latus rectum*.

For such an ellipse, the axes are two mean proportionals between [*AB*] and [*CD*]. In other words:

$$\frac{|AB|}{|CF|} = \frac{|CF|}{|BE|} = \frac{|BE|}{|CD|}$$

or

$$\frac{p}{2b} = \frac{2b}{2a} = \frac{2a}{p'}.$$

---

23 Put $A(a, 0)$, $B(x, 0)$, $C(x, y)$, and $D(x, mx)$. Then, the continuous proportion yields $(x-a)/y = y/(mx-y)$, which reduces to the equation of a hyperbola.

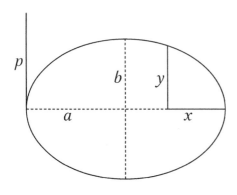

FIGURE 5.30
The construction of two mean proportionals using an ellipse

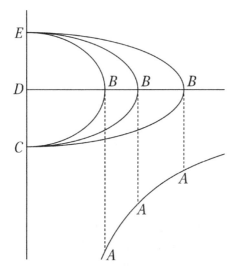

FIGURE 5.31
The extremities A of the *latera recta* [BA] of ellipses with a common axis EC are on a hyperbola

Hence $p.a = 2b^2$, which is a direct consequence of Apollonius's *Konika* 1.21.[24]

This lemma allows San Vicente to formulate a transformation theorem (MS 2.2, proposition 205, fol. 248^v; fig. 5.31): if BA are the *latera recta* of the ellipses on the common axis EC, then the points A are on a hyperbola. The proof is short: all rectangles (i.e., |DB|.|BA|) are equal to twice the square on DC.

From fol. 246 onward, San Vicente lets an unknown student write down his ideas. In this forty-page part, there are two main subjects, the first of which is inscribing the largest quadrilateral in a conic section and the second is comparing the ellipse with the circle.

---

24   Heath, *Apollonius of Perga*, 19; Apollonius Pergae, *Les coniques d'Apollonius de Perge*, 43.

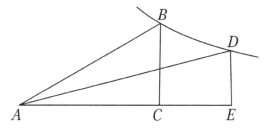

FIGURE 5.32
The vertices of right-angled triangles with a common vertex A and with one of their perpendiculars on a common line are on a hyperbola

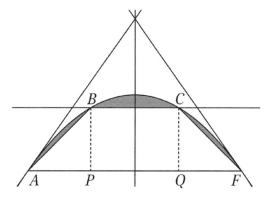

FIGURE 5.33
How can the largest inscribed quadrilateral in a parabola be constructed?

In a new section, San Vicente begins with relatively easy theorems such as: if the triangles △ABC and △AED are equal, then B and D are on a hyperbola with vertex A (MS 2.2, proposition 216, fol. 244ʳ; fig. 5.32). This is nothing new, as this result was used by San Vicente in the second transformation theorem. On the next page, we find the condition for the ellipse: all isoperimetric triangles on the same base have vertices on the ellipse.

For a parabola, San Vicente finds that one will obtain the largest inscribed quadrilateral on the chord [AF] if one constructs equal segments AB, BC, CF (MS 2.2, propositions 235–36, fol. 235ᵛ; fig. 5.33; an analogue proposition for the hyperbola is proposition 231, fol. 237ᵛ).

On the same page, the following problem is posed: "To find the largest inscribed polygon with a given number of sides in a parabolic segment." This time, San Vicente has written down the solution: "Divide [AF] in as many equal parts as the number of sides added to [AF] and draw parallels with the axis through these points."[25] This is applied by the unknown scribe (MS 2.2,

---

25   Consider the parabola $y = x^2$, the fixed points $A(a, a^2)$ and $F(a, a^2)$, and the variable points $B(t, t^2)$ and $C(t, -t^2)$. Then the trapezoid AFCB will have its maximum area for $t = \frac{1}{3}a$. By trisecting the line segment [AB], one can construct the inscribed trapezoid with maximum area. It can easily be checked that the parabolic segments are equal.

proposition 247, fol. 229ᵛ) who trisects *AF* by *P* and *Q* and finds *B* and *C*. Again, a proof is lacking, but there is a reference to Archimedes's *De quadratura parabolae* (On the quadrature of the parabola). Archimedes applied the quadrature of the parabola by inscribing triangles,[26] and now San Vicente tries to generalize this procedure.

---

26   Archimedes, *Les œuvres complètes d'Archimède*, 377–404.

CHAPTER 6

Exhaustion: The Road to Infinitesimals

1 Sequences and Series[1]

A part of MS 2.1 was written by Moretus after San Vicente had left for Rome. This manuscript holds the key to San Vicente's ideas and accomplishments: the treatment of geometric series. Two theorems, independent of the other theorems in this part, merit some attention.

Theorem 460 (MS 2.1, fol. 202^v) states a property nowadays known to every secondary school pupil: "In any triangle, the perpendiculars are concurrent."

The second theorem is even more important because it introduces properties of *infinite* sequences. Proposition 479 (MS 2.1, fol. 207^r) states that:

> If one takes away from [AB] the half [AC], of the remainder [CB] again the half [DB], of what then remains again the half [CE], of the remainder [ED] again half [DF], of [AB] again half [EG] and so on. I say that *the end of this sequence* will be where [AB] is trisected [fig. 6.1].[2]

By successively removing half of the segment, which was obtained in the previous step by the same procedure, and by alternating between sides, one ultimately reaches the point where the line is trisected. San Vicente's proposition amounts to $\sum_{i=0}^{n}\left(\frac{-1}{2}\right)^i = \frac{2}{3}$.

In other propositions, San Vicente will use the word "terminus" (meaning the limit of a sequence) for this "end of the sequence." Unfortunately, he never defines this concept in his manuscripts, though he does attempt to clarify its meaning in *Problema Austriacum*: "Terminus progressionis est seriei finis, ad quem nulla progressio pertinget, licet in infinitum continuetur, sed quovis intervallo propius ad eum acedre poterit" (The terminus of a progression is the end of the series, which no progression can reach, not even if it is continued to infinity, but which it can approach nearer than any given segment).[3] This is as

---

1 This chapter is co-authored with Herman van Looy.
2 "Auferatur ex AB dimidia AC ex reliqua CB dimidia DB et reliqua DC dimidia CE et ex reliqua ED auferatur DF et ex reliqua EF auferatur dimidia EG et sic in infinitum. Dico *finem progressionis* esse ubi dividitur linea AB triseriam."
3 San Vicente, *Problema Avstriacvm*, 2:55, definition 3.

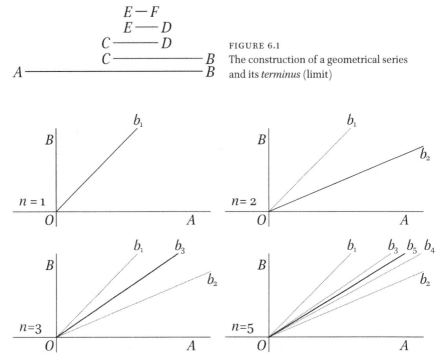

FIGURE 6.1 The construction of a geometrical series and its *terminus* (limit)

FIGURE 6.2 Trisecting a right angle by infinitely many bisections. Begin with an angle ∠AOC (in this figure, we have used a right angle to be trisected) and construct the interior bisector b_1. For the angle with sides OA and b_1, construct the interior bisector b_2. Now construct the interior bisector b_3 of b_1 and b_2, the interior bisector b_4 of b_2 and b_3, the interior bisector b_5 of b_3 and b_4, and so on

close to the definition of a limit one can get, given the tools at San Vicente's disposal. It is the first time that San Vicente, or anyone else for that matter, formulates a proposition for infinite series and *de facto* accepts the limit of the series. This is nothing other than a giant step forward, even if, for now, the assertion is limited to one geometric series. What San Vicente does is nothing less than accepting that this infinite sum *is equal to* a number. Or, in his frame of mind, that the result of removing an infinity of lines from a segment is a point.

San Vicente immediately recognizes that the same procedure can be applied to establish the trisection of an angle. By doing so, he can trisect an angle using only ruler and compass methods, unfortunately not in a finite number of construction steps (fig. 6.2).[4]

After a couple of paragraphs on the properties of conic sections, San Vicente returns to the subject of infinite sequences and series (MS 2.1, fols. 276–309).

---

4 San Vicente, *Problema Avstriacvm*, 2:111–12, proposition 107.

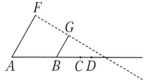

FIGURE 6.3
The determination of the *terminus* (limit) of a geometric series

Among other propositions, we find the equivalent of $\lim_{n\to+\infty}\sum_{i=0}^{n}\frac{1}{2^i}=2$ (MS 2.1, proposition 720, fol. 300^r).

Another interesting theorem (MS 2.1, proposition 722, fol. 301^r) is: if [AB] is divided at any point C, [CB] by D in the same proportion, [CD] by E and [ED] by F and so on then the position of the terminus X of the sequence will be such that $\frac{|AX|}{|XB|}=\frac{|AB|}{|CB|}$.

This theorem is the equivalent of $\lim_{n\to+\infty}\sum_{i=0}^{n}(-x)^i=\frac{1}{1+x}$, the generalization of earlier results.

In MS 2.3, Boelmans describes twenty-nine properties of sequences, including the exhaustion theorem.

Proposition 5 (MS 2.3, fol. 220^r; fig. 6.3) gives the construction of the terminus of a (decreasing) geometric sequence.[5] If AB, BC, CD, and so on are a geometric sequence, the terminus is found by constructing the parallel line segments [AF] and [BG], equal in length to [AB] and [BC] respectively. The intersection of the straight line FG is the terminus of the sequence.

The importance of this theorem cannot be underestimated. In constructing the terminus, San Vicente gives it a physical reality. This explains why he accepts the terminus as an admissible mathematical entity.[6]

After a couple of theorems on the division of a line segment into a sequence, Boelmans tries to generalize a well-known property of a geometric sequence.

In proposition 12 (MS 2.3, fols. 223^v–224^r; fig. 6.4), Boelmans proves that if [AB], [BC], [CD], and so on are a geometric sequence with ratio 1/2 and terminus E, then the square on [AE] (i.e., the square AF) is triple the sum of the squares on [AB], [BC], [CD]. Boelmans actually states:

$$3\left(a^2+\left(\frac{a}{2}\right)^2+\left(\frac{a}{4}\right)^2+\left(\frac{a}{8}\right)^2+\ldots\right)=(2a)^2$$

$$\text{or } a^2\left(1+\frac{1}{4}+\frac{1}{16}+\ldots\right)=\frac{4}{3}a^2$$

---

5 The construction is repeated in MS 3.2, fol. 467^r and MS 7.5, proposition 5 (40), fol. 105^r.
6 San Vicente, *Problema Avstriacvm*, 2:98.

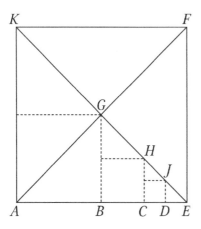

FIGURE 6.4
The determination of the limit of $\sum_{i=0}^{n} \dfrac{a^{2}}{2^{2n}}$

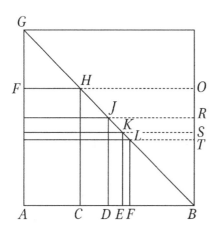

FIGURE 6.5
The relation between certain lines and certain areas in a square, from which it follows that
$$\sum_{i=0}^{\infty} q^{i} = \frac{1}{1-q} \,(q<1)$$

a result that was already known to Archimedes.[7]

In proposition 13 (MS 2.3, fols. 224^{v}–225^{r}; fig. 6.5), he proves that if $|AC|$, $|CD|$, $|DE|$, and so on are a geometric sequence with terminus B then[8]
$$\frac{|AB|^{2}}{\square HA + \square JC + \square KD + \cdots} = \frac{|GA|+|PA|}{|PA|}.$$

---

7 *De quadratura parabolae*, proposition 23, see Archimedes, *Les œuvres complètes d'Archimède*, 401.
8 If we put $|AC| = a < 1$, then $|AB| = |GA| = \dfrac{a}{1-a}$ and $|PA| = \dfrac{a^{2}}{1-a}$. Then $\square HA = S_{0} = \dfrac{a^{3}}{1-a}$. The other rectangles are similar to $\square HA$ by a ratio a^{2}. Therefore $S = \square HA + \square JC + \square KD + \ldots = \sum_{i=0}^{\infty} S_{i} = \dfrac{a^{3}}{1-a} \sum_{i=0}^{\infty} a^{2i} = \dfrac{a^{3}}{(1-a)^{2}(1+a)}$, and also $|AB|^{2} = \dfrac{a^{2}}{(1-a)^{2}}$ whence $\dfrac{|AB|^{2}}{\sum_{i=0}^{\infty} S_{i}} = \dfrac{1+a}{a}$. On the other hand, $\dfrac{|GA|+|PA|}{|PA|} = \dfrac{a+a^{2}}{a^{2}} = \dfrac{1+a}{a}$.

EXHAUSTION: THE ROAD TO INFINITESIMALS

This property opens up many perspectives, including the possibility of calculating the sum of the squares □GH, □HJ, □JK, □KL, and so on.[9] This also allows us to deduce that $1+q+q^2+q^3+\ldots=\dfrac{1}{1-q}(q<1)$,[10] a corollary that Boelmans does not mention here.

The last theorem, 29 (MS 2.3, fol. 238r), is written by San Vicente himself: "[AB] is divided by C, between [AB] and [AC] the mean proportional [AD] is constructed, between [AD] and [AC] the mean proportional [AE], and so on. The terminus of this sequence will be thus that [CB] is divided into three continuous proportionals." This theorem was already mentioned as proposition 696 in MS 2.1.

In his book *Problema Austriacum*, as a corollary to proposition 87, San Vicente resolves Achilles's paradox using geometric progressions.[11] Aristotle had posed this paradox in his book *Physics*.[12] It states that if Achilles gives a tortoise a head start, he will never be able to overtake the tortoise. To do so, he would first have to reach the position where the tortoise started. When he reaches this point, the tortoise has moved on, so Achilles will now have to reach this new position of the tortoise. And so on. Aristotle's conclusion therefore is that Achilles will never overtake the tortoise. Using decreasing geometric progressions, San Vicente is not only able to contradict Aristotle's conclusion but also to determine the point where the tortoise is overtaken by Achilles.[13]

2 The Exhaustion Method

In his treatment of infinite sequences and series, San Vicente defines a new and important concept: *exhaustion*. The method as such was not new, as Archimedes had previously put it to work. However, for the first time in the history of mathematics, the method got its name in proposition 727 of San

---

9   It is now easy to calculate the sum of the rectangles $S' = \square GC + \square HD + \ldots = \sum_{i=1}^{\infty} S'_i = \dfrac{a^2}{(1-a)^2(1+a)}$ while S was calculated in footnote 8. It is then obvious that $S_\square = S' - S = \dfrac{a^2}{1-a^2}$ and therefore $\sum_{i=0}^{\infty} a^{2i} = \dfrac{1}{1-a^2}$. The ratio of the squares to the rectangles beneath is $\dfrac{1-a}{a}$, therefore $S_\square = \dfrac{1-a}{a}S$, which obviously yields the same result.

10   Putting $q = a^2$ in the result obtained in the previous footnote, $\sum_{i=0}^{\infty} a^{2i} = \dfrac{1}{1-a^2}$, we find the familiar formula $1+q+q^2+q^3+\ldots=\dfrac{1}{1-q}$ $(q<1)$.

11   San Vicente, *Problema Avstriacvm*, 2:101–3.

12   "The second is the so-called 'Achilles,' and it amounts to this, that in a race the quickest runner can never overtake the slowest, since the pursuer must first reach the point whence the pursued started, so that the slower must always hold a lead" (Aristotle, *Physics*, Book VI, 9, 239b15).

13   Dhombres, "L' innovation comme produit captive," 143–45.

Vicente's MS 2.1 (fol. 303ᵛ): *Sit quantitas AB et ab eadem auferatur aliqua quantitas nempe AC et loco AC dematur maior AD et loco AD dematur maior AE et loco AE dematur maior nempe AF et hoc semper fiat. Dico quod non exhaurietur in omne casu haec linea* (If a magnitude *AB* is diminished by a magnitude *AC*, then instead of *AC* a larger part *AD*, then instead of *AD* a larger part *AE*, and instead of *AE* a larger part *AF*, and this is repeated, then *AB* will not be exhausted in all cases). He adds the remark: "We have to investigate in which ways it can be exhausted."[14]

One way of exhausting a figure was already described in Euclid 10.1, which states that by continuously removing more than half of a magnitude, a magnitude that is smaller than any given magnitude can be reached. San Vicente, however, does not confine himself to an approximation but boldly accepts the limiting case and states "exhaurietur quantitatem" (the quantity is exhausted). One may wonder whether Stevin's use of infinitesimals inspired him in this move.

San Vicente uses Euclid's method in proposition 730 (MS 2.1, fol. 305ʳ):

> If from three given magnitudes more than half is removed and in such a way that the removed parts are in continuous proportion, and from the remaining pieces one again removes more than half in such a way that the removed parts are in continuous proportion and one keeps on doing this, then the problem is to choose the proportion between the removed parts if one supposes that the magnitudes are exhausted.

The answer is given in proposition 736 (MS 2.1, fol. 307ʳ): "If one keeps removing parts larger than half, and these pieces have (the same) continuous proportion, then the three given magnitudes will be in the same continuous proportion." This is an important proposition because now he deduces the proportion of the wholes *given* the proportion of the parts. It is the basis for his infinitesimal method: one can compare the wholes by comparing the parts. Moreover, the double *reductio ad absurdum* is no longer necessary. It suffices to check whether the conditions for exhaustion are met.

This will also be the basis for his method in *Problema Austriacum*.[15]

Here, we find:

> Let *AB* and *CD* be two magnitudes. Let *AB* be divided at *E* and *G* such that *AE* is not smaller than half *AB* and *EG* not smaller than half of *EB*. If one

---

14 "Oporteat autem determinare quibus modis exhaurietur."
15 San Vicente, *Problema Avstriacvm*, 2:119, proposition 116.

EXHAUSTION: THE ROAD TO INFINITESIMALS 123

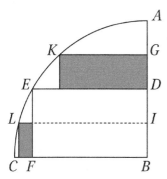

FIGURE 6.6
Euclid's exhaustion method

divides *CD* in the same fashion at *F* and *H* and let *AE*, *EC*, and *CF*, *FH* be proportional. One can proceed indefinitely. I say that the whole of *AB* is to the whole of *CD* as *AE* is to *CF* [fig. 6.6].

A couple of pages later, we see how San Vicente, for the first time, uses exhaustion in a geometric figure, in this case a quarter of a circle (MS 2.1, propositions 737–38, fols. 307^v–308^r): "If the radius *AB* is divided into two parts by *D*, then the rectangle *DEFB* is larger than half of the quarter circle. If one subdivides again at *G* and *I*, then the rectangles *DK* and *FL* are larger than what remained," which leads San Vicente to the conclusion: "Exhauriet ergo haec progressio totum quadrantem" (This progression exhausts the whole quadrant). One can again ascertain that this exhaustion entails taking a limit in our modern sense.

San Vicente adapts the theorem slightly to make it useable for ductus figures, and he will use the exhaustion method again and again in the calculation of the volume and the area of ductus figures:

> If from four magnitudes more than half is removed in such a way that the ductus of the removed part of the first, with the removed part of the fourth, is equal to the removed part of the second plus the removed part of the third, and if one takes from the remainders again more than half, under the same conditions as in the first procedure, and if one continues to repeat this procedure, then the ductus of the first magnitude in the fourth will have a ratio to the second magnitude in the third as the ductus of the removed parts of the first with the fourth has to the removed parts of the second with the third [proposition 743, MS 2.1, fol. 310^r].

Despite having a good command of the method, he sometimes makes mistakes, mainly because he fails to check whether all conditions necessary to apply the method have been met.

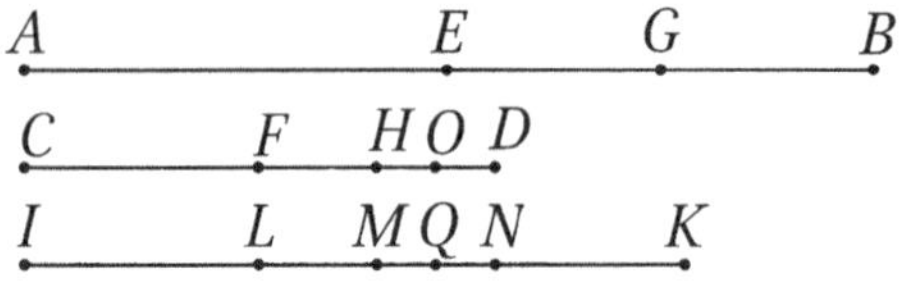

```
A              E    G     B
•──────────────•────•─────•
C        F  H O D
•────────•──•─•─•
I        L  M Q N      K
•────────•──•─•─•──────•
```

FIGURE 6.7
Saint-Vincent's version of the exhaustion theorem

In MS 3.2, one of the missives to Rome, this most important exhaustion theorem is repeated:

> Given two magnitudes *AB* and *CD*; from *AB*, half or more than half is deducted and from *CD* half or more than half of *CD*; then of what remains of the first is again taken away half or more than half *EG*, and also from what remains of the second, this has to be done in such a fashion that the first removed is to the second removed *EG* as the first removed *CF* of the second is to the second removed *FH*; and also for the second removed to the third removed of *AB* as the second removed is to the third removed of *CD*, and this will always be the case. I say that in this fashion *AB* is to *CD* as *AE* is to *CF* or *EG* is to *FH* [proposition 4, MS 3.2, fol. 466ᵛ; fig. 6.7].

The proof now rests on a double *reductio ad absurdum*: one first proves that $\frac{|AB|}{|CD|} > \frac{|AE|}{|CF|}$ leads to an absurdity and then that $\frac{|AB|}{|CD|} < \frac{|AE|}{|CF|}$ also leads to an absurdity.

Proposition 13 of MS 3.2 (fol. 468ʳ) marks the first time that it is stated that a geometric sequence with a ratio smaller than 1 is convergent: "If *AB*, *BC*, *CD*, etc. are in continuous proportion, such that the proportion of *AB* to *BC* is larger than [the proportion of] *BC* to *AB*, I say that the series of the proportions will not constitute an infinite quantity but will end somewhere."

The final sentence of this manuscript reads: "And what we selected from the theory of sequences suffices to prove what we shall undertake in the context of the quadrature of the circle." Clearly, he had already set his mind on a program he would pursue for the rest of his life: the quadrature of the circle.

3 San Vicente's Use of Infinitesimals

The full force of San Vicente's method becomes clear in propositions 36 and 37 of book 7 of *Problema Austriacum* (fig. 6.8):

> If *ABCD* is a figure bounded by two parallel lines *AD* and *BC* on the one hand and by a line *AB* and a curved line *DFC* on the other. After having

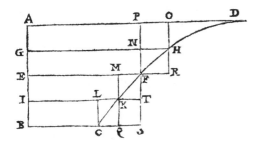

FIGURE 6.8
The application of the exhaustion theorem to the area under a curve. Gregory of Saint-Vincent, Problema avstriacvm plvs vltra qvadratvra circvli (Antwerp: Joannes et Jacobus Meursius, 1647), 7:730
© EHC G 4869

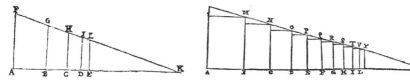

FIGURE 6.9 The application of Saint-Vincent's infinitesimal method to determine the area of a triangle. Gregory of Saint-Vincent, *Problema avstriacvm plvs vltra qvadratvra circvli* (Antwerp: Joannes et Jacobus Meursius, 1647), 2:125 and 138
© EHC G 4869

divided *AB* in four equal parts in *G*, *E*, and *I*, and having drawn parallel lines *GH*, *EF*, and *IK*, the lines *HO*, *FN*, *KM*, and *CL* are drawn equidistant [parallel] to *AB*. If *AD* is smaller than twice the second line *GH*, which is smaller than twice the third and so on. I say that taken together the figures *AHO*, *GFN*, *EKM*, *ICL* are larger than half the figure *ABCD*.[16]

In a scholion, San Vicente adds that the same holds for the residual figures *OHD*, *FNH*, and therefore the figure *ABCD* will be exhausted and thus the area is determined.

An implicit assumption San Vicente makes is that the curve at hand is monotonously decreasing or increasing. In the following theorems and corollaries, San Vicente does introduce the monotony of the function.[17]

San Vicente also uses the subdivision of a segment into segments that are in continuous proportion. He would do so in the case of the hyperbola but also in more elementary cases. For instance, in a lemma to proposition 141, he states that if in a rectangular triangle the side *AK* is divided such that *AK*, *BK*, *CK*, and so on are in continuous proportion, then the trapezoids *FB*, *GC*, *HD*

---

16   San Vicente, *Problema Avstriacvm*, 7:730.
17   This theorem and its corollaries are as close as one can get to a Riemann sum, again given the tools at San Vicente's disposal in the first quarter of the seventeenth century.

will be similar (fig. 6.9, left).[18] This is a generalization of propositions 137 and 138, in which the triangle △AKG is circumscribed about a geometric proportion of rectangles and squares (fig. 6.9, right). According to San Vicente, the triangles above the squares are also in continuous progression, and therefore the trapezoids consisting of the squares and the triangles exhaust the triangle. Moreover, San Vicente knew how to inscribe trapezoids in continuous proportion into a triangle. Remarkably, in this method none of the trapezoids are "infinitely small," as all have a calculable area.

In the case of the hyperbola, San Vicente uses proposition 736 and its corollaries:

> Let AB and CD be two magnitudes. Let AB be divided at E and G such that AE is not smaller than half AB and EG not smaller than half of EB. If one divides CD in the same fashion at F and H and let AE, EC, and CF, FH be proportional. One can proceed indefinitely. I say that the whole of AB is to the whole of CD as AE is to CF [fig. 7.7].

The proposition allows San Vicente to decide that the wholes are in a certain proportion, given that the parts are in that proportion (and meet a number of conditions). In the case of a curve, the wholes are the areas between the curve and a given straight line, and the parts are the areas of the inscribed (or circumscribed) parallelograms.

The following paragraphs are an anachronistic explanation of San Vicente's method with the aim of demonstrating its power as well as its shortcomings. San Vicente himself used geometrical language, rather than the algebraic one used here. Moreover, as San Vicente was one of the pioneers of calculus, some of the concepts used below and treated as self-evident had yet to be conceived.

The property that lies at the basis of San Vicente's ability to find the quadrature of certain curves is the expansion of a series by insertion: *If A, B, C, D are magnitudes in continuous progression, (i.e., $\frac{A}{B} = \frac{B}{C} = \frac{C}{D}$), then it is possible to find numbers E_1, E_2, E_3 such that $\frac{A}{E_1} = \frac{E_1}{B} = \frac{B}{E_2} = \frac{E_2}{C} = \frac{C}{E_3} = \frac{E_3}{D}$.*

---

18   San Vicente, *Problema Avstriacvm*, 2:138. Similar propositions can be found in MS 5789, fols. 116^r, 156^v, and 297^r, where San Vicente constructs a series of inscribed rectangles in continuous proportion into a triangle.

EXHAUSTION: THE ROAD TO INFINITESIMALS 127

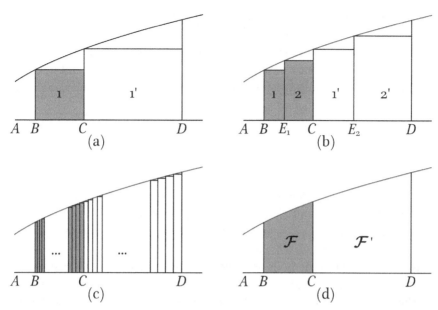

FIGURE 6.10 Saint-Vincent's infinitesimal method to determine the area under a curve

Now suppose that a curve and a straight line are given. Choose four points A, B, C, and D, such that AB, AC, and AD are in continuous proportion. Draw parallel lines through these points and construct the inscribed parallelograms (fig. 6.10). Consider the line segments $[BC]$ and $[CD]$. Suppose the ratio of the inscribed parallelograms is α $\left((a) \dfrac{\text{area 1}}{\text{area 1}'} = \alpha \right)$. Subdivide the line segments $[BC]$ and $[CD]$ at E_1 and E_2, such that the line segments AB, AE_1, AC, AE_2, AD are in continuous proportion, and suppose the respective inscribed parallelograms above the line segments $[BE_1]$, $[E_1C]$, $[C E_2]$ and $[E_2D]$ are again in the same proportion α $\left((b) \dfrac{\text{area 1}}{\text{area 1}'} = \dfrac{\text{area 2}}{\text{area 2}'} = \alpha \right)$. Repeat this procedure, or in other words expand the series by insertion, and let the areas of all the respective parallelograms have the same proportion α $\left((c) \dfrac{\text{area 1}}{\text{area 1}'} = ... = \dfrac{\text{area n}}{\text{area n}'} = \alpha \right)$. By invoking theorem 736, San Vicente concludes that the segments have the same proportion α $\left((d) \dfrac{\text{area F}}{\text{area F}'} = \alpha \right)$. He would apply this method with great success to the case of an orthogonal hyperbola. De Jonghe would apply a slightly adapted method to higher hyperbolae and parabolae.[19]

The symbolism used above makes it seem that these manipulations are easy and straightforward, but we have to remember that San Vicente did not have any appropriate symbolism at his disposal. Moreover, he was trying to make his way through the still uncharted territory of the calculus.

4 The Cavalieri Dispute

The previous section has shown that San Vicente's integration method is based on the exhaustion method and not on some vague calculation with indivisibles. Plagiarism vis-à-vis Cavalieri, as Marin Mersenne (1588–1648) would have it,[20] is out of the question. This is ten years before the publication of Cavalieri's *Geometria*, and both methods are fundamentally different. San Vicente does not use the concept of indivisible, but of infinitesimal, albeit of a special kind.

The power of San Vicente's method becomes clear when comparing it with Cavalieri's method.

In *Geometria*, theorem 2.19, Cavalieri proves that if a diagonal is drawn in a parallelogram, the parallelogram is twice one of the triangles determined by the diagonal.[21] Let *ACDF* be a parallelogram and [*CF*] one of its diagonals (fig. 6.11, left). The proof then rests on considering an arbitrary pair of corresponding line segments [*BM*] and [*HE*]. The line segments are corresponding if $|CB| = |FE|$. It then follows that $\triangle CBM \cong \triangle FEH$ and $|BM| = |HE|$. Since this is valid for all corresponding line segments, Cavalieri concluded that area $\triangle FAC$ = area $\triangle FCD$.

Using Cavalieri's method, Evangelista Torricelli (1608–47) created a paradox.[22] Consider the rectangle *ACDF*, and let one side *AC* be smaller than

---

19 In the case of higher hyperbolae or parabolae, which have an equation $y = x^r$ ($r \neq$ -1), the method yields inscribed parallelograms for which the areas are in continuous proportion and hence can easily be summated as a geometric series.

20 Marin Mersenne, *F. Marini Mersenni minimi cogitata physico-mathematica, Volumes 1–3* (Paris: Antoni Bertier, 1644), 192. "At verò cùm neque dederit quadraturam eo modo quo solet a Geometris expectari, cùm in ea exhibenda longè, quàm ipsam quadraturam, diffiliora supponat, vel postulet; neque meminerit [San Vicente] *ulla tenus Geometriae per indivisibilia, eruditissimi Bon. Cauallieri, quandoquidem primus illam per indiuisibilia methodum edidit*, quae tamen illi praeluxisse videtur, nostris Geometris displicuit." See also Van Looy, "Chronologie en analyse," 335.

21 Andersen, "Cavalieri's Method of Indivisibles," 322.

22 Tiziani Bascelli, "Torricelli's Indivisibles," in Jullien, *Seventeenth-Century Indivisibles Revisited*, 105–36, here 119. Tacquet would raise similar objections (see chapters 14 and 15).

EXHAUSTION: THE ROAD TO INFINITESIMALS   129

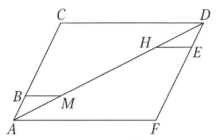 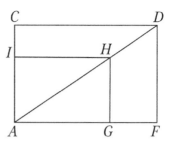

FIGURE 6.11   Cavalieri's application of his *indivisibilia* method to determine the area of a triangle and Evangelista Torricelli's counter-example

the other *CD*, now call two lines corresponding when they intersect the diagonal in the same point (fig. 6.11, right).

It is obvious that $|IH| > |HG|$, and this holds true for all lines parallel to *IH* and *HG*. Therefore, area $\triangle FAC$ = area $\triangle FCD$, which is obviously false.

San Vicente's method does not give rise to these paradoxes, because he either uses physical objects or infinitesimals.

In proposition 155 (*Problema Austriacum* 2, 150, fig. 6.12), San Vicente proves that:

> Consider a geometric progression of squares *ABMP*, *BCNH*, …
> Let $\triangle AKG$ be the triangle circumscribed about these squares.
> Let *AKIG* be a rectangle.
> Let *PM* intersect *AI* in *L*, *HN* intersect *AI* in *S*, …
> Then the ratio of the squares *ABMP* and *BCNH*, … is equal to the ratio of the rectangles *QILM* and *OLSN*, …
> Hence the geometric series of the squares is equal to the series of the rectangles.

The proof is straightforward.

Furthermore (*Problema Austriacum* 2, 151, proposition 157), the rectangles *GQPM*, *MONH*, and so on are in the same proportion. Although San Vicente does not draw the conclusion explicitly, it is clear that one arrives at the equality of both triangles $\triangle AKG$ and $\triangle IGK$. Moreover, the result easily lends itself to generalization.

The same results can be found if we start out with inscribed rectangles instead of squares. If the geometric progression of the bases is expanded, the results will continue to hold.

Because every argument San Vicente makes is about physical objects (rectangles, squares, triangles, etc.) and then by exhaustion turns to a limiting process,

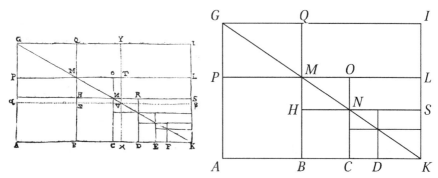

FIGURE 6.12   The proof that Saint-Vincent's infinitesimal method does not have the flaws of Cavalieri's method of indivisibles. Gregory of Saint-Vincent, *Problema avstriacvm plvs vltra qvadratvra circvli* (Antwerp: Joannes et Jacobus Meursius, 1647), 1:150
© EHC G 4869

he does not encounter the paradoxes associated with Cavalieri's method. Here, the elongation of the horizontal side is compensated for by the diminution of the vertical side.

San Vicente was aware of Mersenne's criticism, and he also showed that he had read Cavalieri's book in depth. For instance, he writes (MS 2.1, fol. 134ᵛ): "Oporteat ostendere Cavalerij methodum per indivisibilia demonstrandi esse *paralogisticam* [my italics]" (It is necessary to show that Cavalieri's method of demonstrating with *indivisibilia* is *a fallacy*). He clearly knows the flaws of Cavalieri's theory and, like Torricelli, uses *indivisibilia* to arrive at absurdities.

For instance (fig. 6.13):[23]

> Let [EB] be to [FG] as [BC] is to [GH] and so on.
> Therefore, as [EB] is to [BC], so all [lines] [FG] are to all [lines] [GH].
> Therefore, as [EB] is to [BC] so the area ADB is to the area DBCA, which is clearly false.
> Therefore, Cavalieri's method is not valid.

Another example[24] is (fig. 6.14):

> If [AC] and [AE] are divided in the same proportion, then the *indivisibilia* of ABDC and ABFE (i.e. the lines parallel to [AB]) are equal.

Therefore, ABDC and ABFE have an equal area, which is clearly false.

---

23   MS 9.1, fols. 186ʳ–187ᵛ.
24   MS 9.1, fols. 186ʳ–187ᵛ.

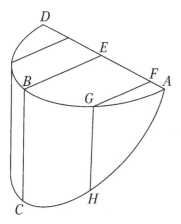

FIGURE 6.13
One of Saint-Vincent's counter-examples to Cavalieri's *indivisibilia* method

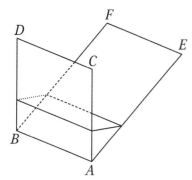

FIGURE 6.14
One of Saint-Vincent's counter-examples to Cavalieri's *indivisibilia* method

Alphonse Antonius de Sarasa (1617–67) put up a defense of San Vicente against Mersenne in his book *Solutio problematis a R.P. Marino Mersenno minimo propositi* (Solution to a problem proposed by, Friar Marin Mersenne of the Order of Minims [1649]), which was graciously acknowledged by Christian Huygens.[25]

---

25 "Father de Sarasa, who has successfully freed you from the criticism of Mersenne"; Huygens, *Oeuvres complètes: Tome II*, letter 102.

CHAPTER 7

Infinitesimal Calculus at Work

1 The Hyperbola[1]

The properties San Vicente found for the hyperbola, most of them discovered during his Antwerp period, must count among his greatest achievements. The main theorems on the hyperbola were published in *Problema Austriacum* as theorems 106 to 109 of book 6.

What we find in the manuscripts are San Vicente's various attempts to achieve the quadrature of the hyperbola.

In doing so, he discovered many properties relating to the logarithmic properties of the hyperbola.

Between the propositions about segments of conic sections are two propositions (MS 2.1, propositions 679 and 681, fol. 283^v; fig. 7.2) that are the advent of another breakthrough. A hyperbola *ABC* is given as are its asymptotes.

If the lines *AD*, *BE*, *FG*, and so on, parallel to one of the asymptotes, are drawn and divided into two equal segments, then the midpoints *L, M, N, O, P* are points of a hyperbola with the same asymptotes.

Moreover, the area of *ABCKD* is twice that of *LMPKD*. From the fact that the line segments are halved, San Vicente concludes that the area is halved as well; he does not provide a proof. Again, it is not clear whether San Vicente sees the lines as shorthand for rectangles or whether he implicitly accepts that an infinity of lines constitutes an area.[2] This proposition is not present in *Problema*

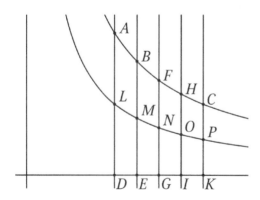

FIGURE 7.1
If *L*, ..., *P* divide [*AD*], ..., [*CK*] respectively into two equal segments and *A*, ..., *C* are on a hyperbola, then so are *L*, ..., *P*, and the area under the hyperbola *AC* is twice that of the area under the hyperbola *LP*

---

1 This chapter is co-authored with Herman van Looy.
2 In the last case, this would mean that, in the 1610s, he entertained ideas similar to Cavalieri's.

INFINITESIMAL CALCULUS AT WORK 133

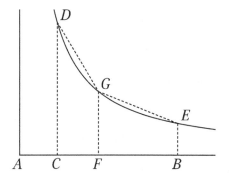

FIGURE 7.2
If [AF] is a mean proportional between [AC] and [AB], then [DG] and [GE] are equal segments

Austriacum. After the theorems about conic sections in MS 2.1, we find a series of theorems about sequences to solve the problem of two mean proportionals (propositions 696–702, fols. 290^r–292^r).

In the first theorem (MS 2.1, proposition 696, fol. 290^r), San Vicente gives a method to find a mean proportional between two line segments [AB] and [CD]: construct the mean proportional [EF] between the two line segments, then construct the mean proportional between the second line segment [CD] and the third segment [EF], which is the mean proportional from the previous step. This procedure is repeated for the third and the fourth, which was found in the previous step. If this construction is repeated over and over again, then the "terminus" of this sequence is one of the two mean proportionals between the original two line segments:[3]

> If [AB] and [CD] are two given line segments and [AE] is the mean proportional between the two, which is constructed on [AB], if one constructs the mean proportional [CF] between [AE] and [CD] on [CD] and if then the mean proportional [AG] between [AE] and [CF] is constructed on [AE] and then the mean proportional [CH] between [AG] and [CF] and if one continues to do this, then the terminus of this ascending or descending sequence will give the mean proportional between the two line segments, because [AI] or [CK] are one of both [MS 2.1, proposition 701, fol. 292^r].

---

3 Putting $|AB| = a$ and $|CD| = 1$, the sequence in theorem 696 is $1, a, \sqrt{a}, \sqrt[4]{a}, \sqrt[8]{a^3}, \sqrt[16]{a^5}, \sqrt[32]{a^{11}}, \ldots$ This sequence actually consists of two sequences, of which the terms are alternately greater or smaller than the sequence's limit. These are the sequences that surface in theorems 701–2: $1, \sqrt{a}, \sqrt[8]{a^3}, \sqrt[32]{a^{11}}, \ldots$ and $1, a, \sqrt[4]{a}, \sqrt[16]{a^5}, \ldots$ It is obvious that these sequences converge on $\sqrt[3]{a}$. For the first sequence, it is easy to prove that the exponent can be written as
$$\frac{2^{2n} - \sum_{i=0}^{n-1}(2^2)^i}{2^{2n+1}} = \frac{2^{2n} - \frac{(2^2)^n - 1}{2^2 - 1}}{2^{2n+1}} = \frac{2 \cdot 2^{2n} + 1}{3 \cdot 2^{2n+1}}$$
for which the limit is $\frac{1}{3}$.

In the next proposition (MS 2.1, proposition 702, fol. 292ʳ), San Vicente states that both mean proportionals are equal. Essentially, these theorems state the same as the first theorem, but now San Vicente constructs the mean proportional on the two segments.

We have to stress that in this theorem San Vicente uses the discovered properties of the hyperbola. This cannot be deduced from a proof, because he does not give one, but from a previous remark: "The terminus of that sequence defines in a hyperbola a point that trisects a hyperbolic arc." When dealing with the properties of hyperbolic segments, San Vicente had claimed that the continuous proportion of the abscissae corresponded with an equality of the segments. Looking for a mean proportional in the abscissae thus corresponds to halving the hyperbolic segments. Because San Vicente had previously noted that a continuing halving of a magnitude can lead to a trisection, this will lead him to the construction of the two mean proportionals between the two given magnitudes.

That San Vicente had this relation in mind becomes clear from the following propositions: "If [AF] is a mean proportional between [AC] and [AB], then [DG] and [GE] are equal segments" (MS 2.1, proposition 698, fol. 291ʳ; fig. 7.2). He confirms this in proposition 700 (MS 2.1, fol. 291ᵛ): if the hyperbolic segments are equal, then the ordinates ("parallelas") are in continuous proportion.

These properties will be important in one of the proofs leading up to theorem 109 of *Problema Austriacum*. Now San Vicente begins with another subject: the exploration of the properties of hyperbolic segments. He no longer explores the logarithmic properties of the hyperbola but tries to determine the proportion between two hyperbolic segments.

The first property San Vicente explores is $\dfrac{\text{hyperbolic segment } ABCDE}{\text{hyperbolic segment } FBCGH} = \dfrac{|AB|}{|FB|}$ (MS 2.2, proposition 393, fol. 150ʳ; fig. 7.3). It seems likely that he obtained the result by comparing lines in both hyperbolic segments.

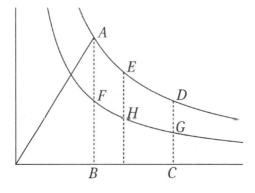

FIGURE 7.3
The determination of the proportion between the hyperbolic segments. ABCDE and FBCGH can be expressed in terms of |AB| and |FB|

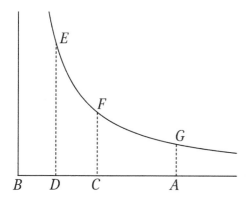

FIGURE 7.4
An exploration of the properties of areas of hyperbolic segments

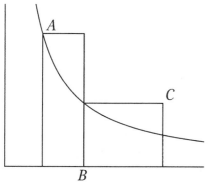

FIGURE 7.5
An exploration of the areas of circumscribed rectangles in a hyperbola

In MS 2.1 (propositions 679 and 681), San Vicente had already concluded that, for the case in which $|AF|$ is equal to $|FB|$, one hyperbolic segment was twice as large as the other.

These results are the starting point for an exploration. San Vicente first asks himself whether the proportion of the hyperbolic segments DEFC and CFGA is always a relation between $\frac{|ED|}{|FC|}$ and $\frac{|GA|}{|FC|}$ (MS 2.2, proposition 397, fol. 148^r; fig. 7.4). He does not yet know the answer, because he writes: "At least, this is generally true if the segments are equal or in continued proportion or not at all," which has to be interpreted as a question: When is this valid? If the segments are equal, proportional, or not?

In the next proposition, San Vicente investigates whether this proportion can perhaps be reduced to the proportion of rectangles AB and BC. Finally, he finds the right restriction: this property is only valid for equal hyperbolic segments (MS 2.2, proposition 399, fol. 147^r; fig. 7.5).

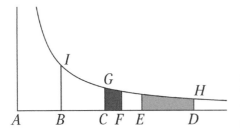

FIGURE 7.6
The construction of a hyperbolic segment with an area twice as large as that of a given hyperbolic segment

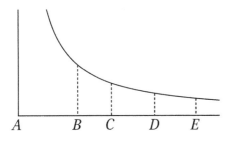

FIGURE 7.7
An exploration of the properties that determine the equality of areas of hyperbolic segments

San Vicente now performs a construction (MS 2.2, proposition 403, fol. 145^r; fig. 7.6): if $|AB|$, $|AC|$, $|AD|$ are in continuous proportion and the mean proportional $[AE]$ between $[AD]$ and $[AC]$, and the mean proportional $[AF]$ between $[AE]$ and $[AC]$ are determined, then the hyperbolic segment EH is twice the segment GF.

This does not solve San Vicente's problem, because he immediately asks himself: Is the question whether the proportion of the segment $[BG]$ to the segment $[CH]$ is composed of $\dfrac{|BI|}{|GC|}, \dfrac{|GC|}{|HD|}$ or $\dfrac{|BC|}{|CD|}$ solved?

His answer is: "R[espondeo] quod non" (I reply that it is not; MS 2.2, fol. 145^r). We do not get an answer to this question. His search continues, because the next question is: "If AB, BC, CD, DE, and so on are equal, which proportion do they have relative to the hyperbolic segments? Are they proportionate?" (MS 2.2, fol. 142^r; fig. 7.7).

A correct statement is found on the following page: if $|AC|$ is twice $|AB|$ and $|AD|$ is twice $|AC|$, and if $[AE]$ is the mean proportional between $[AC]$ and $[AD]$, then the segment $[CF]$ is half the segment $[CG]$ (MS 2.2, proposition 410, fol. 141^v; fig. 7.8). Determining a general proportion between two hyperbolic segments still fails.

Perhaps this is why San Vicente does not proceed along this path for now.

INFINITESIMAL CALCULUS AT WORK

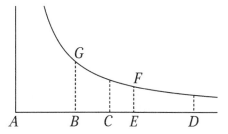

FIGURE 7.8 An exploration of the properties that determine the equality of areas of hyperbolic segments using mean proportionals

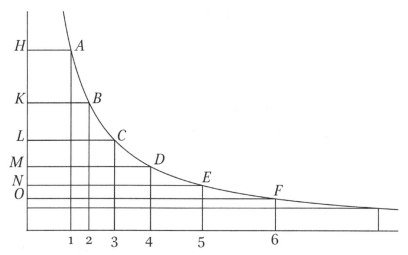

FIGURE 7.9 An exploration of the properties that determine the equality of areas of hyperbolic segments: breakthrough area segments are equal if the ordinates are in continuous proportion

He had, however, in seemingly independent propositions, arrived at important conclusions. In proposition 226 of MS 2.2 (fol. 240^r; fig. 7.9), he states:

> If *ABC* and *DEF* contain equal segments and if *GH* and *GI* are the asymptotes and if *AH, KB, LE, MD, NE, OF* are parallel to the asymptotes, I say that *GO, GN, GM, GL, GK, GH* are in continuous proportion and this is true *ad infinitum*. Likewise, if through *E* and *F* parallels with the other asymptote *GH* are drawn, the same holds true in *GI*.

Proposition 673 of MS 2.1 is undoubtedly San Vicente's most important result: "If *DF, DU, DZ, DY,* and *DX* are in continuous proportion, then the hyperbolic

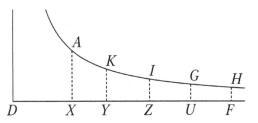

FIGURE 7.10
The logarithmic property of the areas of hyperbolic segments

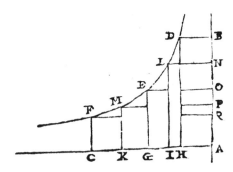

FIGURE 7.11
Rectangles inscribed in a hyperbola. The construction of these rectangles, with bases in continued proportion, is a first step toward proving the equality of the area's hyperbolic segments. Gregory of Saint-Vincent, *Problema avstriacvm plvs vltra qvadratvra circvli* (Antwerp: Joannes et Jacobus Meursius, 1647), 6:585
© EHC G 4869

segments *AXYK*, *KYZI* etc. are equal." San Vicente's research into the properties of a hyperbola will culminate in propositions 106 to 110 of the book *De hyperbola* (On the hyperbola) of *Problema Austriacum*.[4]

In the proof of proposition 108, San Vicente states that if [*DH*], [*LI*], [*EG*], [*MK*], and [*FC*] are parallel to the asymptote along a continuous proportion, then the rectangles *HL*, *IE*, *GM*, and *KF* are equal.

The proof is relatively straightforward.[5] Referring to figure 7.11 and starting with $\frac{|LI|}{|DH|} = \frac{|EG|}{|LI|} = \frac{|MK|}{|EG|} = \frac{|FC|}{|MK|}$ we find that $\frac{|DB|}{|LN|} = \frac{|LN|}{|EO|} = \frac{|EO|}{|MP|} = \frac{|MP|}{|FQ|}$ which derives from the property of the hyperbola that $\frac{|AH|}{|AI|} = \frac{|LI|}{|DH|}$. From which we have $\frac{|AH|}{|AI|} = \frac{|AI|}{|AG|} = \frac{|AG|}{|AK|} = \frac{|AK|}{|AC|}$. This allows us to prove that $\frac{|AH|}{|AI|} = \frac{|AI|}{|AG|} = \frac{|AH|}{AI} = \frac{|HI|}{|IG|}$,[6] but $\frac{|AI|}{|AG|} = \frac{|EG|}{|LI|}$, whence $\frac{|HI|}{|IG|} = \frac{|EG|}{|LI|}$. This last

---

4 San Vicente, *Problema Avstriacvm*, 6:585–86.
5 For a detailed analysis, see Jean Dhombres, "Is One Proof Enough? Travels with a Mathematician of the Baroque Period," *Educational Studies in Mathematics* 24 (1993): 401–19, here 409–14; Charles Naux, "Grégoire de St. Vincent et les propriétés logarithmiques de l'hyperbole équilatère," *Revue des questions scientifiques* 143, no. 2 (1972): 209–21.
6 (|*AI*| − |*AH*|)/(|*AG*| − |*AI*|)=(|*AH*||*AG*|/|*AI*| − |*AH*|)/(|*AK*||*AI*|/|*AG*| − |*AI*|)=|*AH*|(|*AG*||*AI*| − 1)/ (|*AI*||*AK*|/|*AG*| − 1) = |*AH*|/|*AI*| because |*AI*|/|*AG*|=|*AG*|/|*AK*|. The other equality is self-evident.

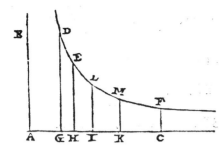

FIGURE 7.12
The logarithmic property of the areas of hyperbolic segments. Gregory of Saint-Vincent, *Problema avstriacvm plvs vltra qvadratvra circvli* (Antwerp: Joannes et Jacobus Meursius, 1647), 6:586
© EHC G 4869

equation immediately reduces to $|HI|.|LI| = |EG|.|IG|$, which can be interpreted as the areas of the rectangles HIL and IGE.

This is generalized in proposition 109 (fig. 7.12):

> Let AB and AC be the asymptotic lines of the hyperbola DEF. Divide AC such that AG, AH, AI, AK, AC are in a continuous proportion. Set GD, EH, LI, MK, FC equidistant from AB [This is "parallel to AB"]. I say that HD, IE, KL, CM are equal segments.[7]

Again, the proof is easy: consider two line segments of which the extremities are in continued proportion. From the previous theorems, we know that the rectangles above them are equal. The number of segments can be augmented by expanding the sequence by insertion. Because the rectangles on them all have the same area and because each segment is subdivided into an equal number of segments, the sum of these rectangles is equal. By implicitly invoking proposition 2.116, the rectangles will exhaust the areas under the hyperbola and above the segment, from which the equality of the hyperbolic areas follows.[8]

These theorems are considered to be the first in which the area of a segment between an axis/asymptote and the orthogonal hyperbola is given, albeit in geometrical language, without the use of logarithms.[9] The relation between

---

7 Translation from Dhombres, "Is One Proof Enough?," 407.
8 Unfortunately, exhaustion is not guaranteed for every choice of a continued proportion. Consider, anachronistically, the hyperbola given by $y = 1/x$ and the geometric proportion $1, a, a^2, \ldots$ The area of the hyperbolic segment above $[1, a]$ and $[a, a^2]$ is $\ln a$, and the area of the rectangle contained in the segment is $(a-1)/a$. The area of the rectangle, however, is not always larger than half of that of the hyperbolic segment. This is only true if $1 < a \lesssim 4.9$. However, if $a > 4.9$, we can expand the sequence and will at one stage find a sequence with ratio $\sqrt[n]{a} < 4.9$ for which we can calculate the area above $\left[1, \sqrt[n]{a}\right]$. The properties of logarithms then allow us to determine the area above $[1, a]$.
9 By that time, the concept of logarithms was known, also by the Jesuits, as is attested by the reference to logarithms in Jan Ciermans et al., *Disciplinae mathematicae traditae anno institutae Societatis Iesu seculari* (Leuven: Everardum de Witte, 1640).

the hyperbolic area and logarithms would be made clear by San Vicente's disciple Sarasa (see chapter 13).

What we have seen here is no mean feat. San Vicente succeeds in finding a quadrature using infinitesimal methods, the first time in modern history this had ever been accomplished. Moreover, his use of the axes as special lines is a first step toward the coordinatization of the plane and thus also toward the algebraization of geometry.

2 Calculation of the Volume of Ductus Figures

In MS 2.1, from fol. 326 onward, written by Boelmans, San Vicente introduces a theory he calls "ducere plani in planum" (multiplication of a plane in a plane) The manuscript was written in Leuven, but the fact that he already has a very good command of the methods indicates that it was developed earlier. Unfortunately, again, a systematic explanation is lacking. It is consequently necessary to revert to his definitions in MS 7.4 (fols. 1^r–2^v), which was written at Ghent and was a preparatory text for *Problema Austriacum*. These definitions are repeated in book 7, *De ductu plani in planum* (On the multiplication of a plane in a plane), of *Problema Austriacum*.

In the first definition,[10] San Vicente explains *ductus plani ACDB in planum EFG* (the multiplication of a plane *ACDB* in a plane *EFG*). To determine the ductus of the plane segment *ACDB* in the plane segment *EFG*, both segments need to have the same height $|AB| = |EF|$. These equal line segments are put onto one another and one of the surfaces is erected at a right angle to the other.

Now San Vicente accepts that all rectangles *HI.KG* create a spatial figure with *ACDB* as base area. This new spatial figure is called the *ductus ACDB in EFG* (fig. 7.13). Note that again this can also be interpreted in a Cavalierian sense, in which a spatial figure is composed of surfaces.[11] In figure 7.13, *ABCD* was taken to be a rectangle. This is not necessary, however. In principle, only [*BA*] and [*FE*] need to be straight line segments of equal length, and [*BD*], [*DC*], and [*CA*] can be any curve (fig. 7.14).

If we take [*DB*] and [*CA*] to be parallel line segments, calculating the volume of the ductus figure can be analytically interpreted as an integral. If we take the axes to be as in figure 7.14, we have that $x = x(y)$ and $z = z(y)$ and the volume is given by $V = \int_0^a \int_0^{x(y)} z(y) dx dy = \int_0^a x(y) z(y) dy$. In other words,

---

10 San Vicente, *Problema Avstriacvm*, 7:703.
11 In *Geometria* (104), Cavalieri constructs the same kind of solid, but he obtains the results by considering "all the rectangles" that make up the solid. Andersen, "Cavalieri's Method of Indivisibles," 312, 352–53.

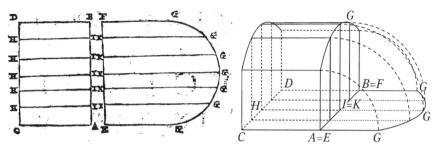

FIGURE 7.13 The construction of a *ductus plani ACDB in planum EFG*. Left: Gregory of Saint-Vincent, *Problema avstriacvm plvs vltra qvadratvra circvli* (Antwerp: Joannes et Jacobus Meursius, 1647), 7:704
© EHC G 4869

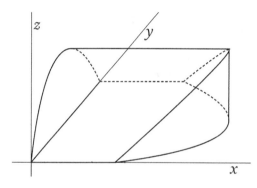

FIGURE 7.14
The construction of a ductus of a parabola in a parabola

San Vicente is performing, limited as the cases he considers may be, the first integrations of functions in two variables.

A second definition[12] defines a special case of the first. This is a ductus of a plane figure in itself. The given surface is mirrored about the height BC, then the mirror image is erected perpendicularly, resulting in a spatial figure *ABCGH* (fig. 7.15).

The third definition[13] is again a special case *ductus plani ABC in se ipsum subalterne* (multiplication of the plane ABC in itself). In this case, two congruent figures are put inversely on the common height and one is erected perpendicularly, resulting in the ductus figure *ABCG* (fig. 7.16).

There is no trace of these general definitions in San Vicente's manuscripts. He only once indicates how a ductus should be performed but uses the word *multiplicare* instead of *ducere*: "Partem circuli perpartem sibi aequalem multiplicare" (To multiply a part of the circle with a part that is equal to it).

---

12 San Vicente, *Problema Avstriacvm*, 7:704.
13 San Vicente, *Problema Avstriacvm*, 7:705.

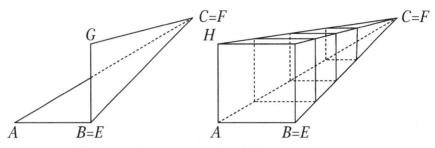

FIGURE 7.15   The construction of a *ductus plani ABC in se*

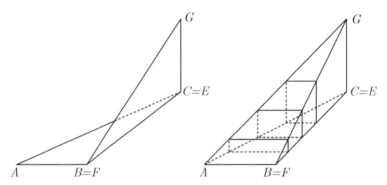

FIGURE 7.16   The construction of a *ductus plani ABC in se ipsum subalterne*

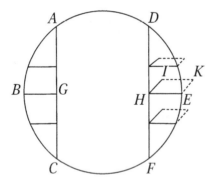

FIGURE 7.17
The ductus of a circle in a part that is equal to it

The geometric construction is performed thus:

> If *ABC* is multiplied by *DEF*, put *ABC* under a right angle by *AC* on the congruent *DF* and *BG* is orthogonal to *HE* and at the same time orthogonal to *HD* such that the plane *DIF* is orthogonal to *DEF*, draw *IK* the perpendicular to the plane *DIF*, and similarly *DEF* are erected perpendicular to the circumference *IK* and these will intersect in the point *K*. And I say that what had to be proven happened [MS 2.1, proposition 744, 314ʳ; fig. 7.17].

INFINITESIMAL CALCULUS AT WORK 143

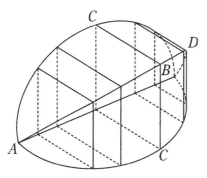

FIGURE 7.18
The *ductus in se* of a circle segment

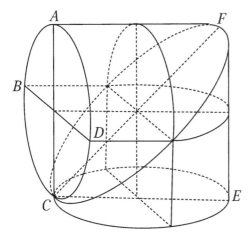

FIGURE 7.19
The *ductus in se* of a circle

Indeed, according to the definitions in *Problema Austriacum*, the spatial figure *DFIK* is the ductus figure.

For the circle sector, San Vicente states that the *ductus in se* will be two equal cylinder segments (MS 2.1, proposition 758, fol. 315^v; fig. 7.18).

After these definitions, the question can be asked of which geometric figures can be obtained by a ductus. For the ductus performed on the circle segment, one reads: "[The ductus of a] part of the circle in itself produces two parts of a cylinder that are equal to one another" (MS 2.1, proposition 756, fol. 314^v; fig. 7.18).

San Vicente states that, for an entire circle, the *ductus in se* will produce two equal cylinder segments (fig. 7.19).

San Vicente already looks at the ductus figure analytically: by drawing dotted lines, he can divide the figure into elementary parts for which he can compute the volume. The volume is calculated either directly or indirectly using an exhaustion method or by comparison with other figures.

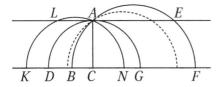

FIGURE 7.20
A property of the ductus

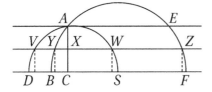

FIGURE 7.21
The proof that the segment *ACB* can be exhausted by constructing inscribed rectangles

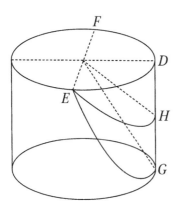

FIGURE 7.22
The calculation of the volume of an *ungula cylindrica* using the exhaustion method

An interesting property resulting from the exhaustion theorem is: "The ductus of *ABC* in *AEFC* is equal to the ductus of *ADC* in *AGC* etc. and therefore the ductus of *ABC* in *AEFC* equals the ductus of *ANC* in *ALKC* [...]" (MS 2.1, proposition 743, fol. 310r; fig. 7.20).

The second corollary shows how San Vicente arrived at this result: "And therefore we conclude that the sequence is useful for the multiplication and for any species of geometry" (MS 2.1, proposition 743, corollary, fol. 310r).[14]

San Vicente has already asserted that if *X* is the middle of *AC* (e.g., *XCBY* is larger than half of the segment *ABC*), and that if *AC* is divided into four equal parts, the additional rectangles are larger than half of what was left over of the segment, and that by continuing to divide *AC*, the segment *ABC* will be exhausted (fig. 7.21).

---

14   Because the segments *ADC* and *AGC* are denoted by three letters, they do not intersect the straight line *AE*, and consequently they are quarter circles.

INFINITESIMAL CALCULUS AT WORK   145

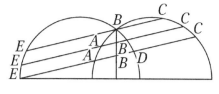

FIGURE 7.23
The *ductus obliquus*

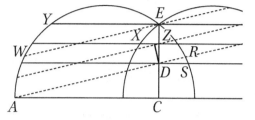

FIGURE 7.24
The exploration of the properties of a *ductus obliquus*

San Vicente now starts thorough research into the nature of these cylinder segments. His attention is drawn to the part of the cylinder that is cut by a plane going through the diameter of the upper base of the cylinder. He and his students will later call this kind of segment an "ungula cylindrica," which roughly translates as a circular hoof. For such an ungula, he calculates the volume using the method of exhaustion. He concludes that for the ungula *EFBH* and *EFBG*, the volumes are in the same ratio as their heights *BH* and *BG* by inscribing pyramids (MS 2.1, propositions 783–84, fols. 330^v–332^r; fig. 7.22). He performs these *ducti* for a multitude of figures.

Then a new subject surfaces: the *ductus obliquus*. The first theorem in which the oblique ductus is used is in a very simple property: "If the parallel lines *EC* are not perpendicular or parallel with respect to the common chord *BB*, then *AB ducta* in *BC* is equal to *DB ducta* in *BE*" (MS 2.1, proposition 846, fol. 345^v; fig. 7.23). This property is an immediate consequence of *BB* being the radical axis of the two circles.

San Vicente's attention is drawn to the relation between the right ductus and the oblique ductus for the same part on the common chord of two circles (MS 2.1, propositions 811–12, 348^r–349^r; fig. 7.24).

$$\frac{\text{ductus of } EDR \text{ in } EWAD}{\text{ductus of } EDR \text{ in } EVTD} = \frac{|DX|}{|DZ|}$$

in which *DX* is the distance between two consecutive parallel lines of the skewed direction *AD* and *DZ* the distance between the corresponding parallel lines of the perpendicular direction *EC*. The proof rests on the fact that there are prisms that satisfy the conditions for exhaustion (e.g., *AD.DR.DX* for the oblique ductus and *TD.DS.DZ* for the right ductus).

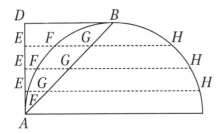

FIGURE 7.25
If in a semi-circle the tangent to the diameter and parallel chords are drawn, then $|EF|.|EH| = |EG|^2$

Since all these prisms have $|AD|.|DR|=|TD|.|DS|$, all the members of the sequence that lead to exhaustion are to one another as DX is to DZ, which according to exhaustion theorem 743 is also the case for the ductus figures themselves.

In later theorems, San Vicente compares two more oblique *ducti* with each other and concludes, analogously, that these are to one another as the distance between the two parallel lines of the oblique direction is to the distance between parallel lines of the other direction.

Remarkably, despite exploring the properties of an oblique ductus in these manuscripts, there is no trace of them in San Vicente's *Problema Austriacum*. One of the reasons may be that the *ductus obliquus* is not necessary for the quadrature of the circle. This usefulness for the quadrature of the circle would in many cases be the decision-maker in keeping or rejecting the subject for the printed version.

Among the last theorems of MS 2.1 are a couple of theorems on segments of the circle. In one of them (MS 2.1, proposition 935, fol. 407ᵛ; fig. 7.25), San Vicente states that the rectangles *FE.EH* are always equal to the squares on [*EG*]. He notes that this is also the case for a parabola.

In the next theorems, the assertion is repeated for the case in which [*DB*] is not parallel to the diameter of the semicircle. In that case, all parallel lines will be at an angle with regard to [*AD*]. The proof is simple because $|EG|^2 = |EA|^2=|EF|.|EH|$.

In the first corollary to this theorem, San Vicente asserts that the ductus of *DAB in se* is equal to the ductus of *DAFB* in *DBHA*, making a curvilinear figure *in se* equal to a rectilinear figure, the pyramid.

In the last theorems of MS 2.1, San Vicente does not explicitly use exhaustion, as knowing the property for lines he immediately draws conclusions about the properties of solids. The previous theorems, however, show that, in each of these cases, there is an exhaustion that is used in the background, but that, perhaps to save time, he does not write it down.

3 Lateral Area of the *Ungula cylindrica* and Relations between Ductus Figures

In the last part of MS 2.2, two subjects are intertwined. The most interesting topic is the calculation of the lateral surface area of the *ungula cylindrica*. To calculate this area, San Vicente uses no fewer than three methods: first with inscribed figures, then with circumscribed figures, and, as an aid to this last method, with an analogy between the sphere and the ungula. The second topic is the relation between different ductus figures.

When calculating the lateral area of the *ungula cylindrica*, San Vicente tries to find a relation to the top surface of the ungula. The first theorem on this subject is: "The area of *ABCD* is to the area of the semicircle *EDA* as the area of *AFCD* is to that of the rectangle *EG*" (MS 2.2, proposition 440, fol. 126^v; fig. 7.26). The proof is very short: construct the largest inscribed curved rectangle *BD* in *ABCD*, which will be to *ABCD* as the rectangle *HI* is to the semicircle *EAD*.

In the course of his research, San Vicente has found an analogy between the sphere and the ungula.[15] A first indication in this sense is given by the following: "If the basis of the semi-ungula through *DE* and *FG* is divided as the quadrant of a sphere through *ML* and *NI*, then the lateral area *AES* will be to the lateral surface area *AGT* as the spherical segment *HPL* is to spherical segment *AQI*" (MS 2.2, fol. 49^v; fig. 7.27).

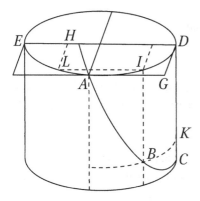

FIGURE 7.26
The calculation of the lateral surface area of the *ungula cylindrica*

---

15 This analogy would later be explored in San Vicente, *Problema Avstriacvm*, 9:4: "Comparatio ungulae cylindricae cum sphaera et aliis corporibus." On this analogy, see Herman van Looy, "De analogie tussen de bol en de ungula bij Gregorius a Sancto Vincentio," *Wiskunde en Onderwijs* 3 (1977): 56–62.

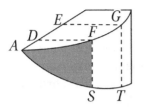 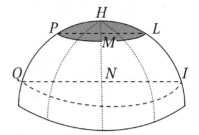

FIGURE 7.27   Exploration of the analogy between the *ungula cylindrica* and the sphere

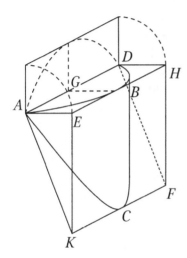

FIGURE 7.28
The calculation of the lateral surface area of an elliptical *ungula*

A proof is lacking, which means that we are unable to see the analogy in detail. The property only becomes clear when San Vicente formulates it for an elliptical ungula, which is more general than a circular ungula (MS 2.2, fol. 31^r; fig. 7.28):

> The lateral surface area of an elliptical ungula is to the area *EF* as the area of the spheroid, generated by the rotation of *ABD* about *AB*, is to the lateral surface area of the cylinder, with the circle, of diameter equal to the minor axis, as a base and with the major axis as height.

Now the analogy becomes clear: while the rotation of the ellipse generates a spheroid, projecting the ellipse on the plane *ADF* generates an elliptical ungula. The line segment *EH* describes respectively the lateral surface of a cylinder with height *AD* and touching the spheroid on the one hand and the tangent plane to the ungula *EHFK* on the other.

If the theorem is formulated for a semicircle we find:

$$\frac{\text{lateral surface area ungula}}{\text{area } EF} = \frac{\text{area sphere}}{\text{lateral surface area cylinder}}$$

This last proportion is equal to 1, which was already known to Archimedes.[16]

Because San Vicente knew this analogy long before he began with the calculation of the lateral surface area using Pappus's theorem, it cannot be excluded that this analogy was the intuitive basis for the more stringent calculation using tangent planes to the ungula.

After a successful calculation of the lateral surface area of the complete ungula, which San Vicente calls *ungula primaria* (primary cylinder hoof) or *sectio primaria* (primary section), he also formulates a theorem for the *sectio secundaria* (secondary section).

This *sectio secundaria* is obtained by intersecting the *ungula primaria* by the plane *ACDF* perpendicular to the basis and with *AC* parallel to *GH*. He claims that the lateral surface area of the section *ABCDEF* of the *sectio secundaria* is to the rectangle *ACDF* as *BE* is to *AF* (MS 2.2, fol. 10ᵛ; fig. 7.29). The proof is self-evident because the lateral surface area of the *sectio secundaria* is $|A'C'|.|BE|$ or $|AC|.|BE|$. San Vicente states that for studying this *sectio secundaria* one best uses the analogy with a spherical segment.

Between these theorems about the lateral surface area of the ungula, we find numerous theorems about the properties of ductus figures.

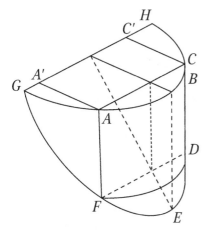

FIGURE 7.29
The calculation of the lateral surface area of a *sectio secundaria*

---

16   Archimedes, *On the Sphere and the Cylinder* 1, corollary to proposition 34. Archimedes, *Les œuvres complètes d'Archimède*, 68–69.

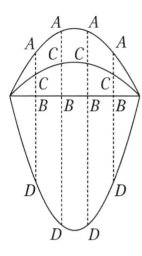

FIGURE 7.30 The calculation of the *ductus in se* of the parabola *AA* using transformations of parabolae

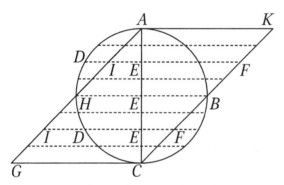

FIGURE 7.31 The transformation of a circular segment into a rectilinear one

Many of these theorems are formulated as transformation theorems for conic sections.

A typical example is: "If *AA* and *CC* are parabolae, and $|AB|^2 = |CB|.|BD|$, then I ask which curve is *DD*? I answer that it also is a parabola although not similar to the first one. Therefore, the ductus of *AB in se* is equal to the ductus of *CB* in *BD*" (MS 2.2, fol. 107ʳ; fig. 7.30).

In these properties of ductus figures, one notices that San Vicente always tries to transform solids that are made up of circular segments into solids that are composed of segments of conic sections or even rectilinear plane figures.

The ductus of the semicircle *ADC in se* is equal to the ductus of *AKC* in *ACG*, in which *HB* is the diameter of the circle (fig. 7.31). The proof is simple: the rectangle *AE.EC* is equal to the square *DE*², but $|AE|=|EI|$ and $|EC|=|EF|$ so $|DE|^2 =$

INFINITESIMAL CALCULUS AT WORK 151

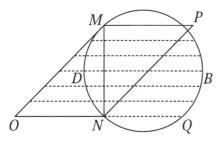

FIGURE 7.32
The calculation of the ductus of *MNO* in *MNP*

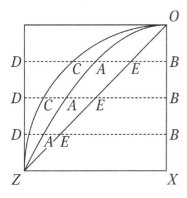

FIGURE 7.33
The transformation of a circle segment into a parabolic segment

$|EI|.|EF|$ (MS 2.2, proposition 473, fol. 111^r). In other words, the parallelepipeds $DE^2.EE$ and $EI.EF.EF$, which exhaust the respective ductus figures, are equal.

In the first corollary, San Vicente states that this theorem is also valid if one takes a chord [*MN*] at random instead of the diameter [*AC*]; in that case, the ductus of *MNO* in *MNP* is equal to the ductus of *MDN* in *MPBQN*, if $|MP|=|MN|=|NO|$ (fig. 7.32).

According to the second corollary, the ductus of a semicircle *in se* is transformed into a prism (actually a pyramid).

This allows for the calculation of the volume of the ungula. The ungula *ACDO* with $|DO| = r$ has a volume $\frac{1}{2}\left(\frac{1}{3}\left(\frac{4r^2}{2}\right)\cdot 2\right) = \frac{1}{6}(4r^3)$ because it is half a pyramid. For an ungula with height $2r$, as San Vicente usually chooses, this is

$$\frac{1}{3}(4r^3)\cdot(1)$$

The transformation of ductus figures on circle segments to ductus figures on parabolic segments is addressed as well (MS 2.2, proposition 505, fol. 94^r; fig. 7.33).

[AB] in se is equal to [DB] in [BE] (the condition for the parabola is $\frac{|OB|}{|OX|} = \frac{|BA|^2}{|XZ|^2}$ or $\frac{|EB|}{r} = \left(\frac{|AB|^2}{r^2}\right)$ so $|AB|^2 = |EB|.|BD|$. This last ductus is half a cube $\left(\frac{1}{2}r^3\right)$, and using (1) San Vicente finds that [CB] in se is $\frac{2}{3}$ of this cube. It is clear that he made an error, as the ratio of $\frac{1}{2}$ to $\frac{2}{3}$ is equal to the ratio of 3 to 4, not 2 to 3.

San Vicente also tries to find other situations in which such relations between solids exist, indicating that he is thinking about transforming solids with circular segments into solids without circular segments, leading him to think that he is within reach of the quadrature of the circle. He therefore asks permission to publish. As has already been noted, the superior general was highly cautious and suggested that his manuscripts be sent to Grienberger in Rome to let him decide. From now on, San Vicente and his students would try to bring order to the results they had found up to now.

CHAPTER 8

Rome and Prague, the Final Discoveries

1 The Missives to Rome[1]

All in all, there were seven consignments to Rome. Most of their contents can be found elsewhere in manuscripts MS 2.1 and MS 2.2. The treatise of November 1624 (MS 3.1) discusses the ungula and its parts as well as the *ductus obliquus*.

The most important consignment from the viewpoint of the history of mathematics is that of January 15, 1625 (MS 3.2). In this treatise, we find the exhaustion theorem, the convergence of geometric series, and the determination of the volume of the ungula and its parts.

The treatise of early 1625 (MS 3.3) contains properties on the ductus of segments of a circle. The first proposition (MS 3.3, fol. 335^r; fig. 8.1) is the basis for a relation between ductus figures: if ABD is a parabola with latus rectum $|AB|=|BD|$, then in the circle AB, $[AF]$ will be equal to $[CE]$. The proof: $\dfrac{|AE|}{|AB|} = \dfrac{|EC|^2}{|BD|^2}$ or $|EC|^2=|AE|.|AB|$ for the parabola, while in the right triangle $\triangle AFB$ we have $|AF|^2 = |AE|.|AB|$.

In the second proposition, the equality of two ductus figures is deduced: the ductus of the segment AKB in ADB is equal to half the cylinder with basis AFB and height $[AB]$ (fig. 9.1). The proof runs as follows: the parallelepiped with the rectangle $|CE|.|EG|$ as base and EE' as height is equal to the parallelepiped with base $|AF|.|FB|$ and height EE' because $|CE|.|EG|=|AF|.|FB|$, according to the previous theorem. Now $|AF|^2=|AB|.|AE|$ and $|BF|^2=|AB|.|EB|$,

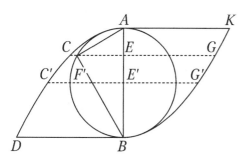

FIGURE 8.1
The properties of the ductus of certain segments of a circle

---

1 Co-authored by Herman van Looy.

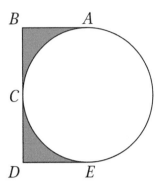

FIGURE 8.2
The mistilineum *ABCDE* (shaded area)

as a consequence $|AF|^2.|BF|^2=|AB|^2 |AE|.|EB|$ or $|AF|^2.|BF|^2 = |AB|^2.|EF|^2$ (i.e., the parallelepiped *AF.FB.EE'*), which is equal to the parallelepiped *AB.EF.EE'*. San Vicente continues: since the conditions for applying the exhaustion theorem hold, the hypothesis also holds. The importance of this proposition in the light of the circle quadrature is that a solid containing a circle is transformed into a ductus figure without a circle.

In the volume sent on May 22, 1625 (MS 3.4), the volumes of several ductus figures that have some connection to the ungula are calculated.

The first part (MS 3.4, fols. 326–30, bis) contains ten lemmas for the theorems of the second part. Half of these ten lemmas are about the *mistilineum ABCDEA* (the shaded figure in fig. 8.2). The *mistilineum* is a plane figure consisting of the part of half a square from which the inscribed semicircle is taken away. More generally, it can refer to a figure of which three of the sides are the sides of a rectangle and the fourth is a curve. The perimeter thus consists of *mistilineum*, mixed lines or lines of a different species (meaning straight and curved lines).

Among other things, the other theorems discuss properties of *corpus cavum primarium* (the primary hollow solid), which is the solid produced by the ductus of a *mistilineum* into a *mistilineum*.

A cubature of the *corpus cavum curvatorum* (the curved hollow solid) is performed in proposition 9 (i.e., the corpus is reduced to a rectilinear solid; MS 3.4, fol. 316^v; fig. 8.3). To do so, in proposition 7 San Vicente had complemented the *corpus cavum primarium* to a figure consisting of the parallelepiped *HC* and the semi-cylinder *KNICFB* (MS 3.4, fol. 314^v).

The proof can be summarized as follows: parallelepiped *HC* = *GAHKLICDEB* + semi-cylinder *HAGIKL* = *GAHKLICDEB* + semi-cylinder *KNICFB* = *GAHKLICDEB* + (*CFD* in *se ductum* + *ICFBKFILK*)

So parallelepiped *HC* - *CF* in *se ductum* = *GAHKLICDEB* + *ICFBKFILK* = *corpus cavum primarium*

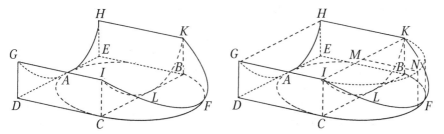

FIGURE 8.3 The calculation of the volume of a *corpus cavum curvatorum*

The last batch of treatises was sent in three consignments of August 22, September 19, and October 17, 1625 (MS 3.5).

In the twenty-five propositions of this part, we find properties of the elliptical cylinder analogous to those of the common circular cylinder. Both exhaustion methods are used on the *ungula elliptica*, on the one hand by inscribing pyramids with their vertex in the midpoint of the basis of the ungula and with a quadrilateral base described on the lateral surface, on the other hand with parallelepipeds constructed on the diameter of the base of the ungula, which is divided into an ever increasing number of equal parts. In proposition 7, one reads that two similar *ungulae ellipticae* will be to each other as the cubes of the corresponding diameter of the ellipse-shaped bases.

In other words, the *corpus cavum primarium* is transformed into a difference of two solids that can be regarded as rectilinear: a parallelepiped and the ductus of a semicircle *in se*, which had earlier been transformed into a pyramid.[2]

All these propositions can also be found in MS 3.2, save for the last five theorems concerning the elliptical cylinder.

San Vicente thought he had made some interesting and intriguing discoveries, which as we have seen was indeed the case. However, none of the treatises sent to Rome shows any indication that squaring the circle was possible.

2   The *Chartae Romanae*

When San Vicente was in Rome, and also when he was in Prague afterward, he continued to look into the properties of solids.[3] He did so by exploring

---

[2] According to Moretus, this is proven in proposition 14 of the previous missive. This missive (MS 3.2, fols. 471–75) ends with proposition 11. It does mention "Sequitur sequenti folio propositio 12a, 13a, 14a." These pages have subsequently been lost.

[3] KBR 5790–91, MS 4.1 and 4.2.

variations of his ductus method. One of the problems he would face was how to "add" ratios. More specifically, if $\frac{A}{B}$ and $\frac{C}{D}$ are known, but not necessarily A, B, C, and D, can $\frac{A+C}{B+D}$ be calculated? It is his failure to solve this question adequately, and the confusion with the common addition, that will be part of his undoing in trying to solve the circle quadrature. Nevertheless, these manuscripts still contain some highly original mathematics.

Although the first traces of San Vicente's transformation method can be found in MS 2.1 (fol. 45), and although he has used it throughout his research, he now explores the transformations in their own right. In the transformation method, one chooses a point A on an axis of the conic section; from this point, draw straight lines that intersect the conic at E_i; through the intersections, draw lines parallel to the other axis, which intersect the first axis at G_i; on these lines, strike off $[G_iH_i]$ such that $|G_iH_i|=|AE_i|$. The resulting figure is again a conic section.[4] With this method, he transforms a line segment into a hyperbola (fig. 8.4) and a circle into both a hyperbola and a parabola.

In this manuscript, San Vicente also introduces a new kind of curve, the *virtual parabola*. These curves are rational curves of the fourth degree with a node and for which the equation is $\sqrt{ax+b}+\sqrt{cx+d}$. San Vicente notes that the diameter of such a curve is a proper parabola (fig. 8.5).

A subject that merits some attention is the analogy between the Archimedean spiral and the parabola. San Vicente notes that the spiral is produced by a point P, which moves uniformly on a radius AC, which is rotated uniformly. In the

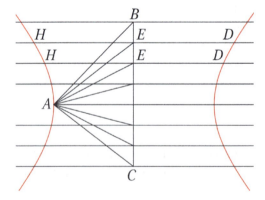

FIGURE 8.4
Transforming a straight line into a hyperbola. In a triangle $\triangle ABC$, draw the lines $[AE_i]$ with E_i on $[BC]$ (i.e., we consider BC to be a degenerate conic). Draw perpendicular lines to BC through E_i and determine the points D_i and H_i on these lines such that $|H_iE_i| = |E_iD_i| = |AE_i|$. Then the points D_i and H_i determine a branch of a hyperbola

---

4   See Karl Bopp, *Die Kegelschnitte des Gregorius a St. Vincentio in vergleichender Bearbeitung* (Leipzig: Teubner Verlag, 1907), 295ff. for a detailed description of this method. Essentially, it is a transformation $x' = x$ and $y'^2 = x^2 + y^2$.

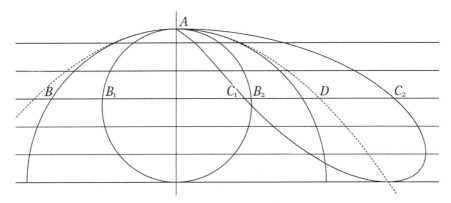

FIGURE 8.5 The construction of the virtual parabola. In a semi-circle, draw a circle with the radius of the semi-circle as diameter. Draw straight lines parallel to the tangent of the pole A of the semi-circle. These lines intersect the semi-circle in B_i and the circle in B_{1i} and B_{2i}. Now construct the points C_{1i} and C_{2i} on the parallel lines such that $|AB_i| = |B_{1i}C_{1i}| = |B_{2i}C_{2i}|$. The points C_{1i} and C_{2i} determine a virtual parabola. The midpoints D_i determine a proper parabola

same way, the parabola is produced by a point P that moves uniformly on the line segment [AC], while C moves uniformly along BD.

San Vicente claims that the parabolic segment between the line segment [AD] and the parabola equals one-third of the triangle △ABD. Equally, when the spiral AC has swept through the circle, the area will be one-third of the circle itself (fig. 8.6). A couple of years later, Cavalieri would publish similar results concerning the areas of the parabola and the spiral.[5] In *Problema Austriacum*, San Vicente would prove that the assertion holds for any parabolic segment and any spiral.[6]

This result would be one of the last of San Vicente's mathematical discoveries. With his attack of apoplexy, his creative period came to an end. All his subsequent work would be a variation on the themes he had explored before his Prague sojourn.

---

5 Alexander, *Infinitesimal*, 100–1.
6 San Vicente, *Problema Avstriacvm*, 6:686–95. The assertion holds for any abscissa α for the parabola and any angle α for the spiral. The equation of the parabola is $y = 2\pi x^2$, and the area of the triangle $\triangle A\alpha P$ is $\pi\alpha^3$. The area under the parabola is $\int_0^\alpha 2\pi x^2 dx = \frac{2\pi\alpha^3}{3}$, and the area of the parabolic segment is $\frac{\pi\alpha^3}{3}$. Analogously, the area swept through by the spiral with equation $r = \theta$ is $\frac{1}{2}\int_0^\alpha \vartheta^2 d\vartheta = \frac{1}{2}\frac{\alpha^3}{3}$. The area of the corresponding circle segment is $\frac{\alpha^3}{2}$.

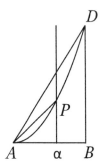 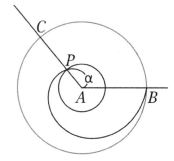

FIGURE 8.6 The analogy between the area under a parabola and the area enclosed by an Archimedean spiral; the area under the parabola is one-third of that of triangle △ABC, and the area swept through by the radius of the spiral is one third of a circle with radius |AB|

3 San Vicente's Legacy

Among the attentive readers of San Vicente's book were Huygens and Leibniz.[7] Both chose not to pursue San Vicente's research program.

Echoes of San Vicente's method can be found in Fermat, but it is not clear whether he was actually influenced in any way by San Vicente's work, even if only from hearsay. In his *Treatise on Quadrature*,[8] Fermat uses a division of the *x*-axis into line segments in continuous proportion for the direct quadrature of segments of higher hyperbolae or parabolae (i.e., curves with an equation $x^p y^q = k$; fig. 8.7).[9] This treatise was written around 1658, although Fermat claims that he had worked on the problems since the mid-1640s.[10]

Fermat's method is based on an approximation of a hyperbolic segment by a rectangle for which the base is "small enough." This is obtained by dividing the *x*-axis into line segments that are in continued proportion. For the higher hyperbolae, the calculation is straightforward, whereas for the higher parabolae he has to use a method that is vaguely similar to expanding a series by

---

7 Hofmann, *Das* Opus geometricum *der Gregorius à Sancto Vincentio.*
8 Pierre de Fermat, *Oeuvres: [Suivi de] observations sur Diophante/De Fermat tome premier, oeuvres mathématiques diverses*, ed. Paul Tannery and Charles Henry (Paris: Gauthier-Villars et fils, 1891), 255ff.
9 Michael Sean Mahoney, *The Mathematical Career of Pierre de Fermat* (Princeton: Princeton University Press, 1973), 244ff.
10 Mahoney, *Mathematical Career of Pierre de Fermat*, 216, 244. In a 1657 letter to Kenelm Digby (1603–65), Fermat claimed to have disclosed his method in a letter to Torricelli in 1646. The correspondence with Torricelli has not been found.

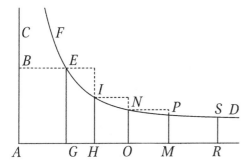

FIGURE 8.7 Fermat's calculation of the area of segments of higher hyperbolas draws its inspiration from Saint-Vincent's methods

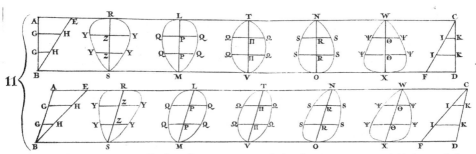

FIGURE 8.8 The construction of higher ellipses in Ignatius de Jonghe, *Geometrica inquisitio in parabolas numero & specie infinitas & iisdem congenitae hyperbolas ac præcipue in quadraturam hyperbolæ Apollonianæ* (Antwerp: n.p., 1688). Figure 11 on foldout page
© MPM B 219
Let *ABDC* be a parallelogram. Let *EB* and *FC* be parallel lines. *RS*, *LM*, ..., *WX* are parallel to *AB*. Let a line *GK*, parallel to *BD*, intersect these lines. Construct the line segments *ZY*, *PQ*, ..., *ΘΨ* on this line, such that *GH*, *ZY*, *PQ*, ..., *ΘΨ*, *IK* are in continuous proportion. Mirror these lines about *Z*, *P*, ..., *Θ* respectively. The locus of the extremities of these line segments (i.e., *Y*, *Q*, ..., *Ψ* and their mirror images) as *GK* moves parallel to itself from *BD* to *AC* are higher ellipses. Higher parabolae and hyperbolae are constructed analogously

insertion.[11] While Fermat's quadratures seem to have all the ingredients of San Vicente's method, the recipe is entirely different. Unlike San Vicente, Fermat has the advantage of using algebraic methods. Because he uses a ratio $\frac{u}{v}$ for a geometric series, he can choose u and v as close to one another as is needed, which eliminates the need for an expansion by insertion.

Nearly ten years after Fermat, in 1688, de Jonghe published *Geometria inquisitio*, in which he treats the quadratures of higher hyperbolae and parabolae,

---

11 For a detailed discussion of Fermat's method, see Mahoney, *Mathematical Career of Pierre de Fermat*, 247–53.

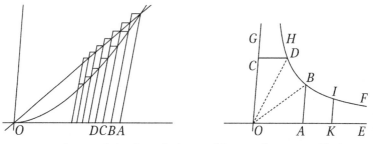

FIGURE 8.9 Ignatius de Jonghe's calculation of the area of segments of higher parabolae and hyperbolae using Saint-Vincent's methods

completely in San Vicente's style.¹² De Jonghe was a student of Tacquet. Essentially, de Jonghe uses San Vicente's method to compare the area of a segment between a curve and a straight line on the one hand and a triangle of which a side is on the straight line.

For a higher parabola $y^n = a^{(n-p)}x^p$ (fig. 8.9, left), he arrives at $\dfrac{\text{area}\triangle OPA}{\text{area parabola } OPA} = \dfrac{n+p}{2n}$, for a higher hyperbola $y^n x^p = a^{(n+p)}$ ($n < p$; fig. 8.9, right), $\dfrac{\text{area } EABF}{\text{area } \triangle OAB} = \dfrac{2n}{p-n}$, and $\dfrac{\text{area } EOCBF}{\text{area } \triangle OCD} = \dfrac{2p}{p-n}$, from which any area $ABIK$ can be determined easily.

De Jonghe also proved, in a fairly cumbersome way, that these curves can be obtained as sections of a plane and a right cone with a higher ellipse as base.

De Jonghe's book came too late to have any influence on the history of mathematics, as San Vicente's methods had already been superseded by differential and integral calculus.

4   Conclusion

What San Vicente did in his first Flemish period was nothing less than lay the foundations of a rigorous infinitesimal calculus. He accepted that an infinite series can generate a finite limit and gave a physical meaning to it. He succeeded in finding a quadrature for a number of curves using a geometric

---

12   Ignatius de Jonghe, *Geometrica inquisitio in parabolas numero & specie infinitas & iisdem congenitae hyperbolas ac præcipue in quadraturam hyperbolæ Apollonianæ* (Antwerp: n.p., 1688). For a detailed analysis, see Bockstaele, "Ignatius de Jonghe"; Bockstaele, "Een vergeten werk."

series of abscissae, generating parallelograms. He established a logically sound method, the exhaustion method, with which he could obtain quadratures of certain curves and cubatures of associated solids in a rigorous manner. In his manuscripts, he sometimes used lines to represent an area. However, the nature of these lines differs from case to case. Sometimes, the lines are used to transform one curve into another, sometimes they are lines in the Cavalierian sense, and sometimes they are shorthand for rectangles and parallelograms. This represents an evolution in San Vicente's thinking. It should be clear that by 1620 San Vicente had developed an integration method for solids (and to a lesser extent for surfaces) that is superior to Cavalieri's *indivisibilia* method. By using exhaustion (i.e., using a limit), San Vicente can, in a strictly deductive way, calculate volumes and areas.

That Grienberger was of the opinion that the quadrature of the circle was not within reach unfortunately postponed San Vicente's publication. Had San Vicente been able to present his results to his confrères in the 1620s, he would have been hailed as one of the greatest ever mathematicians. Now he was not only robbed of his glory but, due to an error, was made the laughing stock of mathematicians, even up to this day. Moreover, that his book was published belatedly made him vulnerable to accusations of plagiarism.

Nevertheless, San Vicente's book has had an influence on the history of mathematics. Fermat pursued a similar path for a while, and Huygens and Leibniz were attentive readers of the book, which essentially showed them that while San Vicente had had some interesting insights, his method was ultimately an intellectual cul-de-sac.

CHAPTER 9

The Erroneous Circle Quadrature

Unfortunately, we cannot bypass San Vicente's circle quadrature entirely. The following paragraphs accordingly give an outline of the first circle quadrature in *Problema Austriacum*,[1] which San Vincente would "prove" a further two times in different ways but while essentially making the same error.

The first traces of this quadrature can be found in MS 4 (fols. 347^{v}ff.; fig. 9.1), which was written either in Rome or Prague. In a figure similar to the one in *Problema Austriacum*, San Vicente claims, without proof, that because

$$\frac{|D_iB_i||B_iE_i|}{|D_jB_j||B_jE_j|} = \left(\frac{|F_iB_i||B_iG_i|}{|F_jB_j||B_jE_j|}\right)^2 \text{ and also } \frac{|D_iB_i||B_iE_i|}{|M_iI_i||I_iL_i|} = \left(\frac{|F_iB_i||B_iG_i|}{|N_iI_i||I_iO_i|}\right)^2 \text{ the ratio}$$

of all triangles $\triangle D_iB_iE_i$ to all triangles $\triangle M_iI_iL_i$ is equal to the square of the ratio of all rectangles $F_iB_iG_i$ to all rectangles $N_iI_iO_i$. We can render his assertion as:

$$\forall i \in \{1,2,\ldots,n\} : \frac{a_i}{b_i} = \frac{c_i^2}{d_i^2} \Rightarrow \frac{\sum_{i=1}^n a_i}{\sum_{i=1}^n b_i} = \frac{\sum_{i=1}^n c_i^2}{\sum_{i=1}^n d_i^2}.$$

San Vicente is now interested in determining the ratio of two solids from the ratio of the corresponding infinitesimals.

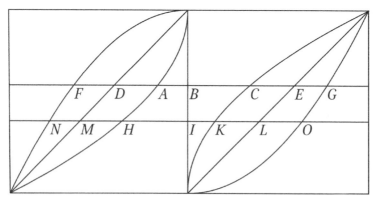

FIGURE 9.1   An attempt to solve the quadrature of the circle using parabolae and a straight line (from Saint-Vincent's manuscripts)

---

1   See also Van Looy, "Chronologie en analyse," 286–94; Hofmann, *Das Opus geometricum der Gregorius à Sancto Vincentio*, 70–71.

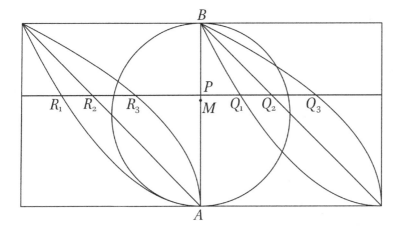

FIGURE 9.2 An attempt to solve the quadrature of the circle using parabolae and a straight line. After Gregory of Saint-Vincent, *Problema avstriacvm plvs vltra qvadratvra circvli* (Antwerp: Joannes et Jacobus Meursius, 1647), 7:704
© EHC G 4869

Knowing $\dfrac{y_i \cdot z_i}{y'_i \cdot z'_i}$, can he calculate $\dfrac{\sum_{i=1}^{n} y_i \cdot z_i}{\sum_{i=1}^{n} y'_i \cdot z'_i}$ or $\dfrac{\sum_{i=1}^{n} y_i \cdot z_i \Delta x}{\sum_{i=1}^{n} y'_i \cdot z'_i \Delta x}$? Which, after application of the exhaustion method, leads to what we write as $\dfrac{\int yz\,dx}{\int y'z'\,dx}$, the ratio of the two solids.

For now, San Vicente does not have an answer, and the problem would haunt him for the years to come. Simply put, his problem amounts to: given $\dfrac{A}{B}$ and $\dfrac{C}{D}$, but not the values of A, B, C, and D, can $\dfrac{A+C}{B+D}$ be calculated? His attempts to solve this problem would end in a complete fiasco.

In *Problema Austriacum*, San Vicente considers two squares with common edge AB, in which parallel diagonals are drawn and in which two arcs of a parabola, with vertices at A and B respectively, are drawn (fig. 9.2). A circle with diameter $[AB]$ is constructed as well.[2] Now a perpendicular to AB is considered. If M is the center of the circle, $a = \dfrac{|AB|}{2}$ its radius, and putting $|MP| = x$ then:

---

2 San Vicente, *Problema Avstriacvm*, 10:1116ff.

$$|PQ_1| = \sqrt{2a(a-x)}, |PQ_2| = a-x, |PQ_3| = \frac{(a-x)^2}{2a}$$

$$|PR_1| = \sqrt{2a(a+x)}, |PR_2| = a+x, |PR_3| = \frac{(a+x)^2}{2a}$$

San Vicente compares the results for two perpendiculars and finds that

$$\frac{\text{Rectangle } Q_3 P \cdot PR_3}{\text{Rectangle } Q'_3 P' \cdot P'R'_3} = \left(\frac{\text{Rectangle } Q_2 P \cdot PR_2}{\text{Rectangle } Q'_2 P' \cdot P'R'_2}\right)^2 = \left(\frac{\text{Rectangle } Q'_1 P' \cdot P'R'_1}{\text{Rectangle } Q'_3 P' \cdot P'R'_3}\right)^4$$

Obviously, the third proportion has the same exponent relative to the second proportion as does the second relative to the first. Now San Vicente performs a ductus by erecting the rectangles on two lines of equal length, P_1P_2 and $P'_1P'_2$. In this way, six solids are formed F_1, F_2, F_3 and F'_1, F'_2, F'_3. Theorem 40 now, erroneously, states that from $\left(\frac{F_2}{F'_2}\right) = \left(\frac{F_1}{F'_1}\right)^t$ it also follows $\frac{F_3}{F''_3} = \left(\frac{F_2}{F'_2}\right)^t$. We consequently have $F_1 = 2a \int_{x_1}^{x_2} \sqrt{a^2 - x^2}\, dx$, $F_2 = \int_{x_1}^{x_2} (a^2 - x^2)\, dx$, $F_3 = \int_{x_1}^{x_2} (a^2 - x^2)^2\, dx$.

The last two integrals can be calculated in a straightforward way, as they could by San Vicente using exhaustion. The first integral can be thought of as the volume of a cylinder with a circle with radius a as base but truncated by two perpendiculars to AB. San Vicente knew how to reduce this kind of figure to a cylinder sector. Therefore, because the proportions $\frac{F_2}{F'_2}, \frac{F_3}{F''_3}$ can be calculated, so can $\frac{F_1}{F'_1}$, using the erroneous theorem 40. Knowing the volume of the cylinder sector implies that the area of the base can be calculated and hence that the area of a circle can be calculated. Unfortunately, San Vicente thought he had solved the problem of the circle quadrature.

CHAPTER 10

Joannes della Faille and the Beginning of Projective Geometry

1   An Itinerant Life

Della Faille is the only student of San Vicente about whom we have a great deal of biographical information, because much of his correspondence with his siblings has been preserved in the family archives.[1]

Della Faille was born on March 1, 1597.[2] He was the son of Jan (van) Karel Sr. (1569–1641), lord of Rijmenam, and of Maria van den Wouwere (1574–1659).[3] In 1606, at the age of nine, della Faille entered the Jesuit school at Antwerp.[4] By 1613, to his father's dismay, he had made clear that he wanted to become a Jesuit himself. He joined the Society on September 12, beginning his novitiate at Mechelen.[5] Two years later, he took his first vows and was sent back to Antwerp to study philosophy.

In 1617, together with twenty-two other students, della Faille took "physics" classes, meaning the second year of his philosophy studies.[6] He defended fifty theses on August 22, 1617.[7] Della Faille and Nuyts were the first two Jesuits to take mathematics classes in 1617.[8] Della Faille seems to have been the only one to have followed the course for three consecutive years.[9] San Vicente was full of praise for his student, as becomes clear from a letter to the Jesuit Remigius Happaert (1608–75):

---

1   For a detailed biography of della Faille, see Meskens, *Joannes della Faille, S.J.*, on which this chapter is based.
2   AFL P12; ARSI Fl. Belg 80, 1613. Schmitz, *Les della Faille*, 3:71.
3   FA PR2, fol. 92ᵛ; PR11, fol. 108ʳ.
4   Droeshout suggests, however, that della Faille did not enroll until 1608, when the college was transferred to Prinsstraat. Droeshout, "Histoire de la Compagnie de Jésus à Anvers," 2:212.
5   AFL P12; ARAA T14/034 486; ARSI Fl. Belg. 80, 1613.
6   ARSI Fl. Belg 44, fol. 8ʳ; fol. 21ʳ; ARAA T14/034 486. Droeshout, "Histoire de la Compagnie de Jésus à Anvers," 3:151–52.
7   AFL 28.4.8.
8   The elogium of della Faille (ARAA T14/034 486) mentions that he first attended classes with San Vicente in 1616. KB MS 20194; ARSI Fl. Belg. 44, fol. 32ʳ. Droeshout, "Histoire de la Compagnie de Jésus à Anvers," 3:199, 248; Van Looy, "Chronologie en analyse," 12–13.
9   ARSI Fl. Belg. 44, fols. 32ʳ, 46ᵛ; KB MS 20194. Droeshout, "Histoire de la Compagnie de Jésus à Anvers," 3:235, 259.

He asked me whether he could come to my room when he had a day off, to study my mathematics books. During that year, I began teaching my mathematics course to the students of second-year philosophy [...]. He set about studying mathematics with great fervor and continued for three years. He successfully defended mechanics theorems more than once. When we said goodbye, he had plenty of scientific notes made during his researches.[10]

San Vicente mentions these theorems again in a letter of 1651 to Huygens, saying that he lost his copy during his time in Prague.[11]

On October 27, 1620, della Faille was ordered to teach mathematics at the Jesuit college of Dôle and to attend the philosophy courses.[12]

Della Faille would teach mathematics at the college until 1626, during which time he also took theology classes at the university.[13] On April 10, 1621, he was ordained priest.[14] In 1625, he defended his *Theses mechanicae*, which were printed at Dôle. Although no copy has thus far surfaced, it seems that the work described the kind of pendulum clock that was also used by Arnold-Floris van Langren (c.1571–1644).[15]

---

10 Henri Bosmans, S.J., "Deux lettres inédits de Grégoire de St-Vincent et les manuscrits de della Faille," *Annales de La Société Scientifique* 26 (1901): 1–19; Bosmans, "Le mathématicien anversois Jean Charles della Faille de la Compagnie de Jésus," *Mathesis* 41 (1927): 1–12.

11 Huygens, *Oeuvres complètes: Tome I*, 158–59, letter 105.

12 ARSI Fl. Belg. 44, fols. 60^v–61^r. See also ARSI Lugd. 14, fols. 97^v, 130^r, 142^r, 150^v, 163^r, 181^v. François de Dainville and Marie-Madeleine Compère, *L'éducation des jésuites (XVIe–XVIIIe siècles)* (Paris: Ed. de Minuit, 1978), 326; de Dainville, *L'éducation des jésuites*, 342. The Jesuit college at Dôle was established in 1582, but a chair of mathematics was not introduced until 1615. Antonella Romano, "Les jésuites et les mathématiques: Le cas des collèges français de la Compagnie de Jésus," in *Christoph Clavius e attività scientifica dei gesuiti nell'età di Galileo*, ed. Ugo Baldini (Rome: Bulzoni Editore, 1995), 243–82, 270; de Dainville, *L'éducation des jésuites*, 326.

13 ARSI Lugd. 14, fols. 97^v, 130^r, 142^r, 150^v, 163^r; ARAA T14/034 486. Romano, *La Contre-Réforme mathématique*, 536, 571.

14 ARSI Fl. Belg. 11, fol. 142^v. Meskens, *Joannes della Faille, S.J.*, 42; Pierre Delattre, S.J., ed., *Les établissements des jésuites en France depuis quatre siècles: Répertoire topo-bibliographique publié à l'occasion du quatrième centenaire de la fondation de la Compagnie de Jésus 1540–1940* (Enghien: Institut supérieur de théologie, 1940–57), col. 152.

15 Letter from della Faille to Arnold Floris van Langren, November 26, 1636, published by Van de Vyver, "L'école de mathématiques des jésuites," 108–9. See also Ribadeneyra and Alegambe, *Bibliotheca scriptorum Societatis Jesu*, 233; Martin Lipen, *M. Martini Lipenii bibliotheca realis philosophica omnium materiarum, rerum, & titulorum, in vniverso totivs philosophiæ ambitu occurrentium: Ordine alphabetico sic disposita, ut primo statim aspectu titvli, et sub titulis autores ordinata velut acie dispo* (Frankfurt am Main: Fridericus, 1682), 903; http://diglib.hab.de/drucke/va-4f-4/start.htm (accessed February 4, 2020); Jan Frans

FIGURE 10.1　Pieter van der Plas, "Joannes della Faille," oil on canvas, c.1620
PRIVATE COLLECTION

Della Faille applied to be sent on missionary work in 1623, but the superior general refused the application, replying that there was for the time being no need for missionaries from France and that he would be called upon when the occasion presented itself.[16]

Della Faille returned to the Low Countries in 1626.[17] In Leuven, he acted as a substitute professor for San Vicente. He taught physics to Jesuit students and mathematics to lay students.[18] Among these seem to have been the sons of the Paltzgraf (count of Palatine) and several sons of Polish noblemen who had come to study at Leuven.[19] It is highly likely he was occupied with the pastoral care of the troops of the Spanish army garrisoned in Leuven as well.[20]

In November 1628, della Faille started his tertianship in Lier, though it was to prove short lived.[21] In February 1628, the Madrilene Jesuits wrote to the superior general to send German mathematicians "muy buenos" (very good [mathematicians])[22] to teach at the Estudios Reales of the Colegio Imperial that had opened in February 1629.[23] As San Vicente had been struck by an attack of apoplexy and was unable to make the journey, Della Faille was sent to Madrid

---

Foppens, *Bibliotheca belgica, sive virorum in Belgio vita, scriptisque illustrium catalogus, librorumque nomenclatura* (Brussels: Petrus Foppens, 1739), 604; Augustin de Backer, S.J., Aloys de Backer, and Carlos Sommervogel, S.J., *Bibliothèque des écrivains de La Compagnie de Jésus* (Brussels: Schepens, 1890), 1785.

16 ARSI Lugd. 5, fol. 443ᵛ, December 11, 1623.
17 ARSI, Fl. Belg. 44, fol. 152ʳ; ARAA T14/034 486.
18 Van de Vyver, "L'école de mathématiques des jésuites," 267; Meskens, *Joannes della Faille, S.J.*, 41–42.
19 ARAA T14/034 486. Schmitz, *Les della Faille*, 3:76; Meskens, *Joannes della Faille, S.J.*, 43–44. Some of these contacts may have lasted a long time. When he was in Spain, della Faille sent information on the political situation in Poland to the Spanish provincial general Francisco Ribera (1537–91). BRAHM CC 9/6788, fol. 372ᵛ.
20 Della Faille is mentioned in the lists of the *Missio Castrensis*, but without dates. See Audenaert, *Prosopographia iesuitica Belgica antiqua*, 1:336, 3:285.
21 ARSI Fl. Belg. 44, fol. 203ᵛ; KB MS 20199. Hubert P. Vanderspeeten, S.J., "Le R.P. Jean-Charles della Faille de la Compagnie de Jésus dans la seconde moitié du XVIᵉ siècle," *Précis historiques 2e série* 3 (1874): 77–117, 132–42, 191–219, and 241–47, 216; Bosmans, "Deux lettres inédits," 38; Brouwers, *Carolus Scribani 1561–1629*, 220–22.
22 ARSI Tolet 9, fol. 29 in margin.
23 The Estudios Reales did not live up to expectations. By 1634, it had only sixty students in the upper college, none of them of high social standing. John H. Elliott, *The Count-Duke of Olivares: The Statesman in an Age of Decline* (New Haven: Yale University Press, 1986), 188. See Mariano Esteban Piñeiro and Mauricio Jalón, "Juan de Herrera and the Royal Academy of Mathematics," in *Scientific Instruments in the Sixteenth Century: The Spanish Court and the Louvain School*, ed. Jacques van Damme, Koenraad van Cleempoel, and Gérard L'Estrange Turner (Madrid: Fundación Carlos de Amberes, 1997), 35–42, here 41. On mathematics and mathematics professors at the college, see Udías, "Los libros y manuscritos."

in his stead.[24] He left the Low Countries on March 23, 1629, never to return again. On May 1, 1630, he took his solemn vows.[25]

While in Madrid, della Faille did not forget that he was first and foremost a priest: he visited the sick in the hospitals and prisoners in jail, but above all he acted as a confessor.[26] He redistributed the money he received from his family among the poor and needy.[27]

At some stage, San Vicente learned of della Faille's *Theses mechanicae*, in which he found an interesting theorem about the center of gravity of a circle sector and an ellipse sector. Upon hearing that Guldin was about to publish on the same subject, San Vicente wrote to della Faille to prepare a publication.[28] After some hesitation, della Faille agreed and sent the manuscript to San Vicente. San Vicente obligingly edited the book, which appeared in 1632. It was the only book della Faille would publish, yet it attained immediate fame (Galileo, for instance, drew Cavalieri's attention to it).[29]

At the behest of the rector of the Madrilene Society of Jesus, della Faille was appointed to the chair of cosmography at the Consejo de Indias (Council of the Indies) on September 30, 1637.[30] Henceforward, he would carry the title of *cosmógrafo mayor* (senior cosmographer), and along with the title he was also supposed to hold the chair in mathematics and teach mathematics at the palace.[31] From the inception of the office until the expulsion of the Jesuits from Spain, the chair of cosmographer would be held by a Jesuit.

In the summer of 1641, during Portugal's war of independence, della Faille was summoned by Philip IV to be a military advisor to the duke of Alba, the commander of an area of Ciudad Rodrigo. Della Faille would remain there for the best part of the next three years.[32]

---

24 Kadoc ABML Bosmans 116 della Faille, referring to letters of the superior generals to the Flandro-Belgian province, fol. 1077, letter of Vitelleschi to della Faille, February 3, 1629.
25 AFL 28.15.3; ARSI Hisp. 6, fols. 570^r, 575^r.
26 AFL 28.15.20, 28.16, Meskens, *Joannes della Faille, S.J.*, 48.
27 Vanderspeeten, "Le R.P. Jean-Charles della Faille," 136, citing a letter by Don Andres de Anduga y Daca.
28 APUG 534 fols. 65^{r-v}, on the permission to publish; see also ARSI Tolet 9, fols. 149^v and 190^v. Bosmans, "Deux lettres inédits," 24.
29 Rocca and Rocca, *Lettere d'uomini illustri del secolo XVII*, 47.
30 Meskens, *Joannes della Faille, S.J.*, 145; Antonia Heredia Herrera, *Catálogo de las consultas del Consejo de Indias*, 8 vols. (Seville: Diputación Provincial Sevilla, n.d.), 7:159, no. 710; Maroto and Esteban Piñeiro, *Aspectos de la ciencia aplicada*, 199–200 (they misread the name as de la Salle).
31 Maroto and Esteban Piñeiro, *Aspectos de la ciencia aplicada*, 168–70.
32 Omer van de Vyver, S.J., "Lettres de Jean Charles della Faille, S.J., cosmographe du roi à Madrid à M. Fl. van Langren, cosmographe du roi à Bruxelles," *Archivum historicum Societatis Iesu* 46, no. 91 (1977): 73–183, letters 44, 56; see also AFL 28.9 and 28.15.34.

FIGURE 10.2 Astronomical compendium by Cornelis Vinckx (*fl.* 1599–1623) bearing the crest of the family della Faille de Leverghem (Musée de la Vie Wallonne, Liège, no. C525)
© KIK-IRPA A118715

An astronomical compendium is a multipurpose, all-in-one instrument that carries numerous devices for telling the time and performing astronomical calculations, carrying as many instruments as possible filling the available space.

In October or November 1646, della Faille became the mathematics teacher of Don Juan of Austria (1629–79),[33] and he soon also became his confidant and advisor, not to say friend. When, in March 1647, Don Juan became admiral of

---

33 In 1642, Philip IV had his bastard son Don Juan of Austria recognized as his legitimate son and accepted as Infante of Spain. In a letter to a brother (AFL 28.15.34), della Faille wrote that he became tutor five months before Don Juan became admiral of the fleet. This puts his arrival at about five months before March 28, 1647 (i.e., October or November 1646 [see also AFL 28.9]). The subjects della Faille taught him were practical ones such as the art of fortification and the art of navigation. "El P. Juan de la Faille, [...], ha estado estos meses enseñando al señor Don Juan de Austria la marinería"; letter of Sebastián González to Rafael Pereyra, March 26, 1647. *Memorial histórico español* (Madrid: Academia Real de

the fleet, his father allowed della Faille to remain in his service.[34] In 1647, the Armada was directed on a punitive expedition toward the rebellious cities of Palermo and Naples.[35] By the time the Armada reached Sicily, the viceroy had been able to restore authority in Palermo.[36] The Armada arrived off Naples on October 1, 1647, and by the spring of 1648 the revolt had been suppressed.[37] Della Faille took the opportunity of a sojourn in Italy to visit Rome's sacred places.[38]

In 1651, Philip IV had decided on crushing the Catalonian revolt, which had begun in 1640 and had been sustained with French help. By 1650, only Barcelona and the Pyrenean foothills kept alive the chimera of secession.[39] Philip put Don Juan in charge of the troops. Della Faille was still in the prince's retinue, but although the prince treated him as an equal to the accompanying noblemen, in his last letter to his mother della Faille wrote that he had begun to grow tired of the traveling life.[40] Don Juan decided to isolate Barcelona, and after a prolonged and horrifying siege, the city capitulated on October 13, 1652.

The plague of 1650–54 caused severe mortality among a people already suffering from wartime malnutrition. Della Faille, now a teacher at the college of Barcelona, had not forgotten his vows and made the fateful decision to care for the sick and dying. On October 24, he became seriously ill, running a high fever[41] to which he succumbed after eleven days. He died with many Jesuits of the college in attendance on November 4, 1652.[42] His body was interred in the collegiate church.

---

la Historia, 1851); https://catalog.hathitrust.org/Record/000506543 (accessed February 4, 2020), 18:464–73, esp. 469.

34   AFL 28.15.34; Meskens, *Joannes della Faille, S.J.*, 57–58.
35   AFL 28.9 fol. 2ʳ; Josefina Castilla Soto, *Don Juan de Austria (hijo bastardo de Felipe IV): Su labor política y militar* (Madrid: Universidad nacional de educación a distancia, n.d.), 56.
36   Robert A. Stradling, *Philip IV and the Government of Spain 1621–1665* (Cambridge: Cambridge University Press, 1988), 197.
37   Stradling, *Philip IV*, 197–98.
38   AFL 28.9. He was especially interested in the relics of St.-Vincent Ferrer (1350–1419), toward whom he seems to have developed a special devotion from early on. Ferrer was a highly influential Dominican friar during the Western Schism (1378–1417) who crossed Spain, France, and Northern Italy preaching about the mystery of Christ and the impending end of the world. He was declared a saint in 1458.
39   Stradling, *Europe and the Decline of Spain*, 118. On this uprising, see John H. Elliott, *The Revolt of the Catalans: A Study in the Decline of Spain* (Cambridge: Cambridge University Press, 1984).
40   AFL 28.15.39. Meskens, *Joannes della Faille, S.J.*, 61–62.
41   It is not clear whether the cause was a flare up of his previous illness or a new infection, most probably bubonic plague. AFL 28.16, 28.22 (citing the *Libro universal de la fundación, rentas y raíces del presente Colegio de la Compañía de Jesús de Bethlem de Barcelona*, fols. 118ᵛ–119). Van Looy, "Chronologie en analyse," 12.
42   ARSI Arag. 20, fol. 191; Meskens, *Joannes della Faille, S.J.*, 62.

As mentioned earlier, della Faille published only one book, *De centro gravitatis* (On the center of gravity), on which his reputation in the history of mathematics rests. San Vicente claims he could have written as many as thirty treatises.[43]

In a letter to Grienberger, della Faille writes about the results he obtained on the determination of the centroid of an ellipse sector, but he also refers to other projects such as the determination of the center of mass of a cylinder. It seems that in 1630 he sent other mathematical treatises to Rome via the "Belgian route" (i.e., with a detour via the Southern Netherlands) and that he asked Grienberger for a reply via the same route.[44] By 1630 or 1631, della Faille was seeking to publish a book, apparently about geography, or an atlas.[45] He had already made his intentions about this project clear in a letter to Grienberger.[46] However, Moretus had got wind that a similar book was to appear in Holland, and he feared that the market would be too small for two books on the same subject. He therefore refrained from publishing della Faille's book.[47]

I have been able to trace some manuscripts, but I am uncertain about their status. It is not clear whether these were preparatory manuscripts for a book or course notes. The della Faille family archives family contain several interesting manuscripts in his hand.[48] Some of these are simply texts he copied, such as the works of Aristotle, but others contain original mathematical work. Among them are a treatise on conic sections and one on methods in geometry. In the Biblioteca de Palacio de Madrid can be found a treatise on architecture,[49] while the Real Academia de la Historia contains some volumes of manuscripts that may be attributed to della Faille.[50] One of these merits some attention because it reveals that della Faille was still pursuing the Aguilón legacy in optics.[51] In another manuscript, *Método de la geometría* (The geometrical method), della Faille is concerned with the nature of geometry in connection with the work of painters and sculptors, among many other things. It is more a text aimed at getting students to look at propositions at a very basic meta-mathematical level.

---

43 Van de Vyver, "Lettres de Jean Charles della Faille, S.J.," 78–79, 99–100.
44 APUG fols. 243^r–244^v.
45 On the Jesuits and geography, see François de Dainville, *La géographie des humanistes* (Paris: Beauchesne, 1940).
46 APUG 567, fol. 243^r–244^v.
47 Van de Vyver, "Lettres de Jean Charles della Faille, S.J.," letter 7.
48 AFL 28, 29.
49 BPM MS 3729.
50 BRAHM CC 9/2125; 9/2732; 9/2745; 9/2751.
51 Meskens, *Joannes della Faille, S.J.*, 71–72.

2   Conic Sections

The manuscript *Lugares* (On loci)[52] is divided into several chapters, all of which deal with one particular locus, that is, circles, straight lines obtained by circles, straight lines, and parabolae. Some of the results della Faille cites were already well known; in fact, he sometimes gives the definitions of the figure in question in the guise of a construction. For instance, his third proposition in the chapter "Lugares circulares" (Circular loci) states that if, with [AB] as a base, triangles △ABC_i are erected for which $\hat{C}_i$ is constant, then all points will lie on a circle, of which [AB] is a chord.[53]

Some theorems in *On Loci* merit further attention. In one of these theorems, on which he would digress further in *On Conics*,[54] della Faille states that as a line revolves around a point A in the interior of a circle, the tangents at its extremities always intersect on the same line l.[55] Moreover, he states that the line through A and the center of the circle will be perpendicular to the line l (fig. 10.3, left).[56]

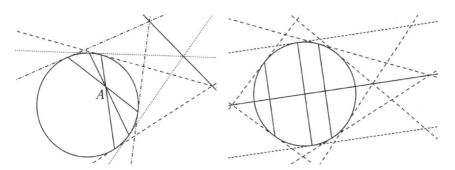

FIGURE 10.3   The definition of the polar line of a pole A in a circle

---

52 AFL 28.2.
53 This is a well-known Euclidean proposition (3.21). See Heath, *Euclid*, 2:49–50.
54 AFL 29.7.
55 This theorem is the converse of de la Hire's theorem: if a point traces out a straight line, then the polars of the point with respect to a conic will rotate around the pole of the straight line (Philippe de la Hire, *Sectiones conicae in novem libros distributæ* [...] [Paris: Stephanum Michallet, 1685], 12, *propositio* 26). See Field and Gray, *Geometrical Work of Girard Desargues*, 37, and Fladt, *Geschichte und Theorie der Kegelschnitte*, 49ff., esp. 52.
56 This is a special case for the circle of the theorem, which states that if A is the pole of a line l with respect to a conic section and if D is the pole of a line k through A and parallel to l, then AD is a diameter. This theorem is also stated in della Faille's *On Conics*.

A special case is one involving the point at infinity: if one takes chords parallel to a given diameter, then the tangents at their extremities will meet on a line through the center of the circle. This line is nothing else than the polar of the point at infinity with respect to a certain direction. If one accepts that the polar of a point is the line that connects the points of tangency of the respective tangents through the given point, then it is immediately clear that a polar of the point at infinity will be perpendicular to the chords of this direction (fig. 10.3, right). Although della Faille clearly had some ideas leading to the concepts of poles and polars, he does not generalize the concept, nor does he exploit all its possibilities. Because his manuscripts only give theorems without proofs, it is unclear whether he realized the prospects his theorems opened up. In *On Conics*, he would further explore some of the properties of polars.

Most of the theorems in the part on rectilinear loci relate to homothetical images and Thales's intercept theorem and are rather straightforward. Only one theorem merits attention, because it is the converse of the cited proposition in the part on circular loci. The theorem states that if a straight line AB intersects the lines CD_i at E_i and the areas of all rectangles D_iCE_i ($|D_iC|.|CE_i|$) are equal, then all D_i lie on the circumference of a circle through C (fig. 10.4). Conversely, if on the lines D_iCE_i, $|D_iC|.|CE_i|$ is constant, then all E_i are on one line.[57] The method is reminiscent of San Vicente's *per subtensas* method.

In the part on parabolic loci, two theorems are interesting because they are equivalent to (a special case of) Pascal's mystic hexagram theorem.

In della Faille's version (fig. 10.5), the theorem reads: "If AD and BC are parallel and $[BC]$ is divided by n lines AC_i, for which $|C_iC_{i+1}|$ = constant, where $B = C_0$, $C = C_n$ and the lines DC_i intersect AC_i at E_i and BE_i intersects AC_i at F_i, then all F_i are on a conic section."

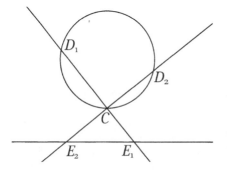

FIGURE 10.4
Defining a circle using one of Saint Vincent's transformation methods. If a straight line intersects the lines CD_i at E_i and $|D_iC|.|CE_i|$ is a constant, then all D_i lie on the circumference of a circle through C

---

57   For proofs of this and following theorems, see Meskens, *Joannes della Faille, S.J.*

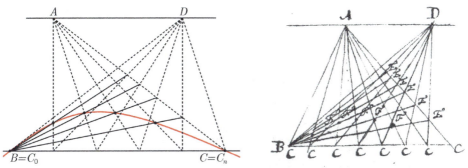

FIGURE 10.5   The construction of a conic section taking inspiration from the construction of a floor tiling in a picture (see figs. 10.7, 10.8)
© AFL 29.7

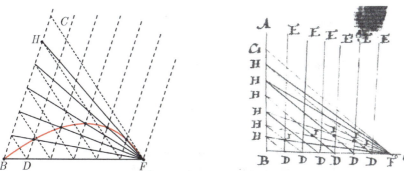

FIGURE 10.6   A special case of the construction of a conic section, using the same mathematical techniques as in figure 10.5. Unwittingly, della Faille uses the line at infinity as the horizon line
© AFL 29.7

A related theorem in which points at infinity are involved is stated again (fig. 10.6). If the line DC moves uniformly along BC, and HF is rotated uniformly about F, then the lines D_iC_i intersect the lines H_iF in a parabola. In Desarguelian terms, this is a special case of the previous theorem in which A and D are chosen as points at infinity. The parallel lines can be seen as a rotation about a point at infinity.

These theorems are nothing else than a special case of the MacLaurin–Braikenridge theorem, which in turn implies Pascal's mystic hexagram theorem.[58] The conditions della Faille imposes seem to indicate the influence of the painting practices of the day and, more particularly, the representation of perspective. Taking this viewpoint, it becomes clear why in figure 10.5 the

---

58   See Meskens, *Joannes della Faille, S.J.*, 84–86.

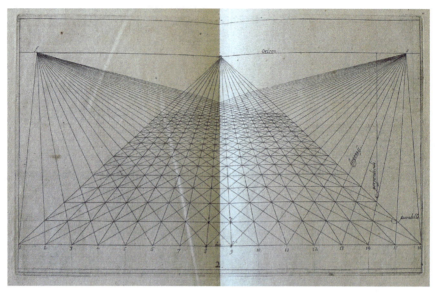

FIGURE 10.7   The construction of floor tiling in Samuel Marolois, Hans Vredeman de Vries, and Albert Girard, *Opera mathematica* [...] (Amsterdam: Ianszoon, 1630). Plate 2
© ARTESIS PLANTIJN HOGESCHOOL CMU P234

line *AD* is required to be parallel to *BC*. *AD* represents the horizon on a canvas of which *BC* is the lower edge (see figs. 3.8, 10.7, and 10.8).

Another manuscript of della Faille's, *Tratado de secciones cónicas* (Treatise on conic sections), treats conic sections.[59] As no proofs are present, it leaves us to wonder whether della Faille was able to prove these theorems for a general conic section, or whether he proved them for each conic section in turn. Nevertheless, della Faille is on the verge of a new concept, namely poles and polars. Because poles and polars are projectively invariant concepts, it is possible to state theorems relating to these concepts for all conic sections, provided all other elements are also projectively invariant.[60]

First, della Faille describes the construction of a tangent from a point to a given conic section. His construction amounts to the theorem that if a complete quadrangle is inscribed in a conic section, then each diagonal point is

---

59 AFL 29.7.
60 See Jan P. Hogendijk, "Desargues's *Brouillon Project* and the *Conics* of Apollonius," *Centaurus* 34 (1991): 1–43, here 18.

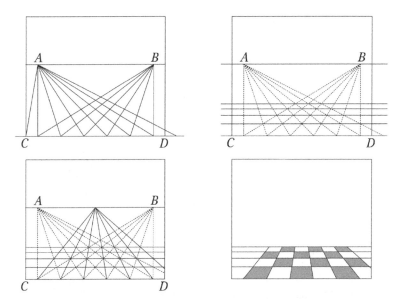

FIGURE 10.8   To draw a tiled floor in two-point perspective, choose two points A and B on a horizontal straight line, either inside or outside the canvas. Divide the horizontal edge CD of the canvas into equal parts (the size of a tile). Connect A and B to these division points. These lines are the diagonals of the tiles. Determine the intersection points E_i of AD with the lines through B. The horizontal lines through E_i determine the perspectival reduction. Now connect, from the division points, the other intersections of the diagonals. This pattern is the desired tile pattern

the pole for the conic section of the line determined by the other two diagonal points (fig. 10.9).[61]

Della Faille's second theorem is actually the same as the first one, if we consider D to be a point at infinity. If three parallel lines intersect a conic section, then the diagonal points of two inscribed quadrangles with vertices in the points of intersection determine a diameter. In modern terminology, the polar of a point at infinity is a diameter (fig. 10.10).

It then follows that all diagonal points of quadrangles formed by the intersection of chords of the same direction lie on the same diagonal. Moreover, the tangents in the points of intersection of chords of the same direction intersect on the same diameter (fig. 10.11).

---

61   Howard Eves, *College Geometry* (Boston: Jones & Bartlett, 1990), 147–48. Many of the theorems on conic sections, such as this one, would now be proven in a circle and then, using invariants under a projection, be generalized to any conic section.

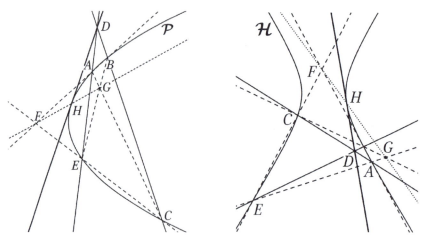

FIGURE 10.9 The construction of the tangent from *D* to a parabola (left) and a hyperbola (right). To find the tangents to a given conic section through a point *D*, draw two lines through *D* that each intersect the conic section in two points. The points of intersection form a complete quadrangle. The drawn lines are thus one of the diagonal pairs and the point *D* is one diagonal point. The other two diagonal points *F* and *G*, which can now be constructed easily, thus determine the polar of *D*. This polar intersects the conic section in the tangent points of the tangents from *D*

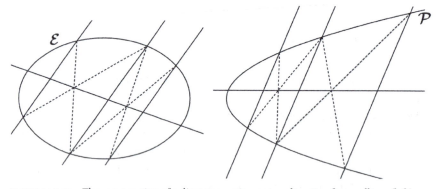

FIGURE 10.10 The construction of a diameter conjugate to a direction for an ellipse (left) and for a parabola (right)

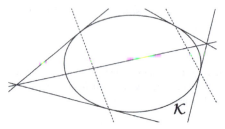

FIGURE 10.11
The construction of a diagonal in a conic section

JOANNES DELLA FAILLE AND THE BEGINNING OF PROJECTIVE GEOMETRY 179

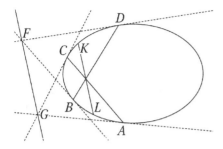

FIGURE 10.12
The exploration of some properties of poles and polars in a conic section

We now find della Faille's first important intermediary result. Let *ABCD* be a conic section 𝒦 and let the intersection of *AC* and *BD* be *E* (fig. 10.12). If *F* is the intersection of the tangents in *B* and *D*, and if *G* is the intersection of the tangents in *A* and *C*; further, if the segment [*KL*], with *K* and *L* on 𝒦, is parallel with *FG* and passes through *E*, then *E* is the midpoint of [*KL*].

He is now able to state that the polars of collinear points, with respect to a conic section, intersect in one point, and that the line through the midpoint of the polar segment of a point and the point itself (the pole) is a diameter.[62] The converse of the first part of this assertion, not mentioned by della Faille, is that as a line *l* rotates about *E*, the tangents at its extremities always meet on *FG*.[63]

Della Faille goes on to state that the angle between the diameter and the tangent in an ellipse is at its maximum when the tangent is parallel to the line between the vertices.[64]

Della Faille needed these theorems for a second important theorem in which he classifies the conic sections according to a certain characteristic. Call the midpoint of the polar segment *E*, the pole *D*, and the intersection of *ED* with the conic section *B*. The relation of the distance of *E* to *B* and the distance of *B* to *D* will determine the kind of conic section 𝒦 one is dealing with (fig. 10.13).

---

62 The polar of a point with respect to a conic section is a straight line. In the following, the phrase "polar segment" is used for the segment between the two intersecting points of the polar and the conic section lying "inside" the conic section. Also note that in a parabola, any line parallel with the axis is a diameter. In Desarguesian geometry, this is a clear consequence of the condition that a diameter has to go through the middle, which lies at infinity. Although della Faille never uses this concept, he clearly accepts these lines as diameters.

63 The more general theorem states that as a line *l* revolves about a fixed point *P*, the tangents at its extremities always meet on a line *m* (see, e.g., Field and Gray, *Geometrical Work of Girard Desargues*, 37).

64 This can be verified easily for an ellipse with equation $x^2/a^2 + y^2/b^2 = 1$. With a little algebra, one finds that the angle between the tangent and the diameter is given by $\theta = \arctan a^2 a^2 - b^2 m^2 a^2 + b^2 m a^2$. It can now be proven that this function is at a maximum for $m = \pm b/a$, exactly the slope of the line between the vertices.

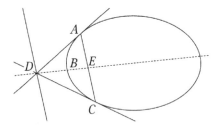

FIGURE 10.13
Given that the curve is a conic section, the relation of the length of [EB] to the length of [BD] determines the kind of conic section the curve is

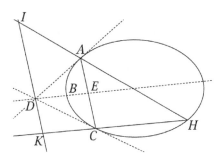

FIGURE 10.14
The proof of della Faille's conic section determination theorem

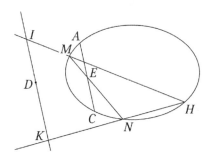

FIGURE 10.15
Generalization of della Faille's conic section determination theorem

1. If $|EB| > |BD|$, then $\mathcal{K}$ is a hyperbola.
2. If $|EB| = |BD|$, then $\mathcal{K}$ is a parabola.
3. If $|EB| < |BD|$, then $\mathcal{K}$ is an ellipse or a circle.

This result is again easily verified (fig. 10.14). Through D, the line parallel to the polar is drawn. If a point H is chosen on the conic section, the lines AH and CH will intersect this parallel in I and K. These points are such that the area of the rectangle with sides ID and DK equals z^2, where $\dfrac{|AE| \cdot |BD|}{|BE|}$.[65]

---

65   See Meskens, *Joannes della Faille, S.J.*, 94–95.

Della Faille is now ready for his first bold generalization (theorem 15; fig. 10.15). The line AC (the polar segment) can be replaced by any line through E. Instead of A and C, the intersections M and N of this line with the conic section are used to project from any point of the conic section onto FG, the polar of E.[66]

Della Faille now states that $|DK|.|DI| = z^2$.[67]

In the case where one of the straight lines passes through D, della Faille notes, the other will be parallel with FG (i.e., with the polar of E).

The following theorems are all corollaries or the inverse assertions of these two theorems. He goes on classifying the conic sections using the length of z, which he compares with that of $[AE]$:

1. If $z < |AE|$, then $\mathcal{K}$ is a hyperbola.
2. If $z = |AE|$, then $\mathcal{K}$ is a parabola.
3. If $z > |AE|$, then $\mathcal{K}$ is an ellipse or a circle.

These results follow directly from the earlier classification.

With these theorems, della Faille proposes a way to construct a conic section when two tangents, and their tangent points, and one other point are given. He fails to state that these conditions define a unique conic section. Nearly all of the following theorems are variations on theorems 14 and 15 and their direct corollaries. The sections on the ellipse, hyperbola, and parabola all find their inspiration in the work of San Vicente.

Della Faille is likely to have proved these theorems by purely synthetic geometric methods. On the other hand, there can be no doubt that he did in fact prove the theorems. The chain of theorems is too complicated to be imagined solely by intuition but not proved. For this, the mathematical concepts probe too deep. However, the proofs are missing, and although we may be able to reconstitute proofs on the basis of early seventeenth-century knowledge of mathematics, it is impossible to retrace how della Faille himself proceeded. In his unified treatment of conic sections, he uses concepts from projective geometry, without, however, ever defining them as such. Yet it is nevertheless appropriate to assign della Faille a place among those early writers on projective geometry and conic sections such as Desargues, Pascal, and Philippe de la Hire (1640–1718).

---

66 This theorem also holds if the point H coincides with either M or N. Suppose H = M, then the tangent at M is taken to be the line HM, while $|HN|=|MN|$.

67 For a proof of this assertion, see Meskens, *Joannes della Faille, S.J.*, 96–99.

3 De centro gravitatis

As mentioned earlier, *De centro gravitatis* is the only book della Faille published. It is a small booklet of fifty-five pages that was published in 1632 by the Antwerp printer Joannes van Meurs (1583–1652). His propositions are formulated in an Archimedean style, which, in view of the disappointing experiences of his teacher San Vicente, must undoubtedly have increased his chances of getting his book published.

Della Faille's book is concerned with a theorem that is in essence a mechanical theorem, yet it is considered as a purely geometrical problem. Closely following the style of his teacher San Vicente, della Faille never uses formulas, nor equations. Each proof is in running text without equations or even fractions. His exposé is purely synthetic. When reading it, it is not always clear what della Faille is aiming at, although with hindsight no theorem is superfluous. On the contrary, the work is of a seldom seen logic with but one aim: proving the grand theorem. Della Faille completely disregards interesting corollaries or side-steps not leading toward his goal. In 1630, he explained to Grienberger that he had already found his theorem as a student of San Vicente. In this letter, della Faille suggests that he did not find his theorem in a geometrical fashion, but with an "analytical method," which is not revealed to the public at large.[68] Della Faille also referred to another result he had obtained, the determination of the centroid of a cylinder-section, indicating that he was using the infinitesimal methods of his teacher.

The first thirty-one theorems are in fact lemmas leading up to theorem 32, which is still taught in any statics course. We know the theorem as: The centroid of a circle sector lies on the bisectrix at a distance d of the center
$$d = \frac{2R \text{ chord } \alpha}{3\alpha} = \frac{4R\sin\frac{\alpha}{2}}{3\alpha},$$
in which R represents the radius of the circle, α the angle of the sector, and chord α the length of the chord, which is subtended by the angle α.

Theorems 33 to 45 allow the determination of the centroids of sectors and segments of circles and ellipses (fig. 10.16). They would allow a construction of the centroids if one can construct a line segment equal to the given arc. Conversely, they would allow one to find the length of an arc if the centroid is known. This condition means della Faille has to draw a clear distinction between the existence of a figure and the possibility of constructing such a figure.

---

68   APUG 567, fols. 243ʳ–244ᵛ.

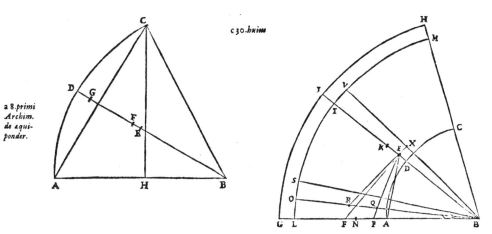

FIGURE 10.16   The determination of the center of mass of a circle sector. Joannes della Faille, *Theoremata de centro gravitatis partium circuli et ellipsis* (Antwerp: Ex officina typographica Joannis Meursii, 1632), 38 (left) and 34 (right)
© EHC G 147861

CHAPTER 11

The Antwerp Students

1   Philip Nuyts

Nuyts was born on March 8, 1597 and baptized in Our Lady's Church in Antwerp two days later.[1] He was the son of Martin Nuyts (1553–1608) and Anna Tempelaers (*c*.1560–?). His father was a printer active in Antwerp in the first decade of the seventeenth century, publishing under the name Martinus Nutius. On February 12, 1613, when he was nearly sixteen, Nuyts entered the Society and was sent to the novitiate in Mechelen.[2] Before joining the Society, he had taken six years of humanities.[3] After his novitiate, he returned to Antwerp to take two years of philosophy classes and two years of mathematics classes,[4] later also taking four years of theology classes.[5]

In 1618, Nuyts is mentioned as being in his second year of the course of mathematics. Before the start of these mathematics classes, Nuyts had asked to be sent to the overseas missions, but this application was apparently rejected.[6] After his studies, he was assigned to the college in Brussels as a teacher of Greek.[7] On March 28, 1626, he was ordained priest in Brussels. He was then sent to Prague, only to have this order revoked and be directed to Madrid.[8] It is not clear whether he embarked on any of these journeys, and in 1627 he was back in the Southern Netherlands to teach Greek at the Mechelen college.[9] He was later sent to Antwerp, where he became the confessor of the count of Feria, the governor of the Antwerp citadel (also known as Zuidkasteel [Southern castle]).[10]

From then on, Nuyts's life could have been that of an ordinary Jesuit: teaching at colleges in Mechelen, Antwerp, Brussels, and Ghent and residing at the

---

1   FA PR2, 188ʳ; PR11, 108ᵛ. He is sometimes referred to as Philip Nutius Sr. because a namesake later joined the Society.
2   ARAA T14/034 403, fol. 2; 3779; Kadoc ABSE 32, 11.
3   ARAA T14/034 31, 24/26.
4   ARAA T14/034 31, 24/26.
5   ARAA T14/034 37, 27/31.
6   ARSI, *Index indepetae*, referring to FG21, fol. 16, May 13, 1617.
7   ARAA T14/034 31, 24/26.
8   ARAA T14/034 403, no. 2.
9   ARAA T14/034 403, no. 2.
10 ARAA T14/034 403, no. 2.

FIGURE 11.1 "P. Philippus Nutius (Nuyts), S.J.," oil on canvas, before 1661. Klooster Jezuïetenhuis Xaverius vzw, Borgerhout
© KIK-IRPA KM008900

professed house in Antwerp,[11] were it not that he was to be sent to the court of Queen Christina of Sweden (1626–89, r.1632–54).

On June 30, 1650, the ambassador of King João IV of Portugal (1604–56, r.1640–56), José Pinto Pereira, arrived in Stockholm to attend Queen Christina's coronation. In his retinue was his chaplain, the Jesuit Antonio Macedo (1612–93), who acted as substitute secretary of the delegation.[12] Macedo's capacity as a priest was known to only a few. By coincidence, having to translate Latin texts, he came into contact with Christina, with whom he entered into discussions. These discussions also concerned the sciences, especially astronomy and the viewpoint of the Catholic Church toward the new world systems. She confessed to him that she would like to meet a Catholic priest, uttering doubts about her Lutheran faith. Macedo advised her to ask for the assistance of priests who were well versed in natural sciences.[13] He obliged and asked the superior general to send two Italian priests, able to enter into discussions about the tenets of faith.[14] From this point on, narratives differ. According to the Jesuit Charles de Manderscheidt (1616–91), the answer came that no suitable Italian candidates were available, but that in the Belgian province candidates could be recruited.[15] In other accounts, the superior general obliged and sent Franciscus de Malines (1612–79), professor of theology at Turin, and Paul Casati (1617–1707), mathematician in Rome.[16] When Nuyts arrived in Stockholm, the Italian priests were already present, although neither knew anything about each other's presence.[17]

---

11   ARAA T14/034 31, 24/26; 32, 62/66; 33; 34, 4/27; 36, 27/31; 37, 27/31; 39, 4/33; 39/2, 3, 55, 75; 41, 57/15; 403, no. 2; 3779.

12   It is suggested by Garstein (*Rome and the Counter-Reformation in Scandinavia*, 624–27) that Macedo's presence was not coincidental. Macedo was one of the priests who were instrumental in converting the Swedish ambassador in Lisbon, Lars Skytte (c.1610–96), to Catholicism. He also suggests that Skytte provided him the information "to smooth his path into the intrigue-ridden corridors of Queen Christina's palace in Stockholm" (627).

13   Garstein, *Rome and the Counter-Reformation in Scandinavia*, 630, referring to letters from Queen Christina to the superior general (1651) and reminiscences of Macedo (1692). In Charles Manderscheidt's account, dated March 14, 1671, we are told that she insisted that the priests be mathematicians. Charles Joseph Nuyts, *Philippe Nutius à la cour de Suède* (Brussels: J. Vandereydt, 1836), 5–6.

14   ARAA T14/034 3675-20, *Narratio de Christina Svecorum Serenissima Regina Gvstavi Adolphi Regis* (Liège: Joannes Mathias Hovius, 1656), 5ff.

15   Account of Charles Manderscheidt, March 14, 1671, in Nuyts, *Philippe Nutius*, 5–6.

16   ARAA T14/034 3675-20, *Narratio de Christina*, 5–6; Johann Wilhelm von Archenholtz, *Mémoires concernant Christine, reine de Suède* (Amsterdam: Pierre Mortier, 1751), 470ff.

17   Casati's narrative states that they arrived on St. Matthew's Day (ARAA T14/034 3675-12). ARAA T14/034 3675-20 and *Narratio de Christina*, 6–7, mention March 6 as their arrival date. See also the account of Manderscheidt, March 14, 1671, in Nuyts, *Philippe*

It seems likely that Christina had asked the Jesuit Gottfried Francken (1595–1654), who was also present in Stockholm at that time,[18] the same questions as she had posed Macedo. Moreover, the delay of an answer from Rome made her anxious to open a different line of communication. She asked for a mathematician who could, also in view of political developments, such as her abdication, instruct her in the Catholic faith.[19] Francken had later returned to the Southern Netherlands,[20] and it is probable that the Flandro-Belgian province responded favorably to his requests, without consulting Rome.

In 1652, Thomas de Bierthe (1601–71), the superior of the Antwerp professed house, proposed sending Nuyts to Sweden. Yet a voyage into Sweden to proclaim Catholicism was to put one's life in peril. Nuyts hesitated, as he feared his ill health would not be able to support his desire to carry out his duty.[21] The provincial Johannes Baptiste Engelgrave (1601–58) nevertheless ordered him to undertake the voyage.[22] During his stay in Sweden, Nuyts would adopt the pseudonym Philip Templars, the maiden name of his mother. According to his own account, he, and his valet,[23] embarked on the journey on March 23, 1652, traveling with Karel Boechelion (1618–52), who was destined for Copenhagen.[24]

Nuyts's voyage seems to have held many dangers, and at one point he was incarcerated in Amsterdam on the basis of false accusations.[25] He eventually arrived in Stockholm on April 11, and by evening time he had been successful in obtaining an audience.[26] He wrote down his account in a diary but later burned his papers to reduce the risk of being exposed.

Nutius, 5–6. On Malines's and Casati's sojourn in Stockholm, see Garstein, *Rome and the Counter-Reformation in Scandinavia*, 637ff., 655ff.

18 On Francken's sojourn in Stockholm, see Garstein, *Rome and the Counter-Reformation in Scandinavia*, 645ff.

19 Nuyts, *Philippe Nutius*, 5. Mathematician, natural scientist, and astronomer should here be understood as synonyms. Most probably she was interested in the world systems of Copernicus (1473–1543) and Tycho Brahe.

20 *Narratio de Christina*, 7.

21 ARAA T14/034 403, no. 2; 3675-74.

22 ARAA T14/034 3675-18, letter of Engelgrave, February 2, 1652, and 3675-4, letter of Nuyts, June 20, 1652.

23 In fact, Boechelion was a Jesuit himself, having entered the Society in 1636. Audenaert, *Prosopographia iesuitica Belgica antiqua*, 1:121.

24 Unless otherwise stated, the following paragraph is based on Nuyts's own account in T14/034 3675-4.

25 The governor of 's-Hertogenbosch had written a letter to the States General in The Hague that Nutius was on a political mission to negotiate the surrender of 's-Hertogenbosch to the Spanish. ARAA T14/034 3675-4.

26 ARAA T14/034 3675-2 and 3675-4.

Nuyts and the queen had candid discussions over the next few days in which he offered her pious books in many languages as well as books about controversial issues in Protestantism. She was also given two relics as a gift, one of which was a relic of Saint Ignatius.

On April 22, Nuyts was summoned to the queen and found her perplexed. Her secretary had opened the letters that had been sent through the usual channels and in one of them it was announced that a Jesuit "Jacobus Nutius" had come to Sweden, but that "he had not come to capture fleas." She decided that he had to leave, and he made preparations for the return journey.

There must have been some procrastination, because on April 27 the queen changed her mind and decided to keep Nuyts prisoner in the palace. If it were to be known that he was a Jesuit, she would no longer be able to contemplate conversion. During the three following weeks, they did not see each other again but sent letters to one another. From the fact that he kept very much to himself, the servants began to have suspicions about him.

On May 17, Francken arrived in Stockholm, carrying with him letters intended for Nuyts.[27] These were opened and read by Christina, and as a result Nuyts was ordered to sail on the first ship to Lübeck, for which he was given a passport for safe passage in Sweden.[28] Francken, on the other hand, was ordered to stay and discuss the dogmas of the faith with her. Nuyts was also instructed to prevent the Italian priests, of whom something was mentioned in the opened letters, from coming to Sweden. The queen feared that the acquaintance with the Jesuits might put her into a politically untenable position. Nevertheless, Nuyts seems to have become an intermediary between Christina and the superior general.

Nuyts's account may not be that accurate, as in fact he was called back by Engelgrave on the orders of Superior General Goswin Nickel (1582–1664, in office 1652–64),[29] and the Italian priests had been in Stockholm since March 1652,[30] when Nuyts was yet to leave Antwerp.

However, after Christina had abdicated and was traveling to Rome in the summer of 1654, she sojourned in Antwerp and Brussels. In Antwerp, she stayed for four months at the mansion of Don García d'Yllan (*fl.* 1650) on Lange Nieuwstraat.[31] Here, it was the Antwerp Jesuits who instructed her in

---

27 ARAA T14/034 3675-4 and 3675-20. Again, accounts differ; according to *Narratio de Christina*, 7, Francken had traveled to Sweden together with Nuyts.
28 ARAA T14/034 3675-3.
29 ARAA T14/034 3675-1.
30 ARAA T14/034 3675-20, *Narratio de Christina*, 6.
31 Carl Johan Reinhold Burenstam, *La Reine Christine de Suède à Anvers et Bruxelles 1654–1655* (Brussels: Alfred Vromant & Cie, 1891), 8. On Christina's visit to the Collegio Romano,

the articles of the Catholic faith,[32] with her former confidant Nuyts undoubtedly among them. When, later that year, she was residing at the archduke's request at his palace on Coudenberg in Brussels, she converted to Catholicism on December 24, 1654.

2 Ignatius Derkennis

Derkennis was born on March 2, 1598, the son of Joannes and Helena(?).[33] He changed his baptismal name to Ignatius when he entered the Society on September 26, 1614.[34] Before joining the Society, he took humanities for six years.[35] After joining, he studied philosophy for two years, followed by four years of theology and two years of mathematics.[36] He followed San Vicente's courses from 1618 to 1620.[37] After his studies, he was first assigned to the college of Mechelen.[38] He taught classical languages for four years, mathematics for two years, and philosophy for three and a half years before becoming prefect of studies and professor of theology at the Leuven college.[39] He also taught a course on fortifications.[40] He became prefect of studies at the college of Antwerp in 1644 and rector of the colleges of Ypres and Leuven. While at Ypres, he cared for the victims of a pestilence epidemic and contracted a fever, which made his brethren fear for his life. According to the eulogy, he was praised for his conduct during this epidemic by Superior General Vincenzo Carafa (1585–1649, in office 1644–49).[41] He may have been active in a dispensary or an apothecary, as is witnessed by his notes on the preparation of some medicines.[42] When Derkennis was at Leuven, one of his students was Juan Caramuel Lobkowitz (1606–82).[43]

---

see Michael John Gorman, "From 'The Eyes of All' to 'Usefull Quarries in Philosophy and Good Literature': Consuming Jesuit Science, 1600–1655," in O'Malley et al., *Jesuits*, 170–89.
32 Burenstam, *La Reine Christine de Suède*, 109.
33 Kadoc ABSE 35, 239.
34 ARAA T14/034 31, 66/12.
35 ARAA T14/034 31, 66/12.
36 ARAA T14/034 35, 48/5.
37 ARAA T14/034 31, 66/12.
38 ARAA T14/034 31, 66/12.
39 ARAA T14/034 37, 52/7.
40 The treatise *Tratado de fortificación militar* (1644) by Juan de Santans y Tapia was dedicated to Derkennis. Alicia Cámara, ed., *Los ingenieros militares de la monarquía Hispánica en los siglos XVII y XVIII* (Madrid: Ministerio de defensa, 2006), 85.
41 ARAA T14/034 398.
42 KBR 3834–45.
43 Jacopo Antonio Tadisi, *Memorie della vita di Monsignore Giovanni Caramuel di Lobkowitz* (Venice: Giovanni Tevernin, 1760), 70. Juan Caramuel y Lobkowitz (1606–82) was a

Derkennis's only mathematical booklet was published as an appendix to the first edition of Tacquet's *Arithmetica theorica et praxis* (Theory and practice of arithmetic [1656]). His importance, however, lies in the theological book *De Deo uno, trino, creatore* (On God, the creator who is one and triune) in which he is the first to use the *consequentia mirabilis* $((\neg A \to A) \to A)$ outside the field of mathematics. He used the theorem to prove that there is a supreme being.[44]

3 Other Students

Alegambe was born on April 1, 1600 in Brussels, the son of Jean, lord of Vertbois (before 1558–1616), and Louise du Blois (1560–after 1602).[45] He entered the Society on October 1, 1615,[46] thereafter studying philosophy for two years, theology for four, and mathematics for one year in 1620.[47] After his studies, he was assigned to the college of Bruges, where he taught basic Latin and syntax.[48] He later taught at the college of Ghent.[49]

Durandus entered the Society on October 1, 1615. He studied mathematics in 1619–20.[50] In 1617, he was sent to Graz to teach mathematics and theology. Not much is known about his career there. We only know that he made measurements of the magnetic declination[51] and that he edited Grienberger's *Euclid*.[52]

      Spanish Catholic Scholastic philosopher and prolific writer. In mathematics, he was one of the early writers on probability. He was also one of the first Spanish writers on logarithms. Juan Navarro-Loidi and José Llombart, "The Introduction of Logarithms into Spain," *Historia mathematica* 35 (2008): 83–101, here 86–90.

44   Gabriel Nuchelmans, "A 17th-Century Debate on the *Consequentia mirabilis*," *History and Philosophy of Logic* 13 (1992): 43–58, here 55–58.

45   M. D. [Jean Charles Joseph de Vegiano (seigneur de Hovel)], *Nobiliaire des Pays-Bas, et du Comté de Bourgogne, Volume 2* (Mechelen: P. J. Hanicq, 1779), 182.

46   ARAA T14/034 31, 20/17.

47   ARAA T14/034 31, 20/17.

48   ARAA T14/034 31, 20/17.

49   ARAA T14/034 34, 36/3.

50   ARAA T14/034 31, 38/19 and 32, 60/48.

51   Measurements at the request of Kircher. Emanuel Swedenborg, *The Principia: Or, The First Principles of Natural Things* (London: W. Newbury, 1846), 119.

52   Foppens, *Bibliotheca Belgica*, 516; Christopher Grienberger, *Euclidis sex primi elementorum geometricorum libri cum parte undecimi ex majoribus Clavii commentariis* [...]: *Contracti per P. Christophorum Grienbergum e Societate Iesu; Brevis trigonometria planorum, cum tabulis sinuum, tangentium, et secantium, ad pantes* (Graz: Haeredes Ernesti Widmanstadii, 1636).

His student Joannes de Swienthohliwicz Kamiensky defended a thesis on the use of a low wall or *fausse braye* of a fortification.[53]

Cox entered the Society on September 24, 1612. He took six years of humanities before joining. After joining the Society, he studied humanities for one year, followed by two years of philosophy and one year of mathematics. He taught syntax at the college of Maastricht. He died of dysentery in Maastricht on October 23, 1622.[54]

---

53   Jacob Durand and Joannes de Swienthohliwicz Kamiensky, *Problema mathematicum ex architectonia militari de Moenibus inferioribus sive falsabraga* [...] (Graz: Apud haeredes Ernesti Widmanstadij, 1636); Foppens, *Bibliotheca Belgica*, 516.

The *fausse braye* is a defensive wall located outside the main walls of a fortification. From this wall, the defenders could fire with muskets against the assailants. In the later seventeenth century, the *fausse braye* was abandoned as a defensive feature because, once it had been captured, it left the main wall vulnerable to mining.

54   ARAA T14/034 31, p. 88/13; 366, no. 20.

CHAPTER 12

The Leuven Students

1   Theodorus Moretus

Theodoor Moerentorff (a name he later Latinized to Moretus, as did his family of printers) was born on February 9, 1602 and baptized two days later in Our Lady's Church in Antwerp.[1] He was the son of Pieter (1544–1616) and Henrica Plantijn (c.1561–1640), a daughter of the famous printer. His father Pieter was a diamond cutter and jeweler.[2] His uncle Jan (I) Moretus (1543–1610) was married to Martina Plantijn (1550–1616), making Balthasar (II) (1615–74), with whom Theodorus corresponded, a nephew along the paternal as well as the maternal line. Jan I Moretus carried on running the printing press in Antwerp,[3] as Plantijn's other sons-in-law would do in Leiden (François van Raphelingen [1539–97]) and Paris (Gilles I Beys [c.1541–95]).

The Antwerp printing house had a long tradition of publishing religious books in close cooperation with clerics. The best-known example is of course the *Biblia Regia* (Royal Bible), a polyglot Bible, published in cooperation with Benito Arias Montano (Montanus [1527–98]) by Christoffel Plantijn (1520–89) between 1568 and 1573.[4]

By the time Theodorus entered the Society, his father had already died. Moretus attended school at the Antwerp Jesuit college for seven years, taking humanities.[5] After his novitiate, he studied philosophy (1620–23) and theology (1623–27) at Leuven. During these years, he also took classes with San Vicente. Like so many of his fellow students, he applied to be sent to the Chinese missions,[6] and as was also the case with his fellow students, his application

---

1   ARAA T14/034 31, p. 62/121; FA PR3 OLV fol. 166; PR11, fol. 198ᵛ.
2   Kadoc ABSE 35, 256. On Moretus, see Hermann Hoffmann, "Der Breslauer Mathematiker Theodor Moretus, S.J.," *Jahresbericht der Schlesischer Gesellschaft für Vaterländische Cultur* 107 (1934): 118–55; Bosmans, "Théodore Moretus de la Compagnie de Jésus"; Schuppener, *Jesuitische Mathematik in Prag*, passim.
3   On Jan Moretus's publications, see Dirk Imhof, *Jan Moretus and the Continuation of the Plantin Press: A Bibliography of the Works Published and Printed by Jan Moretus I in Antwerp (1589–1610)* (Leiden: Brill, 2014).
4   This book had a synoptic version of the Bible in Greek, Latin, Aramic, Syriac, and Hebrew.
5   ARAA T14/034 31, 62/121.
6   Noël Golvers, "The XVIIth-Century Jesuit Mission in China and Its 'Antwerp Connections,'" *De Gulden Passer* 74 (1996): 157–88, here 173–74. The application was signed on July 3, 1626 in Leuven. Golvers refers to ARSI, Fondo Gesuitico 752 (no. 21), fol. 62 for this document.

was turned down. In 1627, he was sent to the college in Bruges, where he taught syntax. During his time in Bruges, he witnessed a violent tempest that led to flooding during high tide. The water could only be stopped after an embankment had been turned into a makeshift dike that had been hastily constructed. The experience, to which he would often refer, later led him to his studies on tides and meteorology.[7]

Between 1634 and 1662, Moretus graduated as doctor of philosophy and theology. In 1628–29, he did his tertianship, after which he was sent to Münster, where he taught mathematics and morals. At about the same time, San Vicente was sent to Prague. San Vicente had asked for an assistant to help him, and as della Faille had been sent to Madrid, Vitelleschi assigned Moretus to the task.[8] For Moretus, this was the beginning of an itinerant life in Bohemia. Part of San Vicente's work in Prague can be attributed to Moretus.

After the Sack of Prague by the Saxons in 1631, Moretus was sent to Olomouc, where he taught philosophy. Some of his students' course notes, written as commentaries on Aristotle, have been preserved in the library of the Klementinum.[9] In 1633, Moretus's own version appeared in print.[10] A year later, he was again directed to Prague, this time as a mathematics teacher. On February 2, 1635, he took his final solemn vows.[11] The importance that was attached to mathematics becomes clear from the fact that Moretus welcomed Emperor Ferdinand II and Archduke Leopold Wilhelm (1614–62) during their visit to the Klementinum in 1638. During this visit, he showed them the large instruments the Klementinum possessed.[12]

Prague was pillaged again, this time by the Swedes, in 1639. In a letter to Niccolò Zucchi (1586–1670), Moretus lamented the Swedish raid on the college library.[13] The Swedes transferred their literary war booty to Uppsala. Together

---

7   Hoffmann, "Der Breslauer Mathematiker Theodor Moretus, S.J.," 122.
8   Kadoc ABME Fonds Bosmans V. 8, 261, referring to letters of Mutius Vitelleschi dated April 7, 1629 and July 21, 1629 in Bohemia Epist. Gen. 1628–37 (277 and 291–95).
9   On physics (second-year philosophy) by Paul Schwabo, and on metaphysics (third-year philosophy) by Johann Bövink. Both were White Canons and active in Strahovice. Hoffmann, "Der Breslauer Mathematiker Theodor Moretus, S.J.," 123–24.
10 De Backer, de Backer, and Sommervogel, *Bibliothèque des écrivains de La Compagnie de Jésus*, 6:1318–21.
11 Kadoc ABSE 47.
12 According to Josef Smolka and René Zandbergen, these were Brahe's instruments. Josef Smolka and René Zandbergen, "Athanasius Kircher und seine ersten Prager Korrespondenten," in Cemus, *Bohemia Jesuitica 1556–2006*, 2:677–704.
13 Cited in Hoffmann, "Der Breslauer Mathematiker Theodor Moretus, S.J.," 125–26 referring to Bibl. Vitt. Emm. MS Gesuiti 371, fol. 86.

with his fellow Jesuits, Moretus fled to Znojmo,[14] where he received the sad tidings that his mother and brother had died. He gave the entire inheritance of his mother and a third of that of his brother to the Klementinum in Prague to restore the library.[15]

In the meantime, Moretus continued to teach mathematics to his students. The next year, 1640, he went to Jihlava, where he held repetitions of rhetoric. In 1641, he was back in Prague only to be sent to Bresnitz (Březnice) as a preacher the next year.[16] He again returned to Prague in 1646, this time as a minister of the college of Saint Nicholas's Church (Kostel svatého Mikuláše).

In 1648, the Swedes attacked the city a second time.[17] In the autumn of 1649, Moretus was transferred from the college to the Klementinum as a professor of biblical studies and mathematics. At the same time, he was also in charge of the university's printing press. He was not allowed to lead a sedentary life, as in the autumn of 1652 he was transferred to Klatovy, where he oversaw the building of the new college. In the following year, he was charged with running the new college.[18] In August 1656, he resigned to go to Neisse,[19] only to be redirected a year later to Glogau (Głogów),[20] where he was a confessor to Johann Franz von Barwitz, Freiherr von Fernemont, lord of Schlawa (1597–1667). He would use the two years spent at Glogau to write books, although during the second year the burden of leading the gymnasium was also put onto his shoulders. The autumn of 1659 saw him leave for Breslau. With the exception of one year when he was in Glogau again, he would remain in Breslau until his death. In Breslau, he was a professor of polemic theology (*controversiae*), about which he published two treatises. He served two years as dean of the faculty of philosophy and, from 1663 until his death, as dean of the faculty of theology. He died unexpectedly on November 6, 1667 of dysentery.[21]

---

14 Hoffmann, "Der Breslauer Mathematiker Theodor Moretus, S.J.," 124.
15 Hoffmann, "Der Breslauer Mathematiker Theodor Moretus, S.J.," 124; Schuppener, "Theodor Moretus," 649.
16 Schuppener, *Jesuitische Mathematik in Prag*, 98.
17 Moretus describes the events at Prague in a letter to Zucchi dated November 21, 1648. Kadoc ABME Fonds Bosmans v.8, 103, referring to Bibl. Vitt. Emm. MS Gesuiti 379, no. 32.
18 Hoffmann, "Der Breslauer Mathematiker Theodor Moretus, S.J.," 127.
19 Kadoc ABME Fonds Bosmans v.8, letter to Joannes Badikoso, dated June 21, 1657, sent from Neisse.
20 Kadoc ABME Fonds Bosmans v.8, referring to KBR 22044, fol. 8, letter to Bolland dated April 26, 1653, sent from Klatovy.
21 Valentin Stansel, *Legatus uranicus ex orbe novo in veterem, hoc est Observationes americanae cometarum factae, conscriptae ac in Europam missae* (Prague: Typis Universitatis Carolo-Ferdinandi, 1683), 142.

During his tenure in Breslau, Moretus acted as a book agent for Otto von Nostitz the Younger (1608–64)[22] and the Polish physician Baltazar Kramer.[23] In four letters to von Nostitz, Moretus informs him of book shipments, both of scientific and of religious books. Moretus also sent von Nostitz mathematical and optical instruments, and it is possible that some of these instruments had been constructed by Moretus himself.[24]

In 1641, public disputations, where theses were defended, were held at the Ferdinandea, the first year that Moretus held the chair of mathematics. He may have had the examples of Antwerp and Leuven in mind. The "mathematical" disputations were on the nature of the winds and on the origin of (water) sources.[25] It seems that when Moretus held the chair of mathematics at the Klementinum, the mathematics department was vibrant and his presence gave a decisive impetus toward the importance it would gain.[26] These theses were also part of the competition with the Carolina, in which mathematics was considered to be a showpiece.

Beginning in 1623, Moretus kept what we could call mathematical diaries, which have unfortunately remained unedited.[27] Moretus had a wide circle of correspondents, among which were Godefred Kinner (c.1610–?), Valentin Stansel (Valentinus Stansel [1621–1705]), Cavalieri, Giovanni Battista Riccioli (1598–1671), Christopher Scheiner (1573/75–1650), Nicolo Zucchi, Athanasius Kircher (1601–80),[28] and Johannes Hevelius (1611–87).[29] Like many of his fellow

---

22 Richard Šípek, *Die Jauerer Schlossbibliothek Ottos des Jüngeren von Nostitz Teil 1 und 2* (Frankfurt am Main: Peter Lang GmbH, 2014), 71–72. Otto der Jüngere von Nostitz (1608–65) was the son of Johann, governor of Wołów. The family was Lutheran. Otto studied in Görlitz, Leipzig, and Strasbourg. At the age of twenty-three, he was made a *Freiherr*, and he soon served on Rudolf II's (1552–1612, r.1576–1612) imperial council. After the sack of Prague, he became the emperor's chamberlain in Vienna. He was later appointed governor of Breslau (1642–50), Schweidnitz (Świdnica) (1651–65), and Jauer (Jawor) (1651–65). For a biography of Otto von Nostitz the Younger, see Šípek, *Die Jauerer Schlossbibliothek*, 15–40.
23 Šípek, *Die Jauerer Schlossbibliothek*, 75.
24 Šípek, *Die Jauerer Schlossbibliothek*, 76–77.
25 Schuppener, *Jesuitische Mathematik in Prag*, 67–68.
26 Schuppener, *Jesuitische Mathematik in Prag*, 68.
27 Theodorus Moretus, "Exercitationes mathematicae, poetice atque sermones" (Manuscript, Národní Knihovna, Prague CR, n.d.); Moretus, "Praelecyiones naturalis Theodori Moreti, S.J." (Manuscript, Národní Knihovna, Prague CR XIV G7, n.d.); Schuppener, "Theodor Moretus," 651–52.
28 Two of their letters are published in Smolka and Zandbergen, "Athanasius Kircher," 681, 690–91.
29 Schuppener, "Theodor Moretus," 652. Unfortunately, like his diaries, his correspondence remains hitherto largely unpublished and unedited.

Jesuits, including San Vicente, Kircher made measurements of the magnetic declination.[30] He also made measurements of the lunar eclipse of April 15, 1642, albeit without an instrument to measure time. He had to revert to measuring the position of the star *Spica*[31] and deducing the time from its position. In 1639, Moretus asked for Kircher's advice on a text written in unfamiliar and mysterious alphabetic characters. Kircher answered that he thought the text was in "Illyrian writing."[32]

Notes of one of Moretus's courses have been preserved.[33] The notes were taken down by Friedrich Füssel (?–1654), who later became abbot at the convent at Teplá. The manuscript is a cross-section of the mathematical practices of the mid-seventeenth century. A mere ten folios are devoted to elementary arithmetic, up to the rule of three. Moretus then pays attention to the construction of sundials and surveying practices, together with some examples on the measurement of heights. The next chapter deals with armillary spheres, with an emphasis on the horizon and the zodiac. As is to be expected, astronomy is the next subject. Here, he deals with three world systems, the Ptolemaic, Copernican, and Tychonic systems. A drawing of the Copernican system is lacking. The last part deals with geography and, again, sundials.

Like his teacher San Vicente, Moretus was a prolific writer, but one who also published quite a lot of books, and if it had depended solely on him, he would have published even more.

---

30 Emanuel Swedenborg, *Principia*, 119; Smolka and Zandbergen, "Athanasius Kircher," 672–73.
    Kircher's project of deducing a theory from measurements from all over Europe, and even from all over the globe if we include the sporadic reports of Jesuits in the colonies, can be seen as the first attempt to gather observational data on a worldwide scale.
31 Smolka and Zandbergen, "Athanasius Kircher," 692–94.
32 About the code, Kircher wrote: "As for the book filled with some sort of mysterious steganography which you enclosed with your letter, I have looked at it and have concluded that it requires application rather than insight in its solver," and also, "Finally, I can let you know that the other sheet which appeared to be written in the same unknown script is printed in the Illyrian language in the script commonly called St. Jerome's, and they use the same script here in Rome to print missals and other holy books in the Illyrian language" (http://philipneal.net/voynichsources/kircher_1639_english/ [accessed February 4, 2020]). History has shown Kircher could not have been further from the truth. Smolka and Zandbergen, "Athanasius Kircher," 683 suggest this text is the so-called Voynich manuscript. The fifteenth-century Voynich manuscript is written in hitherto unidentified characters; nor has the text been deciphered.
33 Theodorus Moretus and Friedrich Füssel, "De mathematica et scientiis naturalibus varii. praecepta medicinalia" (Manuscript, Knihovna kláštera premonstrátu Teplá d23, 1635). The description of this manuscript is based on Schuppener, "Theodor Moretus," 653ff.

In 1635, Moretus's student Gaspar Alexius Francq defended a dissertation, *Propositiones mathematicae de celeri et tardo* (Mathematical propositions on rapid and slow [movements]). Unfortunately, no copy is extant. According to Hermann Hoffmann, the thesis contained twenty-five theorems on ballistics.[34] In 1641, a thesis entitled *De fontibus problema mathematicum* (On the mathematical problems concerning fountains), on hydromechanics, which was defended by Moretus's student Ferdinand Freiherr von Bukau, was published.[35]

With his nephew Balthasar II, Moretus contemplated a book on mathematics and a geographic treatise to complement Ortelius's *Theatrum*. In 1656, he offered Balthasar II his philosophical dissertation, but the latter refused to publish it. Nevertheless, he did publish a number of Moretus's philosophical and ascetic books. In his Breslau period, Moretus published two treatises, one a refutation of the Protestant objections to the cult of Mary and the Immaculate Conception, the other, published anonymously, a catechism compiled from Martin Luther's (1483–1546) writings.

In Breslau, Moretus again had students defend theses on physical subjects.[36] Moretus's theorems on statics (1663) were defended by his student Franz Flüske. One of the theorems deals with Pappus's 8.9 on the equilibrium of a solid on an

---

34   Hoffmann, "Der Breslauer Mathematiker Theodor Moretus, S.J.," 124; Schuppener, "Theodor Moretus," 650.
35   Theodorus Moretus and Ferdinand Ernest L. B. de Bukau, *Virgini Matri fonti sapientiae e terra in caelos exilienti hoc de fontibus problema mathematicum dicabat illustrimus D. Ferdinandus Ernestus L.B. de Buckaw philosophiae et matheseos auditor* (Prague: Typis academicis, 1641); Schuppener, "Theodor Moretus," 650.
36   Theodor Moretus, Joannes Ignatius Stephan, and Balthasar Kirstenio *Propositiones mathematicae, ex statica de raro et denso, demonstartio prooemialis matutina, de aequilibri in liquido natatantium difficultate & modo, propositio a D. Joanne Ignatio Stephan Wansoviensi Silesio, physicae & matheseos auditore: Demonstratio P* (Wrocław: Baumannische Druckerey Johann Christoph Jacob, 1660); Theodor Moretus, Georgius Franciscus Czesch, and Georgus Becke, *Propositiones mathematicae ex optica, de imagine visionis, demonstratio prooemialis matutina tubum opticum conficere ex lentibus sphaericis aeriis, quibus intra aquam [...] Eo modo imagines rerum repraesentantur, sicut passim crystallinis* (Wrocław: Grunder, 1661); Theodorus Moretus and Gottfried Fibig, *Propositiones mathematicae ex harmonica, de soni magnitudine/Propositae [...] a nobili & erudito D. Godefrido Fibig, Silesio Vratislaviensi [...] Anno 1664: Die 4. Septembris, Horis Pomeridianis, In Gymnasio Caesarei Regiiq[Ue] Collegii Vratislaviensis in B* (Wrocław: Baumannische Druckerey Johann Christoph Jacob, 1664); Theodorus Moretus and Johann Heinrich Joseph von Schenckendorff, *Propositiones mathematicae ex geographia de aestu maris/A [...] D. Joanne Henrico Josepho a Schenckendorff & milgast, philosophiae & matheseos auditore: Die & anno DeCIMo QVInto IULII In gymnasio Caesarei Regiiq[Ue] Collegii Vratislaviensis in Burgo Socie* (Wrocław: Baumannische Druckerey Johann Christoph Jacob, 1665); Theodorus Moretus and Caspar Knittel, *Propositiones mathematicae ex astronomia de luna paschali et solis motu* (Wrocław: Jacobi, 1666); Theodorus Moretus and

inclined plane. Moretus, as well as Zucchi and de Gottignies, tried to solve this problem mathematically but they all failed,[37] in part due to their failure to take friction into account, and in part because they did not have the proper mathematical tools (vector calculus) at their disposal. Other theses were defended by his students on the known Jesuit subjects of statics and optics.

Moretus's main scientific work is *De aestu mari* (On the tides), which was published in Antwerp in 1665 and reprinted in Vienna in 1719.[38]

2   Jan Ciermans

Ciermans[39] (fig. 12.1[40]) was born in 's-Hertogenbosch (Den Bosch) on Easter Day, April 7, 1602, the son of Pieter (1550–1617), a cloth merchant, and Elisabeth Thomas (1554–1612).[41] Because of this link with Easter, he would later add Pascasio to his Christian name. From 1613 to 1619, he studied at the Jesuit college of 's-Hertogenbosch, and by the time he joined he had taken eight years

---

Josephus Nicotius, *Propositiones: Mathematicae; Ex hydro-statica de prima; Suppositione; Archimedis* (Brzeg: Christoph Tschorn, 1667).

37   For example, a letter to Zucchi dated November 21, 1648, sent from Prague. Copy by Bosmans (Kadoc ABML Bosmans V 8 Moretus, 103) of an original in the Biblioteca Vittorio Emanuele (MS Gesuitico 379, no. 32).

38   Theodorus Moretus, *Tractatus physico-mathematicus de uestu maris* (Antwerp: Apud Jacobum Meursium, 1665); Moretus, *Tractatus physico-mathematicus de aestu maris* (Vienna: Ignatius Dominicus Voigt, 1719).

39   For a detailed biography, see Omer van de Vyver, S.J., *Jan Ciermans (Pascasio Cosmander) 1602–1648, Wiskundige en Vestingbouwer*, vol. 7, Mededelingen uit het Seminarie voor Geschiedenis van de Wiskunde en de Natuurwetenschappen aan de Katholieke Universiteit te Leuven (Katholieke Universiteit Leuven, Departement wiskunde, 1975); Edwin Paar, "Jan Ciermans: Een Bossche vestingbouwkundige in Portugal," *De Brabantse Leeuw* (2000): 201–16.

40   The sitter is identified as Ciermans (see, e.g., Domingos Bucho and Raul Ladeira, *Cidade-quartel fronteiriça da Elvas e suas fortificações: The Garrison Border Town of Elvas and Its Fortifications* [Elvas: Edições Colibri e Câmara Municipal de Elvas, 2013], 155). I remain very skeptical, however. The text on the painting mentions Pascasius *Brott*. There may be some confusion with Paschasius Broetius, a Netherlandish Jesuit who, shortly after Ignatius had left, joined the Society in Paris. Broetius was to be sent to Abyssinia, at the behest of the Portuguese king, but he succumbed to a contagious disease before he could commence his journey (see *Afbeeldinghe*, 147, 448–49).

41   Paar, "Jan Ciermans," 21–6. www.geneanet.org gives Jonkheer (squire), Mr. Augustijn Peter Augustijn Ciermans (1550–1617), an attorney, as his father, and Elisabeth Godfried Hendrik Teulings (c.1555–1617) as his mother. They had seven children, including Jan and Willem. https://www.stamboomnederland.nl/etalage/Simon_Boom_Project_van_geneal ogie_simon_boom.eu_896/families74.html#f3743 (accessed March 27, 2020) identifies the father as Augustinus (1550–1617) and the mother as Elisabeth Thomas.

of humanities.[42] In the college, he would meet San Vicente for the first time, although it is unlikely that the two became close acquaintances. San Vicente served one year as prefect of studies at the college in 1614–15. On November 6, 1619, Ciermans was admitted to the Society and did his novitiate in Mechelen.[43] From 1621 to 1623, he studied philosophy at the college of Leuven, and in 1624 he attended San Vicente's mathematics classes. On July 29, 1624, together with Van Aelst, he defended statics theses.

Shortly afterward, Ciermans was appointed as a teacher, first in his old school in 's-Hertogenbosch, and later at the college of Ypres.[44] In 1630, he returned to Leuven to take up philosophy. In 1633, he became a member of the Illustre Lieve-Vrouwebroederschap (Illustrious Brotherhood of Our Lady) of 's-Hertogenbosch. Unlike his brother Willem, he agreed to open up membership of the brethren to Protestants.[45] In 1635–36, he was teacher and prefect of studies at Cassel, where he took his four vows and was ordained priest on April 15, 1634.[46]

In 1637, Ciermans succeeded Boelmans as professor at the school of mathematics.[47] During this period, he had a short-lived correspondence with Descartes. At the request of Rector Foppe Plemp (better known as Vopiscus Fortunatus Plempius [1601–71]), he wrote down some remarks on Descartes's *Discours de la méthode* (Discourse on the method).[48] In March 1638, he anonymously wrote a letter to Descartes in which he limits himself to the part on color dispersion and the rainbow. Descartes treats light as an action emanating from a source and propagating rectilinearly in a medium, without transport of matter. However, space is filled with tiny spheres of "subtle matter," which fill the space in translucent matter. It is these tiny particles that through rotation and translation propagate light.

---

42   ARAA T14/034 31, 56/90.
43   Kadoc ABSE 31, 116, and 32, 26; ARAA T14/034 31, 56/90.
44   ARAA T14/034 33.
45   Alexander Frederik Oscar van Sasse van Ysselt, "De transformatie der Illustre Lieve Vrouwe Broederschap te 's-Hertogenbosch," *Taxandria: Tijdschrift voor Noordbrabantsche Geschiedenis en Volkskunde 2de R*. 13 (1906): 237–46, here 237–38; Paar, "Jan Ciermans," 216.
46   ARAA T14/034 35, 23/5.
47   ARAA T14/034 37, 52/11; KBR 20204.
48   For a discussion of this topic, see Van de Vyver, *Jan Ciermans*, 2–3, and Jed Z. Buchwald, "Descartes's Experimental Journey Past the Prism and through the Invisible World to the Rainbow," *Annals of Science* 65, no. 1 (2008): 1–46, here 38–40. The letters can be found in René Descartes, Charles Adam & Paul Tannery (Eds.), *œuvres de Descartes: 2; Correspondance; Mars 1638–Décembre 1639* (Paris: Cerf, 1898), 55–62, letter 116 and 69–81, letter 118. Letters 115 and 117 are letters to Pemplius.

FIGURE 12.1   Portrait allegedly of Jan Ciermans in the Museu Militar Forte de Santa Luzia (Elvas)
WIKIMEDIA COMMONS

To explain dispersion, Descartes assumes that the velocity of light in glass is larger than that in air. From a thought experiment, he then concluded that red light is faster than blue light. Ciermans put forward two objections to this explanation. The first was that, when two different colored beams cross each other, the collisions of the spheres in the beam should bring about a change in color, which is not the case. The second, in which Ciermans assumes that the velocity in glass is smaller than that in air, is that red light is slower than blue light. Ciermans also had doubts about the spheres of subtle matter, which, if emanating from the Sun, would deplete it over time.[49]

Descartes quickly responded, but the answer does not seem to have satisfied Ciermans, who did not reply. For the tiny particles, Descartes used the analogy of the wine press, in which the grapes do not hamper the fluid particles from flowing through the funnels. For the differences in velocity, the argument goes that a ball will roll faster on a firm tabletop than on a carpet. In his book *Disciplinae mathematicae* (The mathematical sciences),[50] Ciermans hints that it will take more research to be able to describe the exact nature of light.

The subjects in Ciermans's *Disciplinae mathematicae* are divided over the months of the year, beginning in October, the start of the school year (fig. 12.2). The subjects dealt with are geometry, arithmetic, optics, statics, hydrostatics, navigation, fortification, the art of warfare, arms, geography, astronomy, and chronology. With the exception of chronology, these are the subjects dealt with by practical mathematicians.[51] Each month has three weeks, each of which is divided into seven subjects followed by *problemata*, with the exception of September, which only has two weeks. All 1,639 theses are reprinted in the book. The work reveals that Ciermans was knowledgeable about developments in engineering, such as the submarine or the sailing cart, although he does not explicitly mention their inventors (Cornelis Drebbel [1572–1633] and Stevin[52] respectively). With regard to military architecture, Ciermans shows himself to be an adept of Stevin and Samuel Marolois (1572–before 1627).[53]

In the *problemata*, Ciermans discusses applications that he thought were feasible, such as a log to dig ditches. However, the most interesting application

---

49 Apparently, here there is some confusion about the different explanations Descartes gave in his treatises *Météores* on the one hand and *Dioptrique* on the other.
50 Ciermans et al., *Disciplinae mathematicae*, December, week 3.
51 On practical mathematics, see, e.g., J. [Jim] A. Bennett, *The Measurers: A Flemish Image of Mathematics in the Sixteenth Century* (Oxford: Museum of the History of Science, 1996); Bennett and Stephen Johnston, *The Geometry of War 1500–1750* (Oxford: Museum of the History of Science, 1996); Meskens, *Practical Mathematics*.
52 Stevin is only referred to as a "Belgian" (i.e., Netherlandish) mathematician.
53 Paar, "Jan Ciermans," 205.

FIGURE 12.2 Two putti demonstrating motion on an inclined plane. Ciermans, *Disciplinae Mathematicae*, January, second week
© MPM PK.OP.02380

he describes is a calculating machine. Unlike other devices, such as Coignet's sector, which are only calculation aids, his machine could actually perform multiplications and divisions. Ciermans begins by stating that such machines already exist, but that they are so slow in operation that it is faster to perform the multiplication or division with pen and paper. He then goes on to say that he is building a device using cog-wheels, which by turning the clock hands can perform multiplication and division flawlessly. Because Ciermans uses the present tense, it is not clear whether the machine had actually been finished.[54]

Ciermans keeps aloof on the subject of geocentrism, although when discussing the orbit of the moon he prefers the elliptical orbit. He is of the opinion that the center of the planetary system can be chosen anywhere, still saving the phenomena. But this choice is a hypothesis, not a physical reality. The vignettes, however, show that he may have been inclined toward the heliocentric system: in one vignette, a *putto* studies the transit of Venus.[55] In light

---

54 Ciermans et al., *Disciplinae mathematicae*, November, week 1 (*problemata*).
55 Patricia Radelet-de Grave, "Guarini et la structure de l'Univers," *Nexus Network Journal* 11 (2009): 393–414, here 394–97.

of the conclusions San Vicente shared with Huygens on the observations at the Collegio Romano, this does not come as a surprise.

When discussing hydrostatics, there is a curious passage in which Ciermans claims that fluids act as machines, and without the explicit presence of machines, everything is performed by nature.[56] Because the book only consists of short statements, it is not possible to gauge the philosophical background to this assertion.

From the theses Ciermans had defended, it becomes clear that his interest lay in practical mathematics and its applications. These can be general subjects, such as the art of warfare, or subjects as mundane as the horse's bridle.

On June 25, 1641, Ciermans had theses defended at Antwerp by Jan Antoon Tucher (before 1619–77), son of alderman Robrecht Tucher (1587–1646). Jan Antoon would become mayor of Antwerp in 1662. I have found a copy of the thesis print entitled *De architectura militari positiones* (Propositions on military architecture),[57] which shows an allegorical representation of Antwerp as a woman on a throne, surrounded by *putti* studying the arts, painting, arithmetic, and geometry. In the background, we can see Poseidon. This central illustration was drawn by Theodoor van Thulden (1606–69) and engraved by Jacques Neeffs (1610–after 1660). The names of these artists alone betray the vast wealth of the Tucher family. At the edge of the print and round the illustration and the theses are eighteen other illustrations.[58] These are the illustrations that can also be found in *Disciplinae mathematicae*. Although Jan Antoon was a lawyer, his theses deal with the fortifications of Antwerp, war being ever present in the Low Countries.

Other students who defended theses include Krzysztof Aleksander Rozdrażewski (*fl.* 1640), Leo Carolus Sapieha,[59] Jakub Hieronim Rozdrażewski (c.1621–62),[60] and Jan Antoon Tucher together with Jacob Hoens.[61]

---

56   Geert Vanpaemel, "Mechanics and Mechanical Philosophy in Some Jesuit Mathematical Textbooks of the Early 17th Century," in *Mechanics and Natural Philosophy before the Scientific Revolution*, ed. Walter Roy Laird and Sophie Roux (Dordrecht: Springer Netherlands, 2008), 259–74, here 269–71.
57   Rijksmuseum Amsterdam RP-P-OB-67.969.
58   Daniel Papebroch (Papebrochius [1628–1714]) describes them as having *fourteen* figures as illustrations.
59   The Sapieha family was an influential Polish noble and magnate family of Lithuanian and Ruthenian origin.
60   Jan Ciermans et al., *Repetitio menstrua quam de geometricis, astronomicis, staticis instituent* [...] (Leuven: Typis Everardi de Witte, 1639).
61   Jan Ciermans, Johann Anton Tucher, and Jacob Hoens, *Repetitio menstrua quam de cosmographicis et geographicis instituent* (Leuven: Everardus de Witte, 1639).

FIGURE 12.3   Thesis print of Jan-Antoon Tucher, *Antverpia Surrounded by Allegorical Representations* (Antwerp: n.p., n.d. [1641?])
© RIJKSMUSEUM AMSTERDAM, RP-P-OB-67.969A

In 1641, Ciermans left for Portugal, hoping to embark for the Chinese missions in Lisbon. He arrived in Lisbon in late 1641, accompanied by one of his students, Hendrik Uwens (1618–67).[62] Ciermans, and later Uwens, taught at the Colégio de Santo Antão, occupying the chair of the Aula da Esfera.[63] Ciermans later tutored Prince Dom Teodósio (1634–53).[64] The chair was set up in 1590, with Clavius's student João Delgado (1553–1612) being the first mathematician to occupy it.[65] The college was the only one in Portugal where military techniques including fortification were taught. From Stafford's college notes, entitled *Tratado de milicía* (Treatise on the militia [1638]), we know that the authors that were studied include Marolois, Jean Errard (1554–1610), and Antoine de Ville (1596–1656).[66] During the war of independence, Ciermans would put his theoretical knowledge to practical use, as he participated in several battles.[67] It was one of the ironies of fate that one of his adversaries and thus technically an enemy was another student of San Vicente, della Faille (see chapter 10).

On April 6, 1644, Ciermans wrote a letter urging Vitelleschi to grant him permission to go to China. Vitelleschi ordered him, while his application was pending, not to participate in any operation that would compromise his neutrality. The king, however, appointed him to the rank of colonel and chief engineer, which led to his expulsion from the Society.[68]

In 1647, Ciermans was taken prisoner by the Spanish while en route from Lisbon to Élvas. The Spanish immediately realized the importance of their

---

62 Uwens was born on April 23, 1618 in Nijmegen, the son of Hendrik and Margareta Buisala. He entered the Society on September 22, 1634 (Kadoc ABSE 31, 145). In 1638, he followed mathematics classes with Ciermans (KBR Hs 20204). In the catalog of 1641, the note "in Sinas" (in China) appears next to his name (KBR Hs 20206). He actually became a missionary in India and died as the rector of the college of Agra on April 6, 1667. Delée, "Liste d'élèves," 58. Norbert Everard van Couwerven, *Sermoon Ter Eeren Vanden H. Ignatius, Fondateur Vande Societeyt Iesu [...] Tot Antwerpen, Op Den XXXI. Iulij, M.DC.LVI.* (Antwerp: Balthasar Moretus, 1656), 25, mentions that Uwens is in "the land of the Moguls."
63 On the Aula, see, e.g., Carlos Fiolhais and José Eduardo Franco, "Portuguese Jesuits and Science: Continuities and Ruptures (16th–18th Centuries)," *Antiguos jesuitas en Iberoamérica* 5, no. 1 (2017): 163–78; Natália de Oliveira and Sezinando Menezes, "Ciência moderna em Portugal: A 'Aula da Esfera' no Colégio de Santo Antão," *Acta scientiarum: Education* 39, no. 3 (2017): 243–53.
64 Henrique Leitão, "Jesuit Mathematical Practice in Portugal 1540–1759," in *The New Science and Jesuit Science: Seventeenth-Century Perspectives*, vol. 4, Archimedes, ed. Mordechai Feingold (Dordrecht: Springer Netherlands, 2003), 229–47, here 237.
65 Denis De Lucca, *Jesuits and Fortifications: The Contribution of the Jesuits to Military Architecture in the Baroque Age* (Leiden: Brill, 2012), 125.
66 De Lucca, *Jesuits and Fortifications*, 123–24.
67 Fernando de Meneses, *Historiarum Lusitanarum ab anno 1640 usque ad 1657 libri decem*, vol. 2 (Lisbon: Joseph Anthony da Sylva, 1734), 499, wrote that Ciermans looked more like an intrepid commander than a Jesuit.
68 Kadoc ABSE 31, 116; 32, 26; 1447, letter of Edmond Lamalle (1900–89) to Omer van de Vyver.

prisoner and refused all Portuguese offers of trading him with high-ranking Spanish captives in Portuguese custody.[69] Ciermans was taken to Madrid, where the Spanish tried every trick in the book, including female seduction, to make him talk. After suffering a blow to the head, he finally gave in. We do not know how many Portuguese secrets he revealed. On June 18, 1648, he participated in the Spanish attack on Olivença and was mortally wounded by a bullet through the head.[70]

3   Willem Boelmans

Boelmans was born in Maastricht in 1603, the son of Aegidius (Gillis) and Liesbeth Liesmans.[71] He had already followed seven years of humanities by the time he joined the Society in 1617, thereafter taking two years of philosophy and four years of theology; he attended courses with San Vicente at Leuven for one year.[72] He taught at the colleges of Maastricht, Ghent, and Leuven.[73] In Leuven, he taught mathematics for at least one year.[74] On August 8 and 9, 1634, he had theses defended on applied mathematics, with Tacquet as one of the defendants,[75] and San Vicente, who was now residing at Ghent, as guest

---

[69] Paar, "Jan Ciermans," 212–13.

[70] Dauril Alden, *The Making of an Enterprise: The Society of Jesus in Portugal, Its Empire, and Beyond 1540–1750* (Stanford: Stanford University Press, 1996), 106; Van Looy, "Chronologie en analyse," 19–20; de Vyver, *Jan Ciermans*; Dhombres and Radelet-de-Grave, *Une mécanique donnée à voir*; Paar, "Jan Ciermans," 214.

[71] Kadoc 35, 397. In his handwritten entry letter, he mentions "Traiectum" as his place of birth. Usually, this translates as Utrecht. However, Utrecht was in the rebellious northern provinces that turned into the republic. As Boelmans himself writes, he attended classes at the Jesuit college of his hometown. Traiectum therefore should read as "Mosa Traiectum," Maastricht, which was in the Southern Netherlands and did have a college.

[72] ARAA T14/034 34, 48/8; 35, 48/13.

[73] ARAA T14/034 32, 96/11.

[74] KBR 20203.

[75] Willem Boelmans and Johannes Groll, *Theses mathematicæ: Geometricæ, arithmeticæ, opticæ, catoptricæ, dioptricæ, mvsicæ, architectonicæ, stereo-staticæ, hygro-staticæ, qvas præside [...] Gvilielmo Boelmans [...] Demonstrabit ac defendet Ioannes Groll [...] Lovanii, in Collegio Societatis I* (Leuven: Viduam Henrici Hastenii, 1634); Boelmans and Laurens van Schoone, *Theses mathematicæ: Geometricæ, arithmeticæ, opticæ, catoptricæ, dioptricæ, mvsicæ, architectonicæ, stereo-staticæ, hygro-staticæ, qvas præside [...] Gvilielmo Boelmans [...] Demonstrabit ac defendet Lavrentivs van Schoone [...] Lovanii, in Collegio Soc* (Leuven: Viduam Henrici Hastenii, 1634); [Boelmans and Philip Jacob], *Theses mathematicæ: Geometricæ, arithmeticæ, opticæ, catoptricæ, dioptricæ, mvsicæ, architectonicæ, stereo-staticæ, hygro-staticæ, qvas præside [...] Gvilielmo Boelmans [...] Demonstrabit*

THE LEUVEN STUDENTS 207

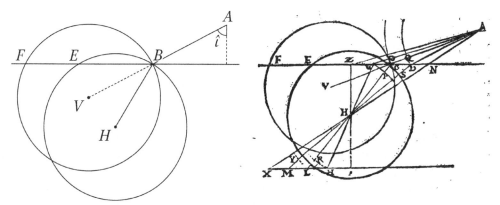

FIGURE 12.4 Let a light ray from A to H cross the boundary at B. Elongate AB and determine V such that |VB| = |HB|. Draw circles with radius R = |HB| and centers V and H respectively, which intersect the boundary in F and E respectively. It is clear that |FB| = 2R sin i and |BE| = 2R sin r yielding $\frac{|FB|}{|BE|} = \frac{\sin i}{\sin r} = n$. Willem Boelmans and Laurens van Schoone, *Theses mathematicæ: Geometricæ, arithmeticæ, opticæ, catoptricæ, dioptricæ, mvsicæ, architectonicæ, stereo-staticæ, hygro-staticæ, qvas præside* [...] *Gvilielmo Boelmans* [...] *Demonstrabit ac defendet Lavrentivs van Schoone* [...] *Lovanii, in Collegio Soc* (Leuven: Viduam Henrici Hastenii, 1634), 19
© KULEUVEN GBIB P 08/16/Q/4-2

of honor at the defenses.[76] Among the optics theses is a noteworthy one: to determine geometrically "which is the path on which the least distance is traveled in both the less dense as the more dense medium."[77] To determine this path, Boelmans starts out with the postulate that nature (in this case light) will pass through as little as necessary in both the less dense and the more dense medium together.[78]

Boelmans argues that the shortest path will cross the boundary at a point B, for which $\frac{\sin i}{\sin r}$, the ratio of the densities (fig. 12.4).

At the time the theses were defended, Descartes had not yet published the sine law. Willebrord Snel van Royen (1580–1626), on the other hand, had already discovered the sine law and this was known in the Low Countries.

---

ac defendet Philippus Iacobi [...] *Lovanii, in Collegio Societati* (Leuven: Viduam Henrici Hastenii, 1634).
76 KBR MS 19337–38, fol. 69.
77 Boelmans and Groll, *Theses mathematicæ*, 18–19. For a detailed analysis, see August Ziggelaar, S.J., "The Sine Law of Refraction Derived from the Principle of Fermat: Prior to Fermat? The Theses of Wilhelm Boelmans, S.J., in 1634," *Centaurus* 24 (1980): 246–62.
78 Boelmans and Groll, *Theses mathematicæ*, 17.

The formulation, including the lemmas preceding the proposition, suggests that Boelmans discovered the sine law independently. More importantly, it is derived from a principle that would later be called Fermat's principle.

4 Willem Hesius

Hesius[79] was born between May 6 and 25, 1601,[80] and he was baptized in St. Andrew's Church in Antwerp on May 25, 1601.[81] He entered the Society on September 22, 1617, after he had studied eight years of humanities. After joining the Society, he studied two years of philosophy, four years of theology, and one year of mathematics.[82] He would later attend courses in Scholasticism.[83] He taught humanities, philosophy, and physics "in accordance with the philosophy of the Society" at the college of Leuven.[84] In 1634, he was transferred to Antwerp, where he was ordained priest. The next year, he went to Ghent, where he became prefect of studies.[85] In 1642, he was attached to the college of Aalst.[86] In 1646, he was in Brussels, where he frequented the court of Archduke Leopold of Austria.[87] He is best remembered for his work in architecture (see chapter 14), especially for his design of the Jesuit church at Leuven. In 1665, he was assigned to the professed house at Antwerp.

Hesius is the author of the emblem book *Emblemata de fide, spe, charitate* (Emblems of faith, hope, and charity [1636]). The book contains 116 wood cuts by Erasmus Quellin the Younger (1607–78). Every section in the book has the same structure: first there is a Bible quotation with a picture, preferably with putti. This is followed by a short summarizing motto, with a short poem revealing, still in enigmatic terms, the meaning of the picture. Finally, there is an extensive poetic meditation in elegiac distichs, which deciphers the picture and from which a meaning for the picture is distilled. Hesius finished his manuscript in 1624, but it only appeared in print in 1636. The pictures reveal that the book was written by a scientist. In many cases, we find that musical

---

79 He is often cited as Willem van Hees, yet the parish registers give Hezius as family name.
80 There is no consensus in the archival sources about his day of birth, ARAA T14/034 31. The *Catalogus personarum* of 1622 even mentions June 11, 1601.
81 FA PR 92, fol. 37ʳ; PR100, fol. 220ʳ.
82 ARAA T14/034 31, 50/39; 35, 38/8.
83 ARAA T14/034 34, 48/9.
84 ARAA T14/034 34, 48/9; 3808.
85 ARAA T14/034 35, 38/8.
86 ARAA T14/034 3808.
87 ARAA T14/034 3808.

FIGURE 12.5 Willem Hesius, "Emblemata de fide, spe, charitate." Manuscript, Erfgoedbibliotheek Hendrik Conscience, Antwerp B 129141, 1624
© EHC B 129141

instruments, anchors, and optical phenomena are used as analogies. The beautiful drawings in the manuscript, which Hesius probably drew himself, make this designation all the more clear (fig. 12.5).

5   Other Students

Van Rasseghem was born in Grimbergen on April 5, 1603 and entered the Society on August 17, 1618.[88] Before joining the Society, he had taken six years of humanities.[89] He studied mathematics for one year. After his studies, he was sent to the college of Kortrijk, where he taught syntaxis,[90] before being assigned to the college of Bruges as prefect of studies.[91] In 1651, he was prefect of studies at the college of Ypres.[92]

88   ARAA T14/034 31, 56/86; 35, 21/13.
89   ARAA T14/034 31, 56/86.
90   ARAA T14/034 32, 42/16.
91   ARAA T14/034 35, 21/13.
92   ARAA T14/034 39, 43/7.

Van Aelst was born on January 30, 1603[93] and entered the Society on July 27, 1619.[94] Before joining the Society, he had studied eight years of humanities.[95] After two years of novitiate and two years of philosophy studies, he studied mathematics with San Vicente in 1624.[96] Together with Ciermans, he defended mechanics theses on July 29, 1624. He taught at the college of Aalst for four years,[97] then at the college of Oudenaarde.[98]

Le Vray was born on September 21, 1604 and entered the Society on October 8, 1620.[99] He had taken six years of humanities before joining the Society.[100] He left the Jesuit order in the early 1620s.

Pynappel was born on December 6, 1605 and entered the Society on September 28, 1620.[101] Before joining the Society, he took seven years of humanities at the college of 's-Hertogenbosch.[102] After joining the Society, he took two years of philosophy and four years of theology.[103] In 1624, he was called *mathematicus* (mathematician). In 1636, he was active at the college of Dunkirk and was also involved in the Missio Navalis (the naval mission), the service responsible for the pastoral care of navy sailors.[104]

---

93   ARAA T14/034 31, 62/125.
94   Kadoc ABSE 36, fol. 33; ARAA T14/034 31, 62/125.
95   ARAA T14/034 31, 62/125.
96   Kadoc ABSE 36, fol. 33.
97   ARAA T14/034 31, 62/125.
98   ARAA T14/034 35, 38/7.
99   ARAA T14/034 31, 72/50.
100 ARAA T14/034 31, 72/50.
101 Kadoc ABSE 36, fol. 95; ARAA T14/034 31, 70/32.
102 ARAA T14/034 31, 70/32; 34, 49/29.
103 ARAA T14/034 35, 36/9.
104 ARAA T14/034 35, 36/9.

CHAPTER 13

The Later Disciples

1   Andreas Tacquet

Although a full study of Tacquet is beyond the scope of this book, it is nevertheless important that we pay at least some attention to him. In a sense, he is the last of the great Flemish Jesuit mathematicians and the last champion of San Vicente's integration methods. It was Tacquet's books that made San Vicente's exhaustion method known to mathematicians throughout Europe.

Tacquet was born in June 1612 in Antwerp and was baptized in St. Walburga Church on June 23, 1612.[1] He joined the Society on October 31, 1631.[2] Before joining, he had studied seven years of humanities at Antwerp, later reading philosophy and mathematics each for two years at the college of Leuven.[3] His teacher of mathematics was Boelmans, the ailing but brilliant student of San Vicente. On August 8 and 9, 1634, he defended theses under Boelmans's supervision.

After his studies in 1635, Tacquet was transferred to the college of Bruges,[4] where he taught humanities (including Greek, among other things). He was eventually sent back to Leuven,[5] where he taught mathematics and finished his education by taking four years of theology.[6] He would follow all movements of the school of mathematics between Leuven and Antwerp.[7]

Tacquet was a cautious man and hardly ever took a firm stand on which view he preferred, neither in mathematics nor in physics or astronomy. He was a teacher in the best sense of the word, not a researcher. He is reported to have had French and Central European noblemen among his students.[8]

Tacquet would first step into the limelight in 1650. On January 31, Philippe Eugène, count of Hornes et d'Herlies (?–1677), defended theses under the

---

1   FA PR65 St Walburga Church, 138^r; PR71, 337^v.
2   ABSE 31, 137; ARAA T14/O34 34, 54/79.
3   ARAA T14/O34 35, 22/23.
4   ARAA T14/O34 35, 22/23; 37, 23/17.
5   ARAA T14/O34 39, 47/5.
6   ARAA T14/O34 39, 47/5.
7   ARAA T14/O34 40, 12/3.
8   Henri Bosmans, S.J., "Le jésuite mathématicien anversois André Tacquet (1612–1660)," *De Gulden Passer* 3 (1925): 63–85, here 65, citing a letter of the rector of the Antwerp college, François van der Meersch (1608–61), to the provincial, Joannes van Renterghem (1606–81).

FIGURE 13.1 Thesis print with mathematical theorems by Phillippe Eugène d'Hornes et d'Herlies. Engraving by Nicolaas Lauwers after Abraham van Diepenbeek, 1652
© KU LEUVEN CAMPUSBIBL. ARENBERG WBIB RAADZAAL LOKAAL 02.16

supervision of Tacquet, which following the custom meant that Tacquet had written them.[9] The subjects were the paradox of Aristotle's wheel, which up to that time was deemed insoluble, and the properties of the cycloid.

On March 29, 1651, d'Hornes et d'Herlies would again defend theses, this time in optics, statics, and ballistics. In the ensuing years, Tacquet often had theses defended on subjects we find time and again in the spheres of the Flemish Jesuits' interests: mathematics, and especially geometry, statics, cosmography, and the art of warfare.[10] These theses would be the starting point of a correspondence between Tacquet and Huygens, with many of these letters passing through the hands of Daniel Seghers (1590–1661), the lay-brother and floral painter. In one of these letters, Tacquet touched upon a philosophical question: in geometry, it is sometimes possible to deduce a true proposition from a false proposition $((\neg A \to A) \to A)$.[11] In his reply, Huygens begged for an

---

9 Andreas Tacquet et al., "Thesis Print" (n.p.: n.p., 1650–51); Tacquet and Philippe Eugène, comte de Hornes et de d'Herlies, *Dissertatio physico-mathematica de motv circvli et sphæra* (Leuven: Corn. Coenestenii, 1650).

10 Bosmans, "Le jésuite mathématicien anversois André Tacquet," 71–72.

11 For instance, as Tacquet notes in Theodore d'Immerselle's (c.1630–1654) thesis, an example can be found in Euclid 9, 12 (Heath, *Euclid*, 397–99). Andreas Tacquet and Theodore

example in geometry, because he could find examples in algebra and astronomy, but he doubted whether there were any in geometry.[12]

Tacquet's *Elementa geometria* (Elements of geometry [1654]) broke with the long tradition of following Euclid's text by adding theorems about solids taken from Archimedes.[13] It also has an introductory chapter on the history of mathematics. A part on trigonometry was added to the 1674 edition, in which the text on spherical trigonometry was copied from Kaspar Schott's (1608–66) books. The book was a huge success and was used throughout Europe. William Whiston (1667–1752), Newton's successor at Cambridge, was a supporter of the book and translated it, adding his comments in an italic typeface. The result was that the book was used throughout Britain in Whiston's English translation.[14] Greek and Italian translations are also known, but none in Dutch or French.

In Tacquet's book *Cylindrocorum et annularium* (On cylinders and rings [1651]), he not only uses the methods of San Vicente but also those of Cavalieri.[15] The 1651 edition contains four books, with a fifth being added in 1659. The first four books are concerned with the volume and the surface area of cylinders, cylinder parts (*ungulae*), and rings. The fifth book deals with the center of gravity, for the calculation of which Tacquet often invokes the Pappus Guldin theorem, which is all but absent in the other books. In many ways, including its layout, the fifth book is sloppier than the other ones, perhaps because Tacquet, suffering from consumption, felt his end was drawing near.[16]

---

d'Imerselle, *Theses mathematicæ* [...] *quas serenissimo Archiduci Leopoldo Wilhelmo dicatas* [...] *demonstrabit illustrissimus Dominus Theodorus d'Imerselle* [...] *in Collegio Societatis Jesu lovanii* [...] *septembris* (Antwerp: Andreæ Bouveti, 1652); Andreas Tacquet and Theodore d'Imerselle, *Theodorvs d'Imerselle Comes de Bovchove et S. Rom. Imp. defendet in Collegio Soc. Iesv lovanii Præside R.P. Andrea Tacquet eiusdem Soc. math: Cos prof.; E A° M DC LII tertiâ Sept* (Antwerp, 1652) are single-sheet thesis prints with symbolic and textual references to d'Imerselle's thesis. In comparison with other theses, d'Immerselle's is rather superficial. Huygens, *Oeuvres complètes: Tome I*, 194ff.

12 On this discussion, see Gabriel Nuchelmans, "17th-Century Debate," 46–49.

13 Andreas Tacquet, *Elementa geometriae planae ac solidae: Quibus accedunt selecta ex Archimede theoremata* (Antwerp: Jacobus Meursium, 1654).

14 On Whiston's Euclid, see, e.g., Lies Bens, "De algebraïsering van Euclides: The Elements of Euclid (1714) van William Whiston" (Master's thesis, Katholieke Universiteit Leuven, 2015).

15 A Dutch paraphrased translation can be found in Herman Dooreman, "Andreas Tacquet *Cylindrica et annularia quinque libris comprehensa*" (Master's thesis, Katholieke Universiteit Leuven, 1966).

16 Dominique Descotes, "Pascal's Indivisibles," in Jullien, *Seventeenth-Century Indivisibles Revisited*, 211–48, here 215–16, suggests that the cycloid contest launched by Pascal in 1658 was the direct inducement for publishing, because the Jesuits wanted their part of the glory in solving Pascal's questions. See also Descotes, "Two Jesuits against the Indivisibles," in Jullien, *Seventeenth-Century Indivisibles Revisited*, 249–73.

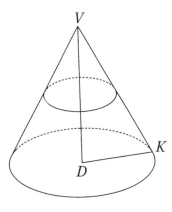

FIGURE 13.2
Tacquet's counter-example to Cavalieri's *indivisibilia* method

Tacquet was highly critical of Cavalieri's method, identifying some apparent paradoxes. Consider, for instance, a right cone generated by the right triangle △VDK. Each plane section of the cone parallel to the base determines a circle. The ratio between the radius and the circumference is 2π. Now, in Cavalieri's method, the triangle △VDK consists of all of these radii. The surfaces of the triangle and of the cone must then, Tacquet argues, have the same ratio as the radii and the circumferences, which is obviously not the case.[17]

Therefore, Tacquet further argues, one can use Cavalieri's method, but only when the results obtained with this method are then rigorously proven. If the result cannot be proven, then it must be considered doubtful.[18] It is as if he views the use of *heterogenia* as a useful experimentation tool for finding areas and volumes.

This contradiction does not arise when using San Vicente's methods. Divide the base circle AB into concentric rings and erect cylinders on them. These cylinders will intersect the lateral area of the cone in circles, which have a known relation to one another. The cylinders thus define rings that are easy to group into a sequence that can be exhausted.

Tacquet's book has two theorems dealing with exhaustion, which are the foundation for nearly all other theorems in the book. The first proposition of the first book is reminiscent of Valerio's theorem: let A and B be two magnitudes, either areas or volumes, and let the ratio of E to F be given.

If one can consecutively inscribe into A and B a sequence of magnitudes that relate to one another such as E to F, and if these magnitudes exhaust A and

---

17 Andreas Tacquet, *Cylindricorum et annularium liber quintus addendus ad quatuor priores anno 1651 editus* (Antwerp: Jacobus Meursius, 1659), 23–24.
18 Tacquet, *Cylindricorum et annularium*, 23–24.

B (i.e., they differ from these by an arbitrarily small amount), then the magnitude A will relate to the magnitude B such as E to F.

Tacquet's general theorem has the advantage that he does not have to repeat a double *reductio ad absurdum* with each proof. It suffices that two sequences exist that exhaust A and B. By introducing the ratio $\frac{E}{F}$ it is sufficient to prove that this ratio is equal to the ratio of the inscribed figures.

The first proposition of the second book introduces another exhaustion method: if a sequence of magnitudes $A_{i,n}$, $B_{i,n}$ can be inscribed in magnitudes A and B, and if likewise a sequence of magnitudes $A_{c,n}$, $B_{c,n}$ can be circumscribed about magnitudes A and B, and if moreover $A_{i,n}$ and $A_{c,n}$ exhaust A and for the corresponding magnitudes, we have that $\frac{A_{i,n}}{B_{i,n}} = \frac{E}{F}$ and $\frac{A_{c,n}}{B_{c,n}} = \frac{E}{F}$ then $\frac{A}{B} = \frac{E}{F}$.

The simplification lies in the fact that it is no longer necessary to exhaust the inscribed and circumscribed magnitudes for each of the magnitudes A and B, as it suffices to calculate the ratio for one or the other.

In Cavalieri's *indivisibilia* method, magnitudes are composed of magnitudes of a different order (e.g., a surface is made up of an infinity of lines, a solid is composed of an infinity of plane figures), and hence Tacquet calls them *heterogenia*. Tacquet argues that this does not meet the standards of geometry and should therefore be rejected. To meet the standards of geometry, one should use *homogenia*, magnitudes of the same kind as the magnitude under investigation. If these *homogenia* can be chosen in such a way that they are a sequence that exhausts the given magnitude, then the sum of arbitrarily many (not infinitely many) *homogenia* can be compared to the given magnitude.

An important new concept is found in Tacquet's definitions of surfaces and solids. He defines a cylinder, for instance, as a solid that is generated by the movement of a circle in such a way that one of the points of the circle segment moves along a straight line. The axis of this cylinder is the straight line joining the center of two of the generated circles. Despite this definition, he does not accept that the cylinder is composed of circles, yet his conception is not all that different from Cavalieri's conception of "all the lines."

An example of Tacquet's method can be found in Eugène's thesis, in which he calculates the area under a cycloid (fig. 13.3):[19] suppose a cycloid is produced by rolling a circle C of radius $|ab|$. Then $|bf|$ equals half the circumference of the circle C. Divide $[bf]$ into four equal line segments. Divide the circumference of

---

19   Tacquet and d'Hornes et d'Herlies, *Dissertatio physico-mathematica*.

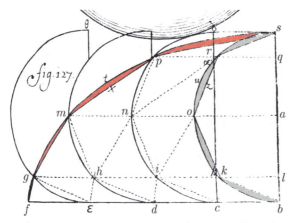

FIGURE 13.3 The determination of the area under a cycloid. Andreas Tacquet and Philippe Eugène Comte d'Hornes et d'Herlies, *Dissertatio physico-mathematica de motv circvli et sphæra* (Leuven: Corn. Coenestenii, 1650) (our shading)
© EHC G 4790

the semicircle into four equal arcs and draw lines through these points parallel to *bf*. In the division, points of [*bf*] draw the generating (semi)circle.

We can immediately deduce that if $\triangle psr \cong \triangle qfe \cong \triangle ghe$ and $\triangle mnp \cong \triangle gmh \cong \triangle mnh$, then the area between *bf*, the cycloid, and the semicircle can be approximated by: $S_4 = 8\frac{|bc|\cdot|lb|}{2} + 8\frac{|bc|\cdot|al|}{2} = 4\cdot|bc|\cdot(|lb|+|al|) = |fb|\cdot|ab|\left[=\pi|ab|^2\right]$.
Now in calculating S_4, we have omitted the shaded curvilinear figures inside the cycloid, but we have also added the shaded figures inside the circle. If [*bf*] is divided into ever more equal line segments, the difference between S_n and the area *fmsob* can be made arbitrarily small. Therefore area *fmsob* + area *bosab* = $\frac{3}{2}$ area generating circle.

This result was not new, as it had already been proven by Gilles Personne de Roberval (1602–75), albeit with the use of indivisibles.[20]

---

20   Léon Auger, *Un savant Méconnu: Gilles Personne de Roberval (1602–1675)* (Paris: Librairie scientifique A. Blanchard, 1962), 43–45; Vincent Jullien, "Roberval's Indivisibles," in Jullien, *Seventeenth-Century Indivisibles Revisited*, 177–210.

In his book *Arithmetica theoria et praxis* (Theory and practice of arithmetic [1656]), Tacquet treats arithmetic in such a way that the ties with geometry are severed altogether.[21]

After Tacquet's death, Henri de Prince (1632–71) collected his manuscripts and published them under the title *Opera mathematica* (Mathematical works [1669]). In this collection, one finds treatises on optics, astronomy, and fortifications. He discusses the heliocentric system, for instance, because it allowed simplifications. He did not see arguments for any of the systems, Ptolemaic and Copernican, as anything more than probable.[22] Ultimately, on the basis of scripture, he chose the Ptolemaic system.[23] On the subject of catoptrics, he was sometimes faced with problems for which he could find no answer within existing optical theories, such as the image formation in concave mirrors.[24]

When the Jesuit Martino Martini (1614–61) visited the Low Countries to gain support for the Chinese missions, Tacquet organized a session in Leuven in which Martini's story was illustrated by means of projected images through the use of a *lanterna magica*.[25] It is the oldest reported lecture based on projected images.[26] The *lanterna magica* is a primitive form of slide projector that could only project very faint images because the light source was limited to candles. The device would be refined to its definitive form by Huygens, and by 1672 it had been introduced to the Germanic countries and was produced on a large scale.[27]

---

21   See Henri Bosmans, S.J., "André Tacquet (S.J.) et son traité d'arithmétique théorique et pratique," *Isis* 9, no. 1 (1927): 66–82.
22   Tacquet, *Opera mathematica*.
23   J. L. [John Lewis] Heilbron, *The Sun in the Church: Cathedrals as Solar Observatories* (Harvard, MA: Harvard University Press, 1999), 188–89.
24   See Thomas M. Lennon, "The Significance of the Barrovian Case," *Studies in History and Philosophy of Science* 38, no. 1 (2007): 36–55.
25   Noël Golvers, "De recruteringstocht van M. Martini, S.J. door de Lage Landen in 1654: Over geomantische kompassen, Chinese verzamelingen, lichtbeelden en R.P. Wilhelm van Aelst, S.J.," *De Zeventiende Eeuw* 10 (1994): 331–44, here 340.
    "And many now use, on the basis of Kircher's advice, this technique to show wonderful things, admired and appreciated by the audience. Among them is the exceptional mathematician Andreas Tacquet of our Society, he showed the whole journey of P. Martino Martini from China all the way to Belgium, as Martini told me himself." Caspar Schott, *Magia universalis naturae et artis, sive recondita naturaliam* [...], *Volume 1* (Bamberg: Joh. Martini Schönwetteri, 1677), 426.
26   Klaus Staubermann, "Making Stars: Projection Culture in Nineteenth-Century German Astronomy," *British Journal for the History of Science* 34, no. 123 (2001): 439–51, here 440.
27   Staubermann, "Making Stars," 440.

One of Tacquet's students, albeit for only half a year, was Ferdinand Verbiest, who shrewdly used his knowledge of science and mathematics to impress the Chinese emperor.[28]

2 Gilles-François de Gottignies

De Gottignies merits some attention as one of San Vicente's students in Ghent. He was born on March 30, 1630 in Brussels, the son of Augustin (?–1656), secretary to the Privy Council, and Margareta Verreycken.[29] The couple had nine children, two of whom would become seigneurs of a lordship. Aegidius's brother Lancelot (1618–73) would become bishop of Roermond and vicar general of the army of Flanders.[30]

De Gottignies went to the Jesuit college of Brussels and then to the Valk college of the University of Leuven. He attended Tacquet's mathematics classes for three years before entering the Society.[31] He was admitted to the Society on November 5, 1653 to begin his novitiate in Mechelen. He would later be assigned to the college of Ghent, where he would study or cooperate with San Vicente.

In 1662, de Gottignies was appointed mathematics professor at the Collegio Romano, a position he held until his death in 1669. As such, he published a substantial number of books, the first being on astronomical observations, in which his correspondence with Cassini about the eclipse of Jupiter can be found.[32] He would remain active as an astronomer with a wide network of correspondents.[33] He made use of a telescope with blackened lenses as well as

---

28 Noël Golvers and Ulrich Libbrecht, *Astronoom van de Keizer: Ferdinand Verbiest en zijn Europese sterrenkunde* (Leuven: Davidsfonds, 1988); Golvers, "F. Verbiest's Mathematical Formation."

29 Kadoc ABSE 31, 175.

30 Stephanie Audenaert, "Aegidius Franciscus de Gottignies (1630–1689), Jezuïet, geleerde en homo universalis," *Het Land van Beveren* 50, no. 4 (2007): 544–60, here 545.

31 Kadoc ABML Bosmans V 3–7 Gottignies, referring to Album Novitiorum Tronchianum. Henri Bosmans, S.J., "La *Logistique* de Gilles-François de Gottignies," *Revue des questions scientifiques 4 Série* 13 (1928): 215–44, here 235.

32 Gilles-François de Gottignies, *Astronomica: Epistolae duae* (Rome: Iacobi Antonij de Lazaris Varesij, 1665).

33 For instance, he was informed of the appearance of a comet in Goa by Giuseppi Candone (1636–1701). Noël Golvers, "Ferdinand Verbiest's 1668 Observation of an Unidentified Celestial Phenomenon in Peking, Its Lost Description, and Some Parallel Observations, Especially in Korea," *Almagest: International Journal for the History of Scientific Ideas* 5, no. 1 (2014): 33–51, here 48.

projections of the solar image on pieces of paper for studying solar eclipses.[34] He tried to improve on existing instruments, with his most important innovation being a mount for the use of larger telescopes.[35]

With his *Elementa geometriae planae* (Elements of plane geometry [1669]), de Gottignies followed in the footsteps of his erstwhile teacher Tacquet, having written a didactical manual instead of a geometrical treatise. According to de Gottignies, pupils needed to become acquainted with geometrical objects before they could develop a deeper understanding of geometry. Geometry has to be offered at a level that is comprehensible for the pupil and in accordance with his development,[36] a piece of advice that is still adhered to in didactics of mathematics.

As professor of mathematics, de Gottignies's work was devoted to logistics, which he argued should be an independent subject within mathematics. At one point, he was audacious enough, together with his friend Michelangelo Fardella (1650–1718),[37] to organize a public session in which he invited all adversaries to logistics. They would either present their case at the meeting or have it sent to de Gottignies in writing. However, because Fardella fell ill, the meeting never materialized.[38]

In much the same way that Tacquet was mistrustful of indivisibles, so too was de Gottignies of Descartes's new mathematics, arguing it did not bring mathematical certainty. The importance of his books on logistics does not lie in the fact that they were innovative but that their arguments and examples force the reader to think about the foundations of arithmetic and mathematics.[39]

De Gottignies was a regular contributor to the *Giornale di letterati* (Journal of literati), an Italian journal modeled after the *Journal des sçavans* and the *Philosophical Transactions of the Royal Society*. Many of his observations were published in this journal next to other contributions and articles.[40]

From 1683 to 1687, de Gottignies was in Naples at the behest of the viceroy, who had ordered a large telescope with a mount. De Gottignies was called in to explain its operation.[41] What was only to be a sojourn of a couple of weeks turned out to be a stay of a couple of years. In Naples, de Gottignies suffered

---

34 Audenaert, "Aegidius Franciscus de Gottignies," 550.
35 Audenaert, "Aegidius Franciscus de Gottignies," 550.
36 Audenaert, "Aegidius Franciscus de Gottignies," 547.
37 See also ARAA T14/034 1876.
38 Bosmans, "La *Logistique* de Gilles-François de Gottignies," 240–41.
39 Audenaert, "Aegidius Franciscus de Gottignies," 549.
40 Audenaert, "Aegidius Franciscus de Gottignies," 553.
41 Audenaert, "Aegidius Franciscus de Gottignies," 556.

two strokes, the second of which impaired his physical and mental abilities. He returned to Rome in 1687 and died in the city on April 6, 1689.[42]

3   Alphonse Antonius de Sarasa

Sarasa was born in Nieuwpoort on October 31, 1617 and entered the Society on September 28, 1632.[43] Before joining the Society, he had taken six years of humanities,[44] and after joining he took another year of humanities, two years of philosophy, and four years of theology.[45] In the 1639 *Catalogus personarum*, he is called *magister*. He was assigned to the college of Ghent,[46] where he cooperated with San Vicente. By 1665, he had been transferred to the professed house at Antwerp.[47]

In a response to San Vicente's *Problema Austriacum*, Mersenne challenged Sarasa to find the third logarithm geometrically if, for three given magnitudes, two logarithms are given. With the aid of San Vicente, he accepted the challenge[48] and published it in *Solutio problematis a R.P. Marino Mersenno minimo propositi* (Solution to the problem proposed by the Minim Reverend Marin Mersenne [1649]). Sarasa's concept of logarithm is closely related to our concept, but it is not entirely the same. He used Henry Briggs's (1561–1630) definition, according to which logarithms are numbers with constant difference matched with numbers in a continued progression. In other words, terms of an arithmetic sequence are matched to a geometric sequence, which is just what San Vicente did in his publications. In this definition, it follows that logarithms with different bases can be applied to these sequences—and Sarasa does give four examples of different logarithms of a geometric sequence.

For his solution of Mersenne's problem, Sarasa based himself on the book on the hyperbola in *Problema Austriacum*. He then proposed a scholion in which he discussed the nature of logarithms and their relationships with hyperbolic areas. In doing so, he took San Vicente's earlier results out of the realm of geometry and put them where they belong: among infinitesimal calculus. There were still a number of steps to be taken before one gets from Sarasa's

---

42   Audenaert, "Aegidius Franciscus de Gottignies," 556.
43   ARAA T14/034 35, 53/70.
44   ARAA T14/034 35, 53/70.
45   ARAA T14/034 37, 41/21; 39, 36/4.
46   ARAA T14/034 37, 41/21.
47   ARAA T14/034 42, 1/4.
48   Robert P. Burn, "Alphonse Antonio de Sarasa and Logarithms," *Historia mathematica* 28 (2001): 1–17, here 7ff.

concept of logarithm to the modern concept of natural logarithm, steps that by the end of the century would be taken by Leibniz, among others, the attentive reader of San Vicente's book.[49]

Sarasa's book would be the last influential work to be published by a member of San Vicente's school.

---

49   Hofmann, *Das* Opus geometricum *der Gregorius à Sancto Vincentio*.

CHAPTER 14

The Jesuit Architects

No history of the Flemish Jesuit mathematicians would be complete without some attention being paid to their work as architects.

1   Ad maiorem Dei gloriam

As with many aspects of Jesuit life, Rome also wanted to have some control over the building of churches and other edifices. The norms that were applied to assess the plans sent to Rome were practical, functional, and financial.[1] Superior General Acquaviva had given clear instructions at the beginning of the seventeenth century on how Jesuit buildings (but not churches) should look: "The building to be erected ought to be adapted to our religious mission, in that it be simple, hygienic, and functional, and in none of its parts be pretentious either in substance or in form. It should be designed for practical living and not for pomp and ornamentation."[2]

These instructions were similar to those contained in the document *De ratione aedificorum* (Building guidelines) of 1558, which states that "it is necessary to establish a norm for building the houses and colleges that depend on us, so as to avoid [...] their seeming to be palaces fit for nobles [...]; they should give visible testimony to poverty."[3]

The First General Congregation (1558) created the office of *consiliarius aedificorum*, the consultant for building plans. The purpose of the position was to oversee the design and construction of the Society's new buildings.[4] The office did not impose a specific style, nor did it issue specific guidelines, although it goes without saying that the very existence of such an office imposes the artistic views of the one who holds it on his decisions vis-à-vis permissions. Thus, the stamp of lay brother Giovanni Tristano (active 1555–1575) and especially of

---

1   Giovanni Sale, S.J., "Architectural Simplicity and Jesuit Architecture," in *The Jesuits and the Arts 1540–1773*, ed. John W. O'Malley, S.J., and Gauvin Alexander Bailey (Philadelphia: Saint Joseph's University Press, 2005), 27–44, here 29.
2   Translation by Sale, "Architectural Simplicity," 33.
3   Translation by Sale, "Architectural Simplicity," 33.
4   Sale, "Architectural Simplicity," 38.

his successor Giovanni de Rosis (1538–1610) can clearly be seen, without, however, imposing something that could be termed a "Jesuit style."[5]

From the mid-seventeenth century onward, this task would be assigned to the mathematics professor at the Collegio Romano, and from 1662 until his death, de Gottignies would carry out these tasks. As can be expected, we find him criticizing designs on practicality, functionality, and simplicity.[6]

For the Netherlands, the result was that sixteenth-century Jesuit churches—as in other northern provinces—were built in the gothic tradition. The main Jesuit architects in this period were lay brothers Hendrik Hoeymaker (1559–1626) and later Joannes du Blocq (1583–1656).

Hoeymaker was born in Bruges on December 22, 1559, the son of a master-mason.[7] He had to abandon his philosophy studies to attend to the family business. On April 25, 1585, he joined the Society,[8] and after two years of novitiate in Tournai he took his first vows in August 1587.[9] He became a coadjutor on February 4, 1596.[10] According to the Jesuit necrology, he built the churches of Ghent, Brussels, Tournai, Valenciennes, Mons, and Ypres. Although his design for the Brussels church was in essence gothic, it was also the first design to incorporate Renaissance stylistic elements. In 1609, the annual reports called it *opus excellens at in hoc genere in Belgio primum* (an excellent work, the first of its kind in Belgium). After a pause in the building process, the church was completely redesigned by Jacob Franckaert (1583–1651) using the already present foundations.[11]

The real watershed from gothic to baroque would be the Antwerp Jesuit church designed by Aguilón, who had worked with Hoeymaker on the church of Mons.[12] In 1607, both men were called to Mons with the assignment of drawing up plans for a church. As professionals, they were also asked to choose between two possible building sites.[13] This indicates that by this time Aguilón was considered to be an experienced architect. At the same time, the Mons Jesuits wanted to build a new college. The plans were approved and work soon

---

5   Gauvin Alexander Bailey, "'Le style jésuite n'existe pas': Jesuit Corporate Culture and the Visual Arts," in *The Jesuits: Cultures, Sciences, and the Arts 1540–1773*, ed. John W. O'Malley, S.J., et al. (Toronto: University of Toronto Press, 1999), 38–89.
6   Audenaert, "Aegidius Franciscus de Gottignies," 551.
7   Kadoc ABSE 33, 2; 34/1, 4–5. Braun, *Die Belgischen Jesuitenkirchen*, 12–18 mentions Tournai as his birthplace.
8   Kadoc ABSE 34/1, 4–5; ARAA T14/034 31, 38/23.
9   Kadoc ABSE 34/1, 4–5.
10 ARAA T14/034 31, 56/90.
11 Braun, *Die Belgischen Jesuitenkirchen*, 14.
12 Braun, *Die Belgischen Jesuitenkirchen*, 14.
13 Braun, *Die Belgischen Jesuitenkirchen*, 29.

began, with the church being finished in 1617.[14] Unfortunately, however, the church was destroyed at the end of the eighteenth century after the expulsion of the Jesuits. Neither Aguilón's nor Hoeymaker's share in the drawing of the plans is clear. Joseph Braun suggests that because of Aguilón's lack of practical knowledge, the ideas and initial plans were his, while Hoeymaker converted them into practical building schemes.[15]

At about the same time, Aguilón also worked with du Blocq, at that time himself a novice, on the plans for a church in Tournai. Four plans for the church are extant, one of which is ascribed to Aguilón.[16] Aguilón's plan has one nave, with the building measuring 90' (26.42 meters) by 40' (11.60 meters). The choir ends in half a decagon. Du Blocq would draw two more plans based on Aguilón's proposals.[17]

The first Netherlandish churches in which the inspiration of Italian examples can be seen are those at Scherpenheuvel (1609–27) and Brussels (Discalced Carmelites' Church; 1607–11). In both cases, the façades, designed by Wenceslas Cobergher (1560–1634), clearly show Italian influence.[18] The façade of the Augustinians' church at Antwerp, also designed by Cobergher, can be seen as an amalgamation of traditional typologies with Italian modernity.[19]

The most magnificent Jesuit church in the Netherlands is undoubtedly the Church of Saint Charles Borromeo in Antwerp, which broke with all gothic traditions. Whereas in a gothic church attention is drawn to the light radiating in through stained-glass windows, baroque churches try to overwhelm the visitor with much pomp and grandeur.

Once the Antwerp college had moved to Engels Huys, the Jesuits began dreaming of a new church with an annexed professed house on the site of Huys van Aeken. They intended their church to represent the newly triumphant Catholicism,[20] and their impressive and richly decorated building would be finished less than fifteen years later. The church was the result of the cooperation of three remarkable men: Aguilón, Peter Huyssens (1578–1637), and Rubens. They were administratively supported by Scribani, then the provincial

---

14   Braun, *Die Belgischen Jesuitenkirchen*, 30.
15   Braun, *Die Belgischen Jesuitenkirchen*, 67–68.
16   Braun, *Die Belgischen Jesuitenkirchen*, 67–68. See http://data.bnf.fr/16240952/fr__francois_d__aguillon/ (accessed February 12, 2020).
17   For the plans of the novitiate and the church, see http://data.bnf.fr/16210838/frere_jean_du_blocq/ (accessed February 12, 2020).
18   Krista de Jonge, "Architectuur ten tijde van de Aartshertogen: Het Hof achterna," in *Bellissimi ingegni, grandissimo splendore*, vol. B-15, *Symbolae*, ed. Krista de Jonge, Annemie de Vos, and Joris Snaet (Leuven: Universitaire Pers Leuven, 2000), 11–42, here 21.
19   De Jonge, "Architectuur ten tijde van de Aartshertogen," 23.
20   Thijs, *Van Geuzenstad tot katholiek bolwerk*, 79–80.

of the Flemish province, and Jacob Tierens (Tirinus [1580–1636]), the rector of the professed house.[21]

The first real plans date back to 1613. In that same year, the Jesuit architect Huyssens was called to Antwerp.

Huyssens was born in Bruges on February 22, 1577, the son of the wealthy Jacob and Cathelijne Doudens.[22] He received training as a master-mason. On October 5, 1596, he entered the Jesuit novitiate at Tournai[23] and became coadjutor on December 8, 1613.[24] Just before ending his novitiate, he was sent to Douai as a technical advisor in the construction of the church.[25] The Antwerp church was his first major undertaking—and also his undoing.[26] For Huyssens, his collaboration with Aguilón marked a watershed. In all his subsequent projects, he adopted specific elements that he first encountered in the Antwerp church. While Aguilón was the designer of the church and needed a practical collaborator, Huyssens turned out to be much more than that, surpassing Aguilón in putting a stamp on the architectural practice of the seventeenth century.[27]

Together with Aguilón, Huyssens worked on the design of the new church, and the plans were soon sent to Rome, only to be rejected several times.[28] On

---

21 On Tirinus, see Delée, "Liste d'élèves," 13.
22 ARAA T14/034 29, the *Catalogus personarum* for 1615 mentions 1576 as date of birth, while that of 1625 (ARAA T14/034 32, 8/52) mentions 1578. Also KBR MS 1016, 306. H. de Smet, "Pieter Huyssens, architect van de Sint-Walburgakerk," in *Sint-Walburga: Een Brugse kerk vol geschiedenis*, ed. Jozef van den Heuvel (Bruges: Jong Kristen Onthaal voor Toerisme, 1982), 77–86, here 77–79.
23 ARAA T14/034 29, the *Catalogus personarum* for 1615 mentions 1598 as entry date, which is repeated in subsequent *catalogi personarum*, also KBR MS 1016, 306. De Smet, "Pieter Huyssens," 79; Bert Daelemans, S.J., "Pieter Huyssens, S.J. (1577–1637), an Underestimated Architect and Engineer," in *Innovation and Experience in the Early Baroque in the Southern Netherlands: The Case of the Jesuit Church in Antwerp*, ed. Piet Lombaerde (Turnhout: Brepols, 2008), 41–52, here 42.
24 ARAA T14/034 31, 8/52.
25 De Smet, "Pieter Huyssens," 79; Daelemans, "Pieter Huyssens, S.J.," 42.
26 The high cost led to Huyssens being temporarily suspended as an architect by Superior General Vitelleschi. Poncelet, *Histoire de la Compagnie de Jésus*, 556; Daelemans, "Pieter Huyssens, S.J.," 44.
27 For a discussion of this subject, see Daelemans, "Pieter Huyssens, S.J."
28 According to de Mayer, it was a fortuitous coincidence for Antwerp and Aguilón that Manare—the visitor of the Belgian province and one of the most influential critics of the projects—and Superior General Acquaviva died soon after one another, in 1614 and 1615 respectively. Marcel de Maeyer, *Albrecht en Isabella en de Schilderkunst: Bijdrage tot de Geschiedenis van de 17e-eeuwse Schilderkunst in de Zuidelijke Nederlanden* (Brussels: Koninklijke Vlaamse Academie voor Wetenschappen, Letteren en Schone Kunsten van België, 1955), 120.

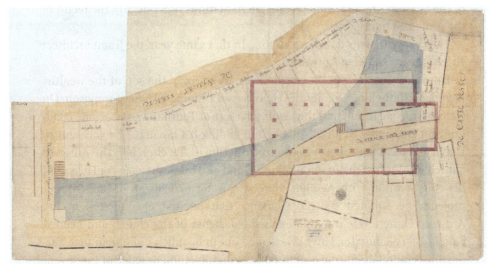

FIGURE 14.1 Project plan for the new Jesuit church in Antwerp, situated across the soon to be arched Korte Rui (a city-canal), c.1615
© FA 12#8442

these plans, probably drawn by Aguilón, the envisaged buildings remained within the confines of the existing building block.[29] Two years later, the Jesuits had acquired a considerable number of houses in the adjacent Katteveste (now Sint-Katelijnevest) and Wijngaertstraat. Add to this a number of expropriations, and the Jesuits were able to create a large square, erasing three streets in the process.[30]

For the Antwerp church, Aguilón introduced the baroque style, possibly influenced by Sebastiano Serlio's (1475–c.1554) and Hans Vredeman de Vries's (1527–c.1607) books on architecture.[31]

---

29 Lombaerde, "Façade and the Towers," 79.
30 Lombaerde, "Façade and the Towers," 80. Also ARAA T14/034, 2028, and Kadoc ABSE 108.
31 Piet Lombaerde, "The Façade and the Towers of the Jesuit Church in the Urban Landscape of Antwerp during the Seventeenth Century," in *Innovation and Experience in the Early Baroque in the Southern Netherlands: The Case of the Jesuit Church in Antwerp*, ed. Piet Lombaerde (Turnhout: Brepols, 2008), 77–96. A Dutch as well as a French translation of Serlio's *Architectura* was published by various Antwerp printers between 1546 and 1553. Meskens, *Wiskunde tussen Renaissance en Barok*, 221–23.

A copy of Vredeman's *Architectura oder Bauung der Antiquen auss dem Vitruvius* was available in the libraries of both the colleges of Antwerp and Mechelen. These copies are preserved in the Royal Library Albert I (KBR VB 5.321 C 3 RP and VB 5.335 C 1 RP respectively).

The ground plan, and hence the concept of the church, was drawn up by Aguilón.[32] The first of his plans was inspired by the plan of the Gesù (Chiesa del Santissimo Nome di Gesù all'Argentina [Church of the most holy name of Jesus at the Argentina]) in Rome, the mother church of the Society of Jesus.[33] It was not the first time that this church had inspired Jesuit architects in the Low Countries, as in 1583 the church of Douai had also been erected along Gesù lines.[34] The church was built from 1583 to 1591, when Aguilón studied and thereafter taught at its college.[35] The first plan for the Antwerp church envisaged a central nave with four side chapels on each side, communicating with each other through small corridors.[36]

Other plans show a centered plan (*Zentralbau*) with a large circular open space and three radiating chapels on each side.[37] Yet another plan shows a ground plan based on a Greek cross. In this plan, a cupola is inscribed in a square with three naves as a side. The cupola is carried by eight pillars. At the corners of the outermost naves, we find smaller cupolas. Several authors recognize the cooperation with Rubens in this plan.[38] None of these plans would be realized.

Sebastiano Serlio, *Des antiquites, le troisiesme livre translaté d'Italien en Franchois* (Antwerp: Pieter Coecke van Aelst, 1550); Serlio, *Den Tweeden Boeck van Architecturen S. Serlii, Tracterende van Perspectyven, Dat Is, Het Insien Duer Toercorten*, ed. Peter van Aelst (Antwerp: Mayken Verhulst, wwe Peter Coecke van Aelst, 1553); Serlio, *Den Eersten Boeck van Architecturen [...] Tracterende van Geometrye* (Antwerp: Mayken Verhulst, 1553); Serlio, *Reglen van Metselrijen, Op de Vijve Manieren van Edificien, Te Wetene, Thuscana, Dorica, Jonica, Corintha, Ende Composita* (Antwerp: Peter Coecke van Aelst, 1549); Serlio, *Den Vijfsten Boeck van Architecturen [...] Inden Welcken van Diversche Formen van Templen Getracteert Wordt*, ed. Peter van Aelst (Antwerp: Mayken Verhulst, wwe Peter Coecke van Aelst, 1553).

32 *Historia domus professa* (1625), cited by Braun, *Die Belgischen Jesuitenkirchen*, 168.
33 For a detailed analysis of the early designs, see Braun, *Die Belgischen Jesuitenkirchen*, 159–63; Simon Brigode, "Les projets de construction de l'église des jésuites à Anvers," *Bulletin de l'Institut historique belge de Rome* 14 (1934): 157–74; Joris Snaet, "De bouwprojecten voor de Antwerpse jezuïetenkerk," in de Jonge, de Vos, and Snaet, *Bellissimi ingegni, grandissimo splendore*, 15:43–66.
34 Braun, *Die Belgischen Jesuitenkirchen*, 116–19; Brigode, "Les projets de construction," 163; Snaet, "De bouwprojecten," 52–53.
35 Ziggelaar, *François de Aguilón, S.J.*, 32–34.
36 Brigode, "Les projets de construction," 169; Snaet, "De bouwprojecten."
37 Braun, *Die Belgischen Jesuitenkirchen*, 159–62; Brigode, "Les projets de construction," 163–64; Snaet, "De bouwprojecten," 54–56.
38 Rubens had made sketches of the similar *Santa Maria de Carignano* in Genoa before 1609. Nicolaes Ryckemans (?–after 1636; *fl.* 1616–36), an engraver who worked in Rubens's studio, made engravings on the basis of these sketches that were published as a book in 1622. Peter Paul Rubens and Nicolaes Ryckemans, *Palazzi di Genova* (n.p.: n.p., n.d.); Brigode,

The plans sent to Rome in 1615 finally met with approval, and construction started in the April of the same year.[39] Aguilón would not live to see the finished church; he died on March 20, 1617. Huyssens and Tirinus supervised the building of the church in the years that followed. Rubens helped with the ornamental parts of the church, designing the apse and the ceiling paintings, which were produced by his apprentices, including Van Dijck.[40]

While the ground plan is traditional, the same can most certainly not be said for the façade,[41] which has a magnificent three-tiered screen with a projecting middle part (fig. 14.3). It symbolizes the idea that it is only by becoming a faithful member of the Catholic community that one can—ultimately— enter the Kingdom of Heaven. Without the cross, the façade can be inscribed in a square.[42] The free-standing Ionic columns initiate a vertical movement toward the volutes, which rapidly bring the eye to the fronton, with its gently sloping sides rising toward the cross. This vertical movement is reinforced by the two towers at the sides of the building.

"Les projets de construction," 166; Snaet, "De bouwprojecten," 61–62; August Ziggelaar, "Peter Paul Rubens and François de Aguilón," in Lombaerde, *Innovation and Experience in the Early Baroque*, 31–40.

39 Brigode, "Les projets de construction," 169.

40 Rubens had to be paid ten thousand guilders, which was amortized by a yearly interest payment of 6.25 percent.

Brigode, "Les projets de construction," 171–74; Delée, "Liste d'élèves," 13; John R. Martin, *The Ceiling Paintings for the Jesuit Church in Antwerp* (Brussels: Arcade Press, 1968), esp. 213–19 (English translation of the contract between Tirinus and Rubens) and 220–21 (extract from the church's account book).

On Rubens and architecture, see Rubens and Ryckemans, *Palazzi di Genova*; Koen A. Ottenheym, "De correspondentie tussen Rubens en Huygens over architectuur," *Bulletin KNOB* 96 (1997): 1–11; Konrad Ottenheym, "Peter Paul Rubens's 'Palazzi di Genova' and Its Influence on Architecture in the Netherlands," in *The Reception of P.P. Rubens's Palazzi di Genova during the 17th Century in Europe: Questions and Problems*, ed. Piet Lombaerde (Turnhout: Brepols, 2002), 81–98; Ottenheym and Krista de Jonge, "Civic Prestige: Building the City 1580–1700," in *Unity and Discontinuity: Architectural Relationships between the Southern and Northern Low Countries (1530–1700)*, ed. Krista de Jonge and Konrad Ottenheym (Turnhout: Brepols, 2007), 209–50; Ziggelaar, "Peter Paul Rubens and François de Aguilón," 31–40.

41 For a detailed study of the façade, see Barbara Haeger, "The Façade of the Jesuit Church in Antwerp: Representing the Church Militant and Triumphant," in Lombaerde, *Innovation and Experience in the Early Baroque*, 97–124, on which this description is based.

42 Due to the elevation of the square for sewer works, among other things, four of the seven stairs are now underground, and the square appearance of the façade is broken.

THE JESUIT ARCHITECTS 229

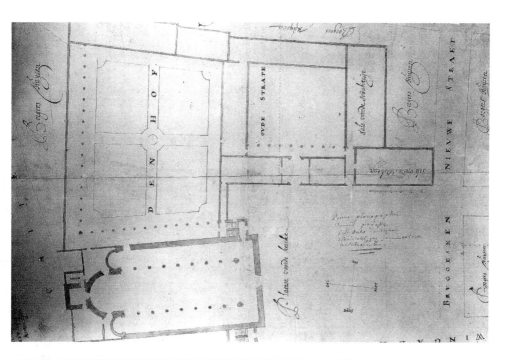

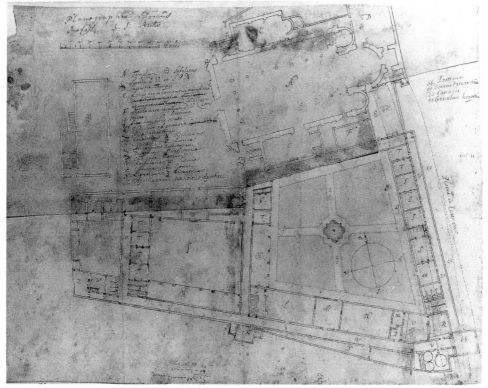

FIGURE 14.2 Designs for the square in front of Saint Charles Borromeo Church
© KIK-IRPA 57430 AND 57577

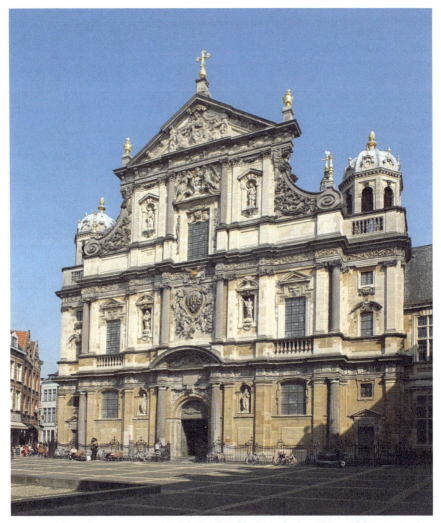

FIGURE 14.3   The Church of Saint Charles Borromeo today
©AD MESKENS

The interior has a three-aisled basilica plan with galleries above the side aisles, similar to the ones used for the earlier gothic Jesuit churches.[43] It has a coffered barrel vault and an apse decorated with antique ornaments. The columns

---

43 Joris Snaet and Krista de Jonge, "The Architecture of the Jesuits in the Southern Low Countries, a State of the Art," in *La arquitectura jesuitica: Actas del simposio internacional*, ed. Isabel Alvaro-Zamora, Ibanez Fernandez, and J. Criado Maínar (Zaragoza: IFC, 2012), 239–76, here 243, 274.

THE JESUIT ARCHITECTS

FIGURE 14.4   The interior of the Church of Saint Charles Borromeo
©AD MESKENS

supporting the round-headed arches are in white marble imported from Italy.[44] To be able to arch a large and spacious church, the ceiling and roofs were made of wood, giving it a lighter but also more vulnerable structure.

The interior is a light-filled and sumptuously adorned space, with the ceiling paintings of the side galleries originally boasting numerous paintings by Rubens (unfortunately lost in the fire of 1718).[45] The altar has an ingenious system of pulleys by which the four paintings can be shown alternately (fig. 14.4).[46]

The church opened in 1621[47] and was consecrated by Bishop Johannes Malderus (1563–1633).[48] The two side chapels were only added in 1622.[49]

In 1651, after the Peace of Westphalia, a visiting Protestant described the church in the following way: "She can hardly be described in words. Inside all is

---

44   Snaet and de Jonge, "Architecture of the Jesuits," 243.
45   Ria Fabri, "Light and Measurement: A Theoretical Approach of the Interior of the Jesuit Church in Antwerp," in Lombaerde, *Innovation and Experience in the Early Baroque*, 125–40; Martin, *Ceiling Paintings*. The design of some of these paintings is known to us through oil sketches by Rubens and watercolors by Jacob de Wit (1695–1754). On this subject, see Knaap, "Meditation, Ministry, and Visual Rhetoric."
46   Mannaerts, *Sint-Carolus Borromeus*, 68–73, esp. 72–73.
47   ARAA T14/034 2/1, fol. 45ᵛ; 3, fol. 3ʳ.
48   ARAA T14/034 3, fol. 15ʳ.
49   Brigode, "Les projets de construction," 171.

of a startling splendor and parade. These Jesuits truly have found their heaven on earth."[50]

Tirinus had to appeal to the city council for help, either financial or material, on a number of occasions. In 1621, he asked for and obtained financial aid for overarching the *ruien* (city canals).[51] In 1622, a couple of houses to the south side of the church needed renovation. Tirinus obtained permission to renovate them on the condition that it would not burden the city treasury and that a thoroughfare had to be maintained so the square could be reached with horse and cart.[52]

The building costs for the church ran to more than five hundred thousand guilders, and the Jesuits had only been able to raise ninety-five thousand from legacies and donations. The archdukes contributed twelve thousand guilders[53] and Philip IV ten thousand guilders,[54] and there were also other sizeable donations from wealthy Antwerp families.[55] In 1616, the city council granted a yearly subsidy of four thousand guilders for a period of five years.[56] However, to build their church, and against the wishes of the superior general, the Jesuits also had to borrow a huge amount of money: nearly three-quarters of the building costs. The total amount exceeded five hundred thousand guilders in debts and arrears by 1625 and nearly bankrupted the Flemish province.[57]

Aguilón's fellow mathematician San Vicente also seems to have been active as an architect. On April 6, 1630, for instance, Vitelleschi wrote to the Bohemian provincial Christopher Grenzino that he preferred San Vicente's plans for the Klementinum, the college at Prague.[58] On April 8, he made his decision

---

50   Jan-Albert Goris, *Lof van Antwerpen: Hoe reizigers Antwerpen zagen, van de XVe tot de XXe eeuw* (Brussels: Standaard, 1940), 79.

51   FA Pk716, fols. 197^v–198^r; request handled by the council on November 5, 1621. The fathers were given twelve thousand guilders as a subsidy.

52   SAA Pk717, fols. 87^r–87^v and MPM R24.3 fols. 5^v–6^r.

53   ARAA T14/034 2027.

54   Brouwers, *Carolus Scribani 1561–1629*, 210; Snaet, "De bouwprojecten," 45.

55   ARAA T14/034 1938.

56   MPM R24.9, fol. 5^r and FA KK591. The money for the subsidy came from a tax on exported beers. In 1619, the Jesuits asked for early payment of the monies. Marinus, "De financiering van de contrareformatie," 242. Also, ARAA T14/034 2026.

57   Marie Juliette Marinus, *De Contrareformatie te Antwerpen (1585–1676): Kerkelijk leven in een grootstad* (Brussels: Paleis der Academiën, 1995), 158–62. Also, ARAA T14/034 1934; 2002–2009; 2026 and 2027; T14/015.01 175–83.

58   Patricia Radelet-de Grave, "Matematica, architettura e meccanica nella scuola di François d'Aguillon e di Grégoire de Saint-Vincent," in *Matematica, arte e tecnica nella storia, in memoria di Tullio Viola*, ed. Livia Giacardi and Clara Silvia Roero (Turin: Ken Williams Books, 2006), 275–92, here 288. In this chapter, the relevant passages from the letters of Vitelleschi are cited in full, after the transcription of Henri Bosmans, S.J. Kadoc ABML

 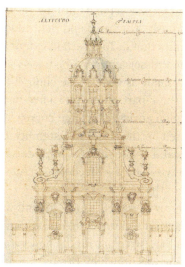

FIGURE 14.5 Left: Saint Michael's Church, Leuven
© AD MESKENS
Right: Hesius's design for the church with a cupola
© KIK-IRPA KM010398

known to San Vicente.[59] Grienberger, San Vicente's nemesis with regard to the ductus method, had also drawn plans for the edifice and tried to have them approved.[60] The stalemate ended when a Czech architect was ultimately chosen instead.

The college at Leuven would not have a church until the second half of the seventeenth century.[61] Its ground plan was drawn by Hesius, one of San Vicente's students in Leuven. He also oversaw the building process.[62]

Hesius was not an architect but had studied architecture in his spare time and acquired extensive knowledge of the subject. His main work would be the Jesuit church at Leuven (fig. 14.5). While the first stone was laid in 1650, the

---

Cahiers Bosmans 125, fol. 243, referring to ARSI Bohem: Epist. Gen., 1623–37. I would like to thank Patricia Radelet-de Grave for making me aware of San Vicente's role as an architect.

59 Radelet-de Grave, "Matematica, architettura e meccanica," 288; Kadoc ABML Cahiers Bosmans 125, fol. 244, referring to ARSI Bohem: Epist. Gen., 1623–37.

60 Radelet-de Grave, "Matematica, architettura e meccanica," 288; Kadoc ABML Cahiers Bosmans 125, fol. 245, referring to ARSI Bohem: Epist. Gen., 1623–37. Grienberger had also drawn (or approved?) plans for the *domus professa* at Antwerp. See http://gallica.bnf.fr/ark:/12148/btv1b84486741.r=grienberger?rk=85837; 2 (accessed February 12, 2020).

61 Braun, *Die Belgischen Jesuitenkirchen*, 141.

62 Braun, *Die Belgischen Jesuitenkirchen*, 142.

church would not be consecrated until 1671, although it was in use by 1666. The façade was vigorously structured and richly decorated and was modeled after that of the Namur church. To a certain extent, the church in Leuven follows the plans for the Brussels church. The nave was elongated with one bay, and the transept was made as wide as the choir.[63] The ceiling of the transept was changed from a three-part ceiling into a five-part ceiling.

The original plan envisaged a large cupola,[64] which was not built because the builders feared for the structure's stability. In 1660, when the work on the cupola began, stability problems were indeed encountered. Repairs on the cross vault, supervised by the Mechelen architect Lucas Fayd'herbe (1617–97), would last six months.

In Ghent, Hesius designed a new gymnasium,[65] and he also redesigned Hoeymaker's plans for a church into one with a baroque design. John Gilissen (1912–98) claims that two plans for the new gymnasium of Ghent were sent to Rome in 1658. One was drawn by Hesius, the other by his former teacher San Vicente, who resided at Ghent. Superior General Oliva approved San Vicente's plans and wrote to Hesius urging him to obey.[66] After building had already started, Hesius became rector of the college. In 1662, Charles de Noyelle (1615–86), acting as substitute for the superior general, approved Hesius's plans.[67]

The plan for the college of Kortrijk, sometimes attributed to Hesius, was actually made by Hesius on the basis of a plan drawn by Huyssens, to which he added comments.[68] The plan for the church of Cassel is in his hand (fig. 14.6). The college and gymnasium had been built in 1634 but lacked a church. In 1666, Hesius sent his plans to the vice-provincial, before sending them to Rome. The design was never realized.[69]

As rector of the college at Mechelen, Hesius designed the new school.[70] He also made drawings for a new high altar at St. Rumbold's Cathedral.[71] When

---

63   Braun, *Die Belgischen Jesuitenkirchen*, 143.
64   Braun, *Die Belgischen Jesuitenkirchen*, 143.
65   ARAA T14/034 3808.
66   Kadoc ABML Cahiers Bosmans 125, fols. 3 and 130, 253, 280ᵛ. Radelet-de Grave, "Matematica, architettura e meccanica," 289.
67   John Gilissen, "Le Père Guillaume Hesius, architecte du XVIIᵉ siècle," *Annales de La Société Royale d'Archéologie de Bruxelles: Mémoires, rapports et documents* 42 (1938): 216–55, here 243–44.
68   ARAA T14/034 2486. Gilissen, "Le Père Guillaume Hesius," 246.
69   ARAA T14/034 2453–55. Gilissen, "Le Père Guillaume Hesius," 247–48.
70   ARAA T14/034 3808.
71   ARAA T14/034 3808. Gilissen, "Le Père Guillaume Hesius," 248; Helena Bussers, "De baroksculptuur en het barok kerkmeubilair in de Zuidelijke Nederlanden," *Openbaar Kunstbezit* 20 (1982): 123–61, here 128.

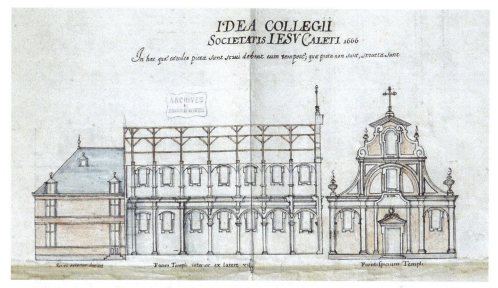

FIGURE 14.6   Design for the church at Cassel by Willem Hesius
© ARAA T14/034 2453

Fayd'herbe and the canons of St. Rumbold's had a dispute about the rood screen, the city council appointed Hesius as an intermediary (fig. 14.7).[72]

In Antwerp, the Jesuits had wanted to build a chapel in their college, but since they did not obtain permission from the cathedral chapter, they had to abandon their plans. Nonetheless, by 1657 the Jesuits were again dreaming of building a church for the college. They had obtained a grant of sixty thousand guilders from the mother of the Losson brothers, all three members of the Society. The college would dedicate the church to Francis Xavier (1506–52), who had been canonized in 1622. Discussions with other orders ran high until, with the consent of the Losson brothers, it was decided that the monies would be used for the construction of a church at Mechelen.[73] The problem of not having a room that was large enough for the spiritual exercises thus remained. In 1677, the baron of Laarne made a large donation with the express purpose of building a church. Hesius, at that time rector of the professed house, drew the plans for this new edifice. The new church was to have a portal in Prinsstraat west of Engels Huys. To the plans, he added a project for a new gymnasium, assuming that the whole block of houses west of the church up to Venusstraat would be acquired. In these plans, the entrance of the college was to be moved

72   Bussers, "De baroksculptuur," 152.
73   Moehlig, "Het Jezuïetencollege te Antwerpen," 43–44.

FIGURE 14.7   The choir at St. Rumbold's Cathedral, Mechelen, designed by Willem Hesius; the plans were executed by Lucas Faydherbe. The gilded arcs and doors were carved by Artus Quellin the Younger
© AD MESKENS

to Venusstraat. Behind the gate was a hall, parallel to the street, which gave access to a courtyard. On the ground floor, he envisaged six classrooms around the yard, and on the first floor he designed a theater and rooms for the priests.[74] However, the superior general, and some of the Jesuits in Antwerp, concluded that they were living in dire times and that the gift should be put to use for another purpose (fig. 14.8).[75]

Della Faille's *Tratado de arquitectura* (Treatise on architecture) of 1636[76] is one of the few treatises about architecture to have been written in seventeenth-century Spain. It is a thirty-six-page booklet in 8°, which, although entitled *Architecture*, largely deals with mathematics, and in particular with projections and intersections of solids. Della Faille describes three kinds of

---

74   ARAA T14/034 1963.
75   Moehlig, "Het Jezuïetencollege te Antwerpen," 44.
76   BPM MS 3729.

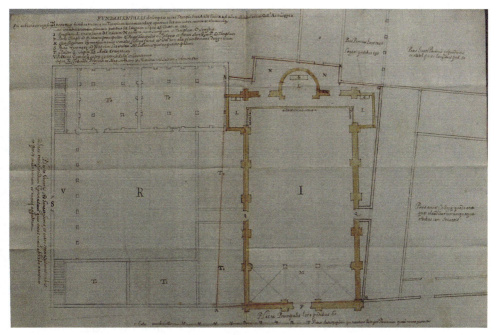

FIGURE 14.8   The plan for a church for the College of Antwerp at Prinsstraat by Willem Hesius
© ARAA T14/034 1963

perspective projections, which are used by architects. *Hignographia*, which shows the building in its plane, *ortographia*, which is the view of the building by rays projecting perpendicularly onto a projection plane, and *superficie*, which shows the building by light rays parallel to one side. The orthographic projection was dealt with by Aguilón in *Opticorum* as a projection in which the eye of the observer is virtually at an infinite distance.[77] Della Faille goes on to discuss some well-known theorems about the orthographic projection of points and lines, about triangles, and about the projection of a circle.

The projections of circles are generally ellipses, and della Faille shows how to obtain the axis of the ellipse and also how to find the projections of points on the circumference of the circle.[78] He repeats these theorems for the projection of ellipses. Della Faille proceeds with stereotomy, or the cutting of stones

---

77   Martin Kemp, *The Science of Art* (New Haven: Yale University Press, 1990), 87, 103 referring to Commandini's *Claudii Ptolomaei liber de analemmate* (Rome: Paolo Manuzio, 1593); and Ziggelaar, *François de Aguilón, S.J.* Aguilón explicitly refers to Commandini as a source of inspiration for this part of his book.

78   Della Faille's arguments are based on what we now call invariants under projection.

(i.e., the intersection of the different shapes that are used in architecture). He identifies the plane, the pyramid, the prism, the sphere, the circular and elliptical cylinder, and the circular and the elliptical cone as possible shapes. Because the surfaces of the pyramid and the prism are planes, he no longer considers them, since they can be reduced to finding the intersection with other shapes of these planes. The seven other shapes result in twelve different combinations of intersections that are found in architecture. For example, a plane will cut a circular or elliptical cylinder in an ellipse. Unfortunately, he does not go so far as to give a description of the construction of these intersections.

2   *Descensus ad inferos*

Although as architects the Jesuits mainly designed and built churches for use by their own brethren, they were not immune from the broader *Zeitgeist*. Ciermans, della Faille, Derkennis, Durandus, and Tacquet, for instance, were all involved in some way or another with the theory or the practice of fortification. A manual on the use of mortars shows that a part of the mathematics course was devoted to ballistics (fig. 14.9).[79] The clearest examples of direct involvement are of course Ciermans and, to a lesser extent, della Faille, who both fought in the Portuguese war of independence, albeit on opposite sides.

The war in the Netherlands consisted primarily of siege warfare, and armies would not necessarily confront each other in the open field. From the mid-sixteenth century onward, Italian-style bastionized fortification began to be introduced to the Low Countries. A defending army could retreat into a fortified town, and the attackers had to lay an often long and costly siege. Towns that could be supplied by river or sea were very hard to subdue for a belligerent army, as the four-year siege of Ostend had shown.[80] By the mid-seventeenth century, Netherlandish fortifications were without peer, their ditches and ramparts giving them an incredible passive resistance, while the outworks, with ravelins, demi-lunes, and horn works, gave a nearly unsurpassable defense in depth.[81] Only when there was sustained frost and the ditches could be crossed would their Achilles' heel be laid bare.

---

79   ARAA T14/034 1880.
80   Charles van den Heuvel, *"Papiere Bolwercken": De introductie van de Italiaanse stede- en vestingbouw in de Nederlanden (1540–1609) en het gebruik van tekeningen* (Alphen aan den Rijn: Canaletto, 1991); Frans Westra, *Nederlandse ingenieurs en de fortificatiewerken in het eerste tijdperk van de Tachtigjarige Oorlog, 1573–1604* (Alphen aan den Rijn: Uitg. Vis, 1992); on the siege of Ostend, see Thomas, *De Val van het Nieuwe Troje*.
81   Christopher Duffy, *Siege Warfare: The Fortress in the Early Modern World 1494–1660* (London: Routledge & Kegan Paul, 1996), 58–105.

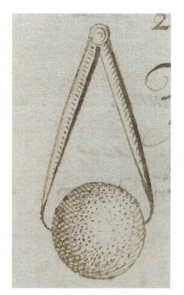 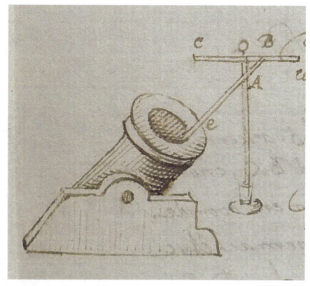

FIGURE 14.9　The use of a caliper (left) and of a *talstock* (tally stick) to aim a mortar (right)
© ARAA T14/034 1880

The renewed war in the Netherlands after the termination of the truce (1621) saw some spectacular victories for both sides. Spinola succeeded in reducing the fortress of Breda in 1624, while stadtholder Frederik Hendrik (1584–1647), nicknamed "Stedendwinger" (Conqueror of towns), reduced 's-Hertogenbosch. Although fortifications called for extensive engineering knowledge, the only engineering school in Europe was to be found in the republic, though some unsuccessful attempts to formalize military education were also made at the court of the archdukes. Small wonder, then, that the nobility sought ways to improve their theoretical skills. By the second quarter of the seventeenth century, the only institution that could accommodate their needs was the Jesuit school of mathematics.[82] In *Disciplinae mathematicae*, Ciermans lamented that his students were only interested in the art of warfare. Nevertheless, three of the topics in his book were military in nature (fortifications, weaponry and war machines, and the properties of gunpowder). Unsurprisingly, Ciermans gives accounts of the reductions of Breda and of his hometown 's-Hertogenbosch. He shows himself knowledgeable of recent developments, and while there is no direct evidence, it seems likely that he had contacts with engineers, if only because they were very active in the Low Countries. Nevertheless, Ciermans's accounts of fortifications remain very superficial.

82　Geert Vanpaemel, "Jesuit Mathematicians, Military Architecture, and the Transmission of Technical Knowledge," in Faesen and Kenis, *Jesuits of the Low Countries*, 109–28.

FIGURE 14.10 Figure in Ciermans, *Disciplinae mathematicae*. April, first week
© EHC G 4827

In 1641, Ciermans left for Portugal, hoping to embark for the Chinese missions in Lisbon. From 1642, he taught at the Collegio de Santo Antão, occupying the chair of the Aula da Esfera, the only one in Portugal where military techniques including fortifications were taught. The Portuguese king showed a particular interest in Dutch fortifications. This does not come as a surprise: having re-found its independence, Portugal was a small nation fighting its neighbor Spain, a situation more or less comparable with the one the Dutch were facing in the Low Countries. Having heard about Ciermans's engineering talent, the crown assigned him, together with the then chair of the Aula da Esfera, Simão Fallon (1604–42), to supervise the construction or rebuilding of the kingdom's fortifications on the border with Spain.[83] The list of fortifications on which Ciermans worked is long but includes the cities of Olivença, Élvas, and Badajoz.[84] In Élvas, he seems to have designed an underground duct for

---

83 De Lucca, *Jesuits and Fortifications*, 128–30.
84 De Lucca, *Jesuits and Fortifications*, 128–30. He participated in reinforcing the fortresses of Praça-Forte (Olivença), Santa Luzia (Élvas), Sant'Iago (Sesimbra), Castel Muralhas (Estremoz), Castelo de Vide, Castel de Campo Maior (Portalegre), Castelo de Alfaiates (Sabugal), Castelo de Amieira, and the fortifications of Vila Viçosa and Vilanueva del Fréno (Badajoz).

the aqueduct. His plan was executed in 1650 by the French engineer Nicolas de Langres (?–1665), and Ciermans's name was all but forgotten.[85]

A French engineer, Charles Lassart, describes Ciermans as if the latter were commander-in-chief of the Portuguese forces. Lassart had a very negative opinion of Ciermans, most likely because Ciermans was a threat to his position as *engenheiro-mor* (chief engineer). It is remarkable that Lassart put Ciermans in Arras during the siege by the French of 1640. Lassart also accuses Ciermans of being a double agent for the Spanish king.[86]

On the Spanish side, we find della Faille, who, from 1639 onward, taught a class of twenty-one royal pages about fortifications as a consequence of his appointment to the Consejo de Indias.[87] Sometimes, della Faille would go with his "student" out into the country in their carriage and perform surveying exercises or make drawings of fortifications.[88] However, with the exception of two or three students, they do not seem to have taken any interest in his lessons. Della Faille complained that it prevented him from solving the navigational problems he was presented by the Consejo de Indias.[89]

As mentioned previously, in the summer of 1641 della Faille was summoned by Philip IV to be a military advisor to the duke of Alba, the commander of an area of Ciudad Rodrigo.[90] He asked Van Langren several times to send manuals with practical information about the use and fabrication of pieces of artillery and about the digging of trenches.[91] Together with the duke of Alba, della Faille inspected the fortifications along the Spanish–Portuguese border.[92] It is remarkable that he kept asking his correspondent to send recent maps of Portugal and Catalonia, suggesting that he could not obtain reliable Spanish maps of these regions. In June, he went to the sector of Badajoz in the service of the count of Santisteban, one of his former students.[93] He was highly critical

---

85 Paar, "Jan Ciermans," 211.
86 "One thing I find dangerous and belittling about the Portuguese, […] is that they would entrust their entire warfare, from locations to campaigns, to a Jesuit born in Den Bosch […], ignorant and without experience, nor hardened in battle, save the siege of Arras, which he helped defend against us." Cited in Paar, "Jan Ciermans," 205–6. It is very unlikely that Ciermans participated in the defence of Arras (1640).
87 Van de Vyver, "Lettres de Jean Charles della Faille, S.J.," letters 33, 35, 36.
88 AFL 28.15.20, letter to his sister Cornelia; 28.16. Vanderspeeten, "Le R.P. Jean-Charles della Faille," 137; Van de Vyver, "L'école de mathématiques des jésuites," 266.
89 Van de Vyver, "Lettres de Jean Charles della Faille, S.J.," letters 35, 36.
90 Van de Vyver, "Lettres de Jean Charles della Faille, S.J.," letter 44, also AFL 28.9.
91 Van de Vyver, "Lettres de Jean Charles della Faille, S.J.," letters 44, 45, 46.
92 Van de Vyver, "Lettres de Jean Charles della Faille, S.J.," letter 49.
93 AFL 28.15.33. Van de Vyver, "Lettres de Jean Charles della Faille, S.J.," letter 54, It is not clear whether the count was a student at the college or one of his private students.

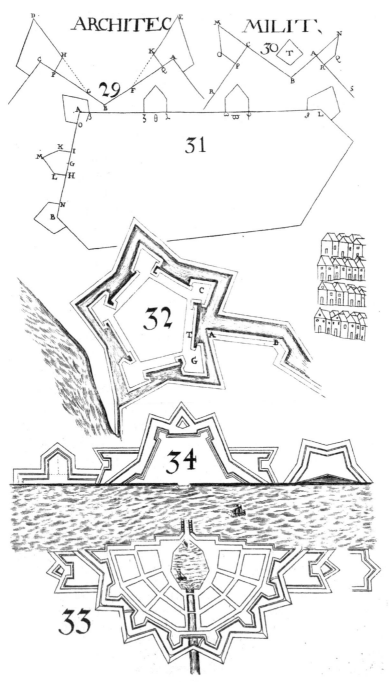

FIGURE 14.11 A page from Tacquet, *Opera mathematica*
© EHC G 4790

of the officers serving in the Spanish army, writing that they "have studied and know nothing of fortifications." Nevertheless, they were adamant about leading every operation themselves and had work done that was costly and useless, according to della Faille. It is clear he had different ideas about strategy. Moreover, he complained, he had not been paid for over six months and had been at the mercy of officers inviting him into their household. He felt utterly useless as a military advisor, not for lack of knowledge but because he was not listened to.[94] He was discharged because he could no longer be paid and was forced to return to Madrid in early 1644.[95]

On a more theoretical level, we find Tacquet's *Opera mathematica*, in which one book is devoted to fortifications.[96] Tacquet wrote the treatise on military architecture over the course of many years, with his first texts probably dating back to 1644 when he was assigned to the mathematics chair. The book is a complete instruction manual and is primarily based on the work of the Dutch engineers Adam Freytach (1608–50), Nicolaus Goldman (1611–65), and Matthias Dögen (1605–72). The book is mathematical in nature, often referring to the companion book *Practical Geometry*. It does not contain any discussion of existing fortified towns.

---

94 Van de Vyver, "Lettres de Jean Charles della Faille, S.J.," letter 55.
95 Van de Vyver, "Lettres de Jean Charles della Faille, S.J.," letter 56.
96 Vanpaemel, "Jesuit Mathematicians," 121–25.

CHAPTER 15

The Influence of the School of Mathematics

The previous pages have described what this book has called the Flemish Jesuit school of mathematics. Whether the word "school" is appropriate depends on one's viewpoint. Angelo de Bruycker, for instance, has taken the view that we can only speak of a mathematics course,[1] as he argues that the mathematics course was always organized in conjunction with the philosophy course and was optional. Moreover, he argues, there was no firm institutional framework and therefore no fixed educational institution, as the course was transferred from Antwerp to Leuven and back again. However, this is a highly materialistic and narrow view of the term "school." If we view the term as a school of thought or intellectual tradition, then it is more than appropriate. What we have here is a well-defined group of people who shared a common opinion about mathematics. San Vicente was undoubtedly the progenitor of this school. He was able, from the time his course was organized, to gather around him a group of very talented mathematicians, whom he influenced with his ideas on infinitesimal calculus, statics, and Aguilón's optics. The demise of the school came when it became clear that San Vicente's methods were not sufficiently general, finding their last application in de Jonghe's book on the quadrature of generalized parabolae and hyperbolae. The end of the school must be placed at the time of Tacquet's death, whose books gave it its final bright glare.

San Vicente's students were very productive, and their manuscripts are dispersed over libraries throughout Europe. The mathematics and mathematical physics that San Vicente and his students produced can only be described as outstanding. Their tale shows that the history of mathematics is not written by a few giants taking enormous leaps forward, but by numerous men advancing step by step in uncharted territory, not even at a continuous pace. The small advances ultimately allow mathematicians to cross a watershed. The Flemish Jesuits, for almost half a century, were walking along such watersheds, in

---

1 See Angelo De Bruycker, "De 'wiskundeschool' van de Vlaamse Jezuïeten in de eerste helft van de zeventiende eeuw: Een herpositionering," *Handelingen van de Koninklijke Zuid-Nederlandse Maatschappij voor Taal- en Letterkunde en Geschiedenis* 58 (2004): 201–20; De Bruycker, "'En todo amar y servir a su divina majestad': Wiskunde in de opleiding en het onderwijs van de Vlaamse Jezuïeten; Een cultuurhistorische benadering," *Scientiarum historia* 32 (2006): 37–58; De Bruycker, "'To the Adornment and Honour of the City': The Mathematics Course of the Flemish Jesuits in the Seventeenth Century," *BSHM Bulletin Journal of the British Society for the History of Mathematics* 24 (2009): 135–46.

calculus and in geometry, as well as in statics, sometimes unwittingly crossing them.

From the study of the manuscripts dating from 1616 to 1625, it is clear that San Vicente was a highly creative mathematician who discovered infinite sequences, transformation theorems for conic sections, the logarithmic properties of the hyperbola, the exhaustion theorems, and could summate geometric series. The first integration method, the ductus figures, and the study of the ungula are the main results of his research. At the end of this period, he thought he would be able to use his methods to solve the quadrature of the circle.

San Vicente sent his principal results to Rome to convince Grienberger of the feasibility of the quadrature of the circle. In the manuscripts dealt with here, there is hardly a trace of the quadrature of the circle. This is the same impression Grienberger had when writing to Vitelleschi: "To confer about the quadrature of the circle and many other beautiful and difficult discoveries he has come to Rome at your request and as upon his arrival he began working with me. But because nothing was in an orderly fashion I could not delve deep into the content." A little further, we read:

> But because neither in the beginning, nor further along, I found a trace of the quadrature, which I expected impatiently, I read on sloppily, and in the end, seeing there were still many difficulties, I have interrupted my work, not to never return to it, but to be able sometime to read it with an open mind. Nevertheless, as I see it now, this will not end right.[2]

Grienberger's judgment was prescient with regard to the quadrature of the circle. Later on, he would write that his negative judgment was partly due to the shortness of the text and also that the quadrature of the circle might have been possible after all. For San Vicente, the decision changed his fate forever. If Grienberger had agreed to let him publish his results up to then, without any quadrature of the circle, he would have been hailed as one of the world's greatest mathematicians. Yet now San Vicente would have to clarify his quadrature of the circle, structure his work more logically, and perhaps explore new ways of obtaining it. He began this work in Prague, helped by Moretus, until he suffered a stroke.

San Vicente has not been dealt with correctly in the history of mathematics. One must ask oneself why. Many reasons can be put forward. His reputation was tarnished by the circle quadrature and more particularly by his

---

2 Van Looy, "Chronologie en analyse," 194, referring to a letter dated October 11, 1627, in ARSI 534, fols. 52–53, see Bosmans, "Lettre inédite de Christophe Grienberger."

unwillingness to admit his failure. As we have seen, his book holds many more exciting new mathematical discoveries. However, unfortunately, his book was out of date by the time it was published. By then, Cavalieri, using a completely different concept, had launched infinitesimal calculus, and Descartes had shown the way to analytical geometry.

On the other hand, a subject that has been all but completely ignored is the influence of these Flemish Jesuits on the history of mechanics. The Jesuit community at large was instrumental in the shaping of the science of statics, and the Flemish Jesuits played a role in it. San Vicente's manuscript for a book on mechanics has been lost, his only tangible contribution being the theses by Ciermans and Van Aelst.[3] Moretus's correspondence, which we know contains letters on this subject, remains unedited.

San Vicente's students were sent to the four corners of Europe. Hardly any of them published a book, although from their manuscripts we can see that they were at the spearhead of mathematical developments in Europe. Della Faille and Moretus, and, from San Vicente's second Flemish period, Tacquet and de Gottignies, merit a lot more attention than they have received up to now. It has been proposed by Bosmans that reading Tacquet's books led Pascal to pay more attention to the way indivisibles were used. In the space of four years between writing *Warnings* in *Letters to Carcavy* and *Lettres de A. Dettonville*, reading Tacquet led him to express his conceptions more meticulously.[4] Huygens recommended San Vicente's book to Leibniz, who acknowledged that he had been inspired by it.[5]

The Antwerp–Leuven school of mathematics had the potential to rival its Roman counterpart. Moreover, its students were as well versed in physics as in mathematics. Boelmans used concepts in optics that would become known as Fermat's principle; Ciermans, Van Aelst, and Moretus tried to explain mechanical phenomena from mathematical principles; Ciermans designed a calculating machine and other mechanical contraptions; and Moretus did research in hydrostatics. All of these Jesuits kept in contact with each other, and they had a wide network of correspondents all over Europe, including Protestant Europe. It is safe to conclude that San Vicente's students were among the

---

3 San Vicente and Van Aelst, *Theoremata mathematica*; San Vicente and Ciermans, *Theoremata mathematica.*

4 Henri Bosmans, S.J., "Sur l'œuvre mathématique de Blaise Pascal," *Revue des questions scientifiques 4S.* 5 (1924): 136–60; 424–50. Descotes, "Pascal's Indivisibles," 213, does not support this view.

5 Hofmann, *Das* Opus geometricum *der Gregorius à Sancto Vincentio*; Dhombres, "Is One Proof Enough?," 403.

leading European scholars and scientists in the second quarter of the seventeenth century.

Why, then, has so little attention been paid to them? We can surmise three principal reasons: they suffered from a belated, and tarnished, publication by San Vicente; they were Jesuits; and they were Southern Netherlandish.

That San Vicente published as late as 1647, and that the content of his book was overshadowed by the controversy on the quadrature of the circle, blemished his reputation. That Mersenne accused him of plagiarism[6] further diminished the appetite to study his book in depth. Yet most of his new mathematics was already developed before 1625. Would it not be fair to give San Vicente his rightful place in the history of mathematics as one of the fathers of infinitesimal calculus? Galileo has—unduly as it turned out—been credited as the inventor of the sector, not because his publication was the first on the subject, but because his manuscript on the sector predates every publication on the subject.[7] Thomas Harriot ($c.$1560–1621) has been rehabilitated on the basis of his manuscripts. Cavalieri is credited with his method as having been conceived in the 1620s on the basis of letters to Galileo. Then why not attribute San Vicente his rightful place in the history of mathematics, the mathematician who, long before Leibniz and Newton, had found a way to calculate areas with rigorous infinitesimal methods? The same remarks can be made for San Vicente's students, who, for a variety of reasons, did not, or were not able to, publish their manuscripts.

The Society of Jesus, since its inception, has had an oracular reputation and was always being accused of consummate Machiavellianism, leading the Jesuits to become the target of hostile comments. Honest criticism by Jesuit scientists was often seen as an attempt to hamper scientific progress. This negative image has persisted up to the present day. Alexander's book on infinitesimals[8] may read as a thriller, but it unfairly reinforces the image of the Jesuits as a scheming society, influencing the powers that be, and, in an obscure way, sidetracking everyone opposing them. In science, careful research time and again has shown that this view is incorrect. One could argue that the Jesuits used Popperian falsification *avant-la-lettre*, and that their criticism was indispensable to putting the scientific revolution on the right track.

---

6 Van Looy, "Chronologie en analyse," 27.
7 The first book on the sector was Thomas Hood's (1556–1620) *The Making and Use of the Geometrical Instrument Called a Sector* (1598), predating Galileo's first publication on the subject by eight years.
8 Alexander, *Infinitesimal*.

On the other hand, the Anglo-Saxon history of science has always had a Protestant, even Puritan, bias, giving the development of science an inescapable direction toward enlightenment and progress. In this view, the Jesuits are usually seen as opposing this development. In France, and to a lesser extent in Catholic Europe, the expulsion of the Jesuits again reinforced the negative bias against them. Although there was no such outspoken anti-Jesuitism in the Southern Netherlands, their European reputation did harm San Vicente's prospects.[9]

Next to these external reasons, there are also internal reasons why being a Jesuit obstructed San Vicente's road to fame. After his visit to Rome, he was sent to Prague. A number of his students, according to him his most talented, were also sent abroad. This dispersal of talent, even if the shortage of mathematics teachers necessitated such moves, prohibited the development of a center of excellence for mathematics outside of Rome. Due to the times in which they were living and the students the Jesuits attracted, or wanted to attract, their focus shifted from pure mathematics and theoretical physics to applied mathematics and military matters. Even if these last subjects remained on a theoretical level, it distracted them from pursuing mathematics proper.

The final reason is that they were Southern Netherlandish. The historiography of the Low Countries has long held the view that, at the end of the sixteenth century, the Spanish re-conquest of the Low Countries initiated a mass migration from the prosperous south to the northern provinces, which were protected by the great rivers. With each town the Spanish took, the south lost more and more of its scientists and artists. In this view, the south became scientifically anemic. Dutch scholars had no reason to oppose this view, as the seventeenth century gave them some of their greatest mathematicians and scientists. Moreover, being a mainly Protestant country, its history of science seemed to agree with the Anglo-Saxon view.

On the other hand, Flemish, and by extension Belgian scholars did not feel an urge to study this period in the history of science, the doctoral thesis of Geert Vanpaemel[10] being an exception. Moreover, the Flemish history of science has been and still is, as Eduard Jan Dijksterhuis (1892–1965) called it,

---

9   On anti-Jesuitism in the Netherlands, see Hendrik Callewier, "Anti-Jezuïtisme in de Zuidelijke Nederlanden (1542–1773)," *Trajecta: Tijdschrift voor de geschiedenis van het katholiek leven in de Nederlanden* 16 (2007): 30–50.

10 Geert Vanpaemel, *Echo's van een wetenschappelijke revolutie: De mechanistische natuurwetenschap aan de Leuvense Artesfaculteit (1650–1797)* (Brussels: Koninklijke Academie voor Wetenschappen, Letteren en Schone Kunsten van België, 1986).

Clio's stepchild[11] and for that matter also *Urania's stepchild*. Flanders has known very few historians of science, and hardly any of them heeded the wakeup call by the inimitable Bosmans, whose historical research focuses on the late sixteenth- and early seventeenth-century Southern Netherlandish mathematicians and has shown the wealth of the material at our disposal. Despite his limited resources, he was able to show that the classical view of a scientifically anemic country was not correct. Father Van de Vyver took up his challenge and has laid the foundations for a thorough history of the Jesuit school of mathematics. However, one of the difficulties he faced, which would also be encountered by later historians of science—including the present one—was and is the fact that the documents of these Jesuits are dispersed over numerous European institutions.

Assessing the influence of the Flemish Jesuit school of mathematics correctly calls for a pan-European project. What this monograph has brought together is only the tip of an iceberg. San Vicente and his students were active in the Low Countries, France, Italy, Spain, Portugal, the Germanic lands, and Sweden—continental Europe, in short, as well as Britain due to the influence of Tacquet's geometry manual. It is hoped that in the foreseeable future a complete history of this potential competitor for the mathematics school of the Collegio Romano can be written.

11 Eduard Jan Dijksterhuis, *Clio's Stiefkind*, ed. Karel van Berkel (Amsterdam: Bert Bakker, 1990).

APPENDIX 1

Chronology of San Vicente's Manuscripts

Manuscript	Chronological MS	Title	Handwriting of	Written at	Date	
1.11	MS 3.1	No title	Willem Boelmans	Leuven	c.1624	sent to Rome November 1624
1.4	MS 3.2	*Lemmata*	Ignatius Derkennis	Leuven	c.1624	sent to Rome January 15, 1625
1.12	MS 3.2	*De progressionibus*	Ignatius Derkennis	Leuven	c.1624	sent to Rome January 15, 1625
1.13	MS 3.2	*Propositiones*	Ignatius Derkennis	Leuven	c.1624	sent to Rome January 15, 1625
1.5	MS 3.3	No title		Leuven	c.1624	sent to Rome beginning of 1625
1.3	MS 3.4	*De involucro*	Theodorus Moretus	Leuven	c.1624	sent to Rome May 22, 1625
1.7	MS 3.5	*Lemmata*			c.1625	sent to Rome autumn 1625
1.8	MS 3.5	*De cylindro elliptico*	Theodorus Moretus	Leuven	c.1624	sent to Rome autumn 1625
1.1	MS 7.4	*Quadraturae circuli liber: Ductu plani in planum*		Ghent	c.1645	preparatory text for *Problema Austriacum*
1.2	MS 7.4	*De ductu plani in planum*		Ghent	c.1645	preparatory text for *Problema Austriacum*
1.9	MS 7.5	*De spirali*		Ghent	c.1645	preparatory text for *Problema Austriacum*
1.10	MS 7.5	*Appendix ad librum De parabola*		Ghent	c.1645	preparatory text for *Problema Austriacum*
1.6		*Quadraturae circuli liber tertius et de ductu superficei in superficiem liber primus*				
2.1	MS 7.1	*De parabola*		Ghent	c.1645	preparatory text for *Problema Austriacum*

(cont.)

Manuscript	Chronological MS	Title	Handwriting of	Written at	Date	
2.2 and 2.3	MS 7	De hyperbola		Ghent	c.1645	preparatory text for Problema Austriacum
3.1 to 3.4	MS 13.1			Ghent	1650–68	preparatory text for Opus geometricum book 3
3.8	MS 13.1	No title		Ghent	1650–68	preparatory text for Opus geometricum Book 3
3.17 to 3.23	MS 14			Ghent	1650–68	preparatory text for Opus geometricum Book 4
3.5 to 3.30	MS 9.3	No title		Ghent	1650–68	preparatory text for Opus geometricum Proaemium
4.1	MS 10.2	Proaemium: De natura rationis geometricae		Ghent	1650–68	preparatory text for Opus geometricum Proaemium
4.2 to 4.22	MS 12			Ghent	1650–68	preparatory text for Opus geometricum book 2
5.1 to 5.26	MS 11			Ghent	1650–68	preparatory text for Opus geometricum books 1 and 6
6.7	MS 1	Lemmata	Francisco de Aguilón	Antwerp	c.1613	
6.2, 6.3, 6.5,	MS 5.1			Prague	1630–31	
6.9 to 6.10	MS 7.10	No title		Ghent	c.1645	preparatory text for Problema Austriacum
6.4	MS 7.11	De potentiis linearum rectarum		Ghent	c.1645	preparatory text for Problema Austriacum
6.1	MS 7.9	Collectiones quaedam mathematicae ex meis chartis		Ghent	c.1645	preparatory text for Problema Austriacum
6.8		No title				
7.1 to 7.6	MS 9.1			Ghent	1650–68	theorems for Opus geometricum
8.6	MS 8.1	Proaemium		Ghent	after 1647	responses to attacks on Problema Austriacum

(cont.)

Manuscript	Chronological MS	Title	Handwriting of	Written at	Date	
8.1 to 8.55	MS 8.10			Ghent	after 1647	responses to attacks on *Problema Austriacum*
8.12	MS 8	*Pro quadraturis*		Ghent	after 1647	responses to attacks on *Problema Austriacum*
9.1 to 9.16	MS 15			Ghent	1650–68	preparatory text for *Opus geometricum* Book 5
10	MS 6.1	Theorems and two letters		Ghent	1632–47	*Chartae Gandenses*
11	MS 6.2	Theorems and two letters		Ghent	1632–47	*Chartae Gandenses*
12.1	MS 5.5	*Quadraturae circuli liber pri[mus]: De progressionibus geometricis*		Prague	1630–31	
12.4 and 12.5	MS 5.6		Theodorus Moretus	Prague	1630–31	
12.2	MS 7.6	*Progressio*				
12.3	MS 7.7	*Liber de progressionibus geometricis*		Ghent	c.1645	preparatory text for *Problema Austriacum*
12.6	MS 7.8	No title				
12.8 to 12.16	MS 7.8	No title				preparatory text for *Problema Austriacum*
12.7		*De progressionibus*	Willem Boelmans	Leuven	c.1624	
13.1	MS 2	*Sectio angulorum*		Antwerp	1617–20	
13.2 to 13.11	MS 2	No title		Antwerp	1617–20	
13.12	MS 2	Index to Amalthea		Antwerp	1617–20	
14	MS 4.1	No title		Rome or Prague	1625–29	*Chartae Romanae*
15	MS 4.2	No title		Rome or Prague	1625–29	*Chartae Romanae*
16.1 to 16.7	MS 9.2			Ghent	1650–68	theorems for *Opus geometricum*

(cont.)

Manuscript	Chronological MS	Title	Handwriting of	Written at	Date
17.1	MS 2.1			Antwerp and Leuven	1617–24
17.2	MS 2.2		Willem Boelmans	Antwerp and Leuven	1617–24
17.3	MS 2.3			Antwerp and Leuven	1617–24
17.4	MS 2.4		unknown	Antwerp and Leuven	1617–24
17.5	MS 2.5			Antwerp and Leuven	1617–24
17.6	MS 2.6		unknown	Antwerp and Leuven	1617–24
17.7	MS 2.7			Antwerp and Leuven	1617–24
17.8	MS 2.8		unknown	Antwerp and Leuven	1617–24
17.9	MS 2.9			Antwerp and Leuven	1617–24
17.10	MS 2.10		unknown	Antwerp and Leuven	1617–24
17.11	MS 2.11			Antwerp and Leuven	1617–24
17.12	MS 2.12		unknown	Antwerp and Leuven	1617–24
17.13	MS 2.13			Antwerp and Leuven	1617–24

APPENDIX 2

Students of the School of Mathematics after 1625

Professor	Name	Birthplace	Birthday	Entry	Father	Mother
Ignatius Derkennis	Sebastianus de Lutiano	Goes	September 30, 1609	June 27, 1626	Frederico	Maria van Coorle
Willem Boelmans	Andreas Tacquet	Antwerp	June 7, 1612	October 31, 1629	Peeter	Agnese Wandelen
	Joannes Groll	Antwerp	July 19, 1613	October 31, 1629	Joannes	Gertrude Gigens
	Laurens van Schoonen	Ghent	July 7, 1613	October 31, 1929	Remaclo (?)	Catharina Thijs
	Philip Jacob	Brussels	March 17, 1612	September 17, 1629		
	Phil. de Roy (?)					
	Hendrik van Munster (?)	Nijmegen	March 2, 1614	September 30, 1631	Bartholomeus	Arnoldina Uwens
Joannes Ciermans	Hendrik Uwens	Nijmegen	April 23, 1618	October 15, 1634	Hendrik	Margareta Busaea
	Joannes Balendonck	Nijmegen	1611	November 7, 1635	Petro	Elizabetha Vijghe
	Joannes Boddaert	Beloeuil	July 26, 1616	September 29, 1635	Daniel	Adriana Aernoudt
	Ludovic du Rieu	Kortrijk	February 10, 1615	September 30, 1634	Joannes	Marina Nullins
	Alexander Barvoets	Bruges	October 15, 1620	September 30, 1636	Willem	Anna Auderan (?)
	Maximiliaan le Dent	Bergues	February 16, 1618	September 30, 1637	Joannes	Jacoba de Meester
Andreas Tacquet	Ignatius Maillot	Mechelen	May 23, 1613	September 27, 1640	Jacob	Susanna del Plano
	Hubert Henschenius	Venray	January 31, 1624	October 5, 1641	Joannes	Beatrice Meus
	Balthasar van Leemputte	Herentals	September 8, 1624	September 22, 1642	Laurentio	Josina Michilsen
	Ignatius Diertiens	Brussels	April 27, 1626	September 18, 1642	Bernardo	? Jacobs
	Lionel Aynscombe	Antwerp	December 30, 1626	October 1, 1642	Arthur	Catharina Anthonis

© KONINKLIJKE BRILL NV, LEIDEN, 2021 | DOI:10.1163/9789004447905_019

(cont.)

Professor	Name	Birthplace	Birthday	Entry	Father	Mother
	Jacob van Weerde	Antwerp	August 8, 1627	October 6, 1643	Joannes	Maria Severdonck
	Henri de Prince	Antwerp	September 14, 1632	September 28, 1650		
	Ignatius de Jonghe	Beveren	November 22, 1632	September 15, 1650	Petro	Anna van der Haeghen
	Aegidius de Gottignies	Brussels	May 10, 1630	November 5, 1653	Augustin	Margareta Verreycken
	Ferdinand Verbiest	Pittem	October 1623	September 2, 1641	Joos	Anna van Hecke

The student's names are from Kadoc ABSE 1447; other data are from Kadoc ABSE 31, 36, 37.

Bibliography

Afbeeldinghe van d'eerste Eeuwe der Societeyt Jesu. Antwerp: Officina Plantiniana, 1640.

Alden, Dauril. *The Making of an Enterprise: The Society of Jesus in Portugal, Its Empire, and Beyond 1540–1750.* Stanford: Stanford University Press, 1996.

Alexander, Amir. *Infinitesimal: How a Dangerous Mathematical Theory Shaped the Modern World.* London: Oneworld, 2015.

Alexander, Amir. "On Indivisibles and Infinitesimals: A Response to David Sherry, 'The Jesuits and the Method of Indivisibles.'" *Foundations of Science* 23, no. 2 (2018): 393–98.

Andersen, Kirsti. "Cavalieri's Method of Indivisibles." *Archive for History of Exact Sciences* 31, no. 4 (December 1985): 291–367.

Andersen, Kirsti, Enrico Giusti, and Vincent Jullien. "Cavalieri's Indivisibles." In *Seventeenth-Century Indivisibles Revisited,* edited by Vincent Jullien, 31–55. Basel: Birkhäuser, 2015.

Andriessen, Jozef. *De Jezuïeten en het samenhorigheidsbesef der Nederlanden 1585–1648.* Antwerp: De Nederlandsche Boekhandel, 1957.

Anselmo, Antonio A. "Album amicorum." Manuscript, Koninklijke Bibliotheek Den Haag 71 J 57, 1594–1602.

Apollonius of Perga. *Les coniques d'Apollonius de Perge: Œuvres trad. pour la première fois du Grec en Français, avec une intr. et des notes, par Paul Ver Eecke.* Edited by Paul Ver Eecke. Bruges: Desclée-De Brouwer, 1923.

Arblaster, Paul. "The Archdukes and the Northern Counter-Reformation." In *Albert and Isabella 1598–1621: Essays,* edited by Werner Thomas and Luc Duerloo, 87–92. Turnhout: Brepols, 1998.

Archenholtz, Johann Wilhelm von. *Mémoires concernant Christine, reine de Suède.* Amsterdam and Leipzig: Pierre Mortier, 1751.

Archimedes. *Archimedis opera omnia cum commentariis eutocii.* Edited by J. L. [Johan Ludvig] Heiberg. Leipzig: Teubner, n.d.

Archimedes. *Les œuvres complètes d'Archimède suivies des commentaires d'Eutocius d'Ascalon: Trad. du Grec en Français avec une introd. et des notes par Paul Ver Eecke.* Edited by Paul Ver Eecke. Liège: Vaillant-Carmanne, 1960.

Aristotle. *Physics, Book VI.* Edited and translated by R. [Robert] P. Hardie and R. [Russell] K. Gaye. Cambridge, MA: MIT, n.d. http://classics.mit.edu/Aristotle/physics.6.vi.html (accessed February 12, 2020).

Audenaert, Stephanie. "Aegidius Franciscus de Gottignies (1630–1689), Jezuïet, geleerde en homo universalis." *Het Land van Beveren* 50, no. 4 (2007): 544–60.

Audenaert, Willem, S.J. *Prosopographia iesuitica Belgica antiqua: A Biographical Dictionary of the Jesuits in the Low Countries 1542–1773; PIBA (4 Vols.).* Heverlee and Leuven: Filosofisch en Theologisch College van de Sociëteit van Jezus, 2000.

Auger, Léon. *Un savant méconnu: Gilles Personne de Roberval (1602–1675)*. Paris: Librairie scientifique A. Blanchard, 1962.

Baetens, Roland. *De Nazomer van Antwerpens Welvaart: De diaspora en het handelshuis De Groote tijdens de eerste helft der zeventiende eeuw* (2 Vols.). Brussels: Gemeentekrediet, 1976.

Bailey, Gauvin Alexander. "'Le style jésuite n'existe pas': Jesuit Corporate Culture and the Visual Arts." In *The Jesuits: Cultures, Sciences, and the Arts 1540–1773*, edited by John W. O'Malley, s.j., Gauvin Alexander Bailey, Steven J. Harris, and T. Frank Kennedy, 38–89. Toronto, Buffalo, and London: University of Toronto Press, 1999.

Bailey, Gauvin Alexander. "Italian Renaissance and Baroque Painting under the Jesuits and Its Legacy throughout Catholic Europe, 1565–1773." In *The Jesuits and the Arts 1540–1773*, edited by John W. O'Malley, s.j., and Gauvin Alexander Bailey, 123–98. Philadelphia: Saint Joseph's University Press, 2005.

Baldini, Ugo. "The Academy of Mathematics of the Collegio Romano from 1553 to 1612." In *Jesuit Science and the Republic of Letters*, edited by Mordechai Feingold, 47–98. Cambridge, MA, and London: MIT Press, 2003.

Baldini, Ugo, and Pier Daniele Napolitani, eds. *Christoph Clavius corrispondenza: Edizione critica; V (1602–1605)*. Vol. 5. Pisa: Dipartimento di Matematica, 1992.

Bangert, William V. *A History of the Society of Jesus*. St. Louis, MO: Institute of Jesuit Sources, 1972.

Barbin, Évelyne, and Anne Boyé. *François Viète, un mathématicien sous la Renaissance*. Paris: Vuibert, 2005.

Bascelli, Tiziani. "Torricelli's Indivisibles." In *Seventeenth-Century Indivisibles Revisited*, edited by Vincent Jullien, 105–36. Basel: Birkhäuser, 2015.

Begheyn, Paul, s.j. "The Jesuits in the Low Countries 1540–1733: Apostles of the Printing Press." In *The Jesuits of the Low Countries: Identity and Impact*, edited by Rob Faesen and Leo Kenis, 129–38. Leuven, Paris, and Walpole, MA: Peeters, 2012.

Bennett, J. [Jim] A. *The Measurers: A Flemish Image of Mathematics in the Sixteenth Century*. Oxford: Museum of the History of Science, 1996.

Bennett, J. [Jim] A., and Stephen Johnston. *The Geometry of War 1500–1750*. Oxford: Museum of the History of Science, 1996.

Bens, Lies. "De algebraïsering van Euclides: The Elements of Euclid (1714) van William Whiston." Master's thesis, Katholieke Universiteit Leuven, 2015.

Bockstaele, Paul. "The Correspondence of Adriaan van Roomen." *Lias* 3 (1976): 85–299.

Bockstaele, Paul. "Ignatius de Jonghe et sa *Geometrica inquisitio in parabolas numero et specie infinitas*." *Janus* 54 (1967): 228–35.

Bockstaele, Paul. "Een vergeten werk over de kwadratuur van parabolen en hyperbolen: Ignatius de Jonghe's *Geometrica inquisitio* (1688)." *Scientiarum historia* 4 (1967): 175–81.

Boelmans, Willem, Willem van Schoone, Laurens van Schoone, Peter Boele, Jan Maarten de la Faille, Frans de la Deuse, Andries de Wulf, and Frans vanden Vivere. *Evstachivs drama como-tragicvm*. Ghent: Ex officina Iudoci Doomsij, 1629.

Boelmans, Willem, and Johannes Groll. *Theses mathematicæ: Geometricæ, arithmeticæ, opticæ, catoptricæ, dioptricæ, mvsicæ, architectonicæ, stereo-staticæ, hygro-staticæ, qvas præside [...] Gvilielmo Boelmans [...] Demonstrabit ac defendet Ioannes Groll [...] Lovanii, in Collegio Societatis I*. Leuven: Viduam Henrici Hastenii, 1634.

Boelmans, Willem, and Philip Jacob. *Theses mathematicæ: Geometricæ, arithmeticæ, opticæ, catoptricæ, dioptricæ, mvsicæ, architectonicæ, stereo-staticæ, hygro-staticæ, qvas præside [...] Gvilielmo Boelmans [...] Demonstrabit ac defendet Philippus Iacobi [...] Lovanii, in Collegio Societati*. Leuven: Viduam Henrici Hastenii, 1634.

Boelmans, Willem, and Laurens van Schoone. *Theses mathematicæ: Geometricæ, arithmeticæ, opticæ, catoptricæ, dioptricæ, mvsicæ, architectonicæ, stereo-staticæ, hygro-staticæ, qvas præside [...] Gvilielmo Boelmans [...] Demonstrabit ac defendet Lavrentivs van Schoone [...] Lovanii, in Collegio Soc*. Leuven: Viduam Henrici Hastenii, 1634.

Bonte, Germain, and François Jongmans. "Sur les origines de Grégoire de Saint Vincent." *Mededelingen van de Koninklijke Academie van België: Klasse der Wetenschappen* (1998): 295–326.

Bopp, Karl. *Die Kegelschnitte des Gregorius a St. Vincentio in vergleichender Bearbeitung*. Leipzig: Teubner Verlag, 1907.

Bos, Henk J. M. *Redefining Geometrical Exactness*. New York: Springer-Verlag, 2001.

Bösel, Richard. "Jesuit Architecture in Europe." In *The Jesuits and the Arts 1540–1773*, edited by John W. O'Malley, s.j., and Gauvin Alexander Bailey, 63–122. Philadelphia: Saint Joseph's University Press, 2005.

Bosmans, Henri, s.j. "André Tacquet (S.J.) et son traité d'arithmétique théorique et pratique." *Isis* 9, no. 1 (1927): 66–82.

Bosmans, Henri, s.j. "Le calcul infinitésimal chez Simon Stevin." *Mathesis* 37 (1923): 12–18; 55–62; 105–9.

Bosmans, Henri, s.j. "La carte lunaire de Van Langren conservée aux Archives Générales du Royaume à Bruxelles." *Revue des questions scientifiques* 54 (1903): 107–39.

Bosmans, Henri, s.j. "Compte rendu de R. Guimarâes: *Les mathématiques en Portugal* (1909)." *Revue des questions scientifiques* 67 (1910): 636–46.

Bosmans, Henri, s.j. "Deux lettres inédits de Grégoire de St-Vincent et les manuscrits de della Faille." *Annales de La Société Scientifique* 26 (1901): 1–19.

Bosmans, Henri, s.j. "Documents inédits sur Grégoire de Saint-Vincent." *Annales de Las Société Scientifique* 27 (1903): 43–44.

Bosmans, Henri, s.j. "Grégoire de Saint-Vincent, note historique." *Mathesis* 38 (1924): 250–56.

Bosmans, Henri, S.J. "Le jésuite mathématicien anversois André Tacquet (1612–1660)." *De Gulden Passer* 3 (1925): 63–85.

Bosmans, Henri, S.J. "Lettre inédite de Christophe Grienberger sur Grégoire de Saint-Vincent." *Annales de La Société d'émulation de Bruges* 63 (1913): 41–50.

Bosmans, Henri, S.J. "La *Logistique* de Gilles-François de Gottignies." *Revue des questions scientifiques 4S*, no. 13 (1928): 215–44.

Bosmans, Henri, S.J. "Le mathématicien anversois Jean Charles della Faille de la Compagnie de Jésus." *Mathesis* 41 (1927): 1–12.

Bosmans, Henri, S.J. "SAINT-VINCENT (Grégoire de)." *Biographie nationale* 141–71. Brussels: Académie royale de Belgique, 1911.

Bosmans, Henri, S.J. "Sur l'œuvre mathématique de Blaise Pascal." *Revue des questions scientifiques, 4 Série.* 5 (1924): 136–60; 424–50.

Bosmans, Henri, S.J. "Sur quelques exemples de la méthode des limites chez Simon Stevin." *Annales de La Société Scientifique de Bruxelles* 37 (1913): 171–99.

Bosmans, Henri, S.J. "Sur les thèses *De cometis* de Grégoire de Saint Vincent." *Bibliotheca mathematica 3F*, no. 4 (1903): 90.

Bosmans, Henri, S.J. "Sur les thèses de statique de Grégoire de Saint Vincent." *Annales de La Société Scientifique de Bruxelles* 44 (1924): 17–22.

Bosmans, Henri, S.J. "Théodore Moretus de la Compagnie de Jésus, mathématicien (1602–1667): D'après sa correspondance et ses manuscrits." *De Gulden Passer* 6 (1928): 57–163.

Braun, Joseph. *Die Belgischen Jesuitenkirchen: Ein Beitrag Zur Geschichte des Kampfes.* Freiburg im Breisgau: Herder, 1907.

Briels, Jan G. C. A. "Zuidnederlandse Onderwijskrachten in Noord-Nederland 1570–1630." *Archief voor de Geschiedenis van de Katholieke Kerk in Nederland* 14 (1972): 89–169.

Brigode, Simon. "Les projets de construction de l'église des jésuites à Anvers." *Bulletin de l'Institut historique belge de Rome* 14 (1934): 157–74.

Brouwers, Lodewijk, S.J. "Aperçu historique sur la province belge de la Compagnie de Jésus." Manuscript, Kadoc ABSE 12429, n.d.

Brouwers, Lodewijk, S.J. *Carolus Scribani 1561–1629.* Antwerp: Ruusbroecgenootschap, 1961.

Brouwers, Lodewijk, S.J. *Het Hof van Liere.* Antwerp: Loyola-vereniging, 1976.

Bucho, Domingos, and Raul Ladeira. *Cidade-quartel fronteiriça Da Elvas e suas fortificações = The Garrison Border Town of Elvas and Its Fortifications.* Elvas: Edições Colibri e Câmara Municipal de Elvas, 2013.

Buchwald, Jed Z. "Descartes's Experimental Journey Past the Prism and through the Invisible World to the Rainbow." *Annals of Science* 65, no. 1 (2008): 1–46.

Buonanno, Roberto. *The Stars of Galileo Galilei and the Universal Knowledge of Athanasius Kircher.* Cham: Springer International, 2014.

Burenstam, Carl Johan Reinhold. *La Reine Christine de Suède à Anvers et Bruxelles 1654–1655*. Brussels: Alfred Vromant & Cie, 1891.

Burn, Robert P. "Alphonse Antonio de Sarasa and Logarithms." *Historia mathematica* 28 (2001): 1–17.

Bussers, Helena. "De baroksculptuur en het barok kerkmeubilair in de Zuidelijke Nederlanden." *Openbaar Kunstbezit* 20 (1982): 123–61.

Callewier, Hendrik. "Anti-Jezuïtisme in de Zuidelijke Nederlanden (1542–1773)." *Trajecta: Tijdschrift voor de Geschiedenis van het Katholiek Leven in de Nederlanden* 16 (2007): 30–50.

Cámara, Alicia, ed. *Los ingenieros militares de la monarquía Hispánica en los siglos XVII y XVIII*. Madrid: Ministerio de defensa, 2006.

Camerota, Filippo. *Il compasso di Fabrizio Mordente, per la storia del compasso di proporzione*. Florence: Leo S. Olschki, 2000.

Casalini, Cristiano, and Claude Pavur, eds. *Jesuit Pedagogy, 1540–1616: A Reader*. Chestnut Hill, MA: Institute of Jesuit Sources, 2016.

Castilla Soto, Josefina. *Don Juan de Austria (hijo bastardo de Felipe IV): Su labor política y militar*. Madrid: Universidad nacional de educación a distancia, n.d.

Catalogue de livres des bibliothèques de la maison professe du college & du couvent des ci devant jésuites d'Anvers. Leuven: J. P. G. Michel, 1779.

Celeyrette, Jean. "From Aristotle to the Classical Age, the Debates around Indivisibilism." In *Seventeenth-Century Indivisibles Revisited*, edited by Vincent Jullien, 19–30. Basel: Birkhäuser, 2015.

Christianidis, Jean. "Archimedes's Quadratures." In *The Genius of Archimedes: 23 Centuries of Influence on Mathematics, Science, and Engineering*, edited by Stephanos A. Paipetis and Marco Ceccarelli, 57–68. Dordrecht, Heidelberg, London, and New York: Springer Science & Business Media B.V., 2010.

Ciermans, Johannes, Krzysztof Aleksander Rozdrażewski, Leo Carolus Sapieha, and Jakub Hieronim Rozdrażewski. *Repetitio menstrua quam de geometricis, astronomicis, staticis instituent [...]*. Leuven: Typis Everardi de Witte, 1639.

Ciermans, Johannes, Jan Antoon Tucher, and Jacob Hoens. *Repetitio menstrua quam de cosmographicis et geographicis instituent*. Leuven: Everardus de Witte, 1639.

Ciermans, Johannes, Jacob Neeffs [illustrator], Maarten Mandekens [illustrator], and Philip Fruytiers [illustrator]. *Disciplinae mathematicae traditae anno institutae Societatis Iesu seculari*. Leuven: Everardum de Witte, 1640.

Ciermans, Johannes, Wolfgang Philips Jakob Unverzagt, Philip Fruytiers [illustrator], and Jacob Neeffs [illustrator]. *Annus positionum mathematicarum*. Leuven: Everardum de Witte, 1640–41.

Claeys-Bouuaert, Ferdinandus. "Une visite canonique des maisons de la Compagnie de Jésus en Belgique (1603–1604); Rapports des visiteurs Olivier Manare et Léonard

Lessius, envoyés au P. Général Claude Acquaviva." *Bulletin de l'Institut historique belge à Rome* 7 (1927): 5–114.

Clavius, Christopher. *Algebra Christophori Clauii barbengensis e Societate Iesu*. Rome: Apud Bartholomaeum Zannettum, 1605.

Clavius, Christopher. *Geometria practica*. Rome: Aloisius Zannetti, 1604.

Clavius, Christopher. *Opera mathematica*. Mainz: Reinhardus Eltz, n.d.

Clavius, Christopher, and Dennis Smolarski, s.j., trans. "Historical Documents Part II: Two Documents on Mathematics." *Science in Context* 15, no. 3 (2002): 465–70.

Codina Mir, Gabriel, s.j. *Aux sources de la pédagogie des jésuites: Le* Modus Parisiensis. Rome: Institutum Historicum s.i., 1968.

Codina Mir, Gabriel, s.j. "The *Modus Parisiensis*." In *The Jesuit* Ratio studiorum *of 1599: 400th Anniversary Perspectives*, edited by Vincent J. Duminuco, 28–49. New York: Fordham University Press, 2000.

Coens, Maurice. "Un panégyrique de S. Ignace prononcé à Anvers en 1656." *Analecta Bollandiana* 74 (1956): 349–61.

[Coignet, Michiel]. "Fábrica y uso del pantometro." Manuscript, Archief della Faille-de Leverghem 28.3, n.d.

[Coignet, Michiel]. "Neunspitziger Passer: Fabricii Mordenti von Salerno Mathematici des Herzogen Alexandrii Fernesi Herzogen zu Parma." Manuscript, University Library Wrocław MS R461, n.d.

Coignet, Michiel. "El uso de las doze diuisiones geometricas." Manuscript, Erfgoedbibliotheek Hendrik Conscience, Antwerp, B264708, 1618.

Coignet, Michiel. "Uso del compasso di Fabricio Mordenti di Salerno Mathematico del. Ser. Mo Principe Alessandro Farnese ducca di Parma composto per Michaelo Coignetto." Manuscript, Biblioteca Nacional Madrid, MS 19.709/32, n.d.

Coignet, Michiel. "Usus regulae pantometrae." Manuscript, MS Riccardiana 956, Florence, n.d.

Coolidge, Julian L. *A History of the Conic Sections and Quadric Surfaces*. Oxford: Clarendon Press, 1945.

Cosentino, Giuseppe. "Mathematics in the Jesuit *Ratio studiorum*." In *Church, Culture, and Curriculum: Theology and Mathematics in the Jesuit* Ratio studiorum, edited by László Lukács, s.j., and Giuseppe Cosentino, 47–80. Philadelphia: Saint Joseph's University Press, 1999.

Courtenay, William J. *Ockham and Ockhamism*. Leiden and Boston: Brill, 2008.

Daelemans, Bert, s.j. "Pieter Huyssens, S.J. (1577–1637), an Underestimated Architect and Engineer." In *Innovation and Experience in the Early Baroque in the Southern Netherlands: The Case of the Jesuit Church in Antwerp*, edited by Piet Lombaerde, 41–52. Turnhout: Brepols, 2008.

Daelemans, Bert, s.j. "Het *Promptuarium pictorum* als spiegel van de ontwerppraktijk der Vlaamse jezuïetenarchitecten in de 17de eeuw." In *Bellissimi Ingegni,*

Grandissimo Splendore, edited by Krista de Jonge, Annemie de Vos, and Joris Snaet, 175–98. Leuven: Universitaire Pers Leuven, 2000.

De Aguilón, Francisco. *Opticorum libri sex philosophis iuxtá ac mathematicis utiles*. Antwerp: Ex officina Plantiniana, apud viduam et filios Jo. Moreti, 1613.

De Backer, Augustin, s.j., Aloys de Backer, and Carlos Sommervogel, s.j. *Bibliothèque des écrivains de la Compagnie de Jésus*. Brussels and Paris: Schepens-Picard, 1890.

De Bruycker, Angelo. "'En todo amar y servir a su Divina Majestad': Wiskunde in de opleiding en het onderwijs van de Vlaamse Jezuïeten; Een cultuurhistorische benadering." *Scientiarum historia* 32 (2006): 37–58.

De Bruycker, Angelo. "'To the Adornment and Honour of the City': The Mathematics Course of the Flemish Jesuits in the Seventeenth Century." *BSHM Bulletin Journal of the British Society for the History of Mathematics* 24 (2009): 135–46.

De Bruycker, Angelo. "De 'wiskundeschool' van de Vlaamse Jezuïeten in de eerste helft van de zeventiende eeuw: Een herpositionering." *Handelingen van de Koninklijke Zuid-Nederlandse Maatschappij Voor Taal: En Letterkunde en Geschiedenis* 58 (2004): 201–20.

De Dainville, François. *L'éducation des jésuites (XVIe–XVIIIe siècles)*. Paris: Les éditions de minuit, 1976.

De Dainville, François. *La géographie des humanistes*. Paris: Beauchesne, 1940.

De Dainville, François, and Marie-Madeleine Compère. *L'éducation des jésuites (XVIe–XVIIIe siècles)*. Paris: Ed. de Minuit, 1978.

De Gottignies, Gilles-François. *Astronomica: Epistolæ duæ*. Rome: Iacobi Antonij de Lazaris Varesij, 1665.

De Gottignies, Gilles-François. *Logistica, sive: Scientia circa quamlibet quantitatem demonstrative discurrens cui mathematicum nullum problema insolubile nullum theorema indemonstrabile*. Rome: De Lazaris, 1675.

De Jong, Michiel A. G. *Staat van Oorlog (1585–1621)*. Hilversum: Verloren, 2005.

De Jonge, Krista. "Architectuur ten tijde van de Aartshertogen: Het Hof achterna." In *Bellissimi Ingegni, Grandissimo Splendore*, edited by Krista de Jonge, Annemie de Vos, and Joris Snaet, 11–42. Leuven: Universitaire Pers Leuven, 2000.

De Jonge, Krista, and Konrad Ottenheym, eds. *Unity and Discontinuity; Architectural Relationships between the Southern and Northern Low Countries (1530–1700)*. Turnhout: Brepols, 2007.

De Jonghe, Ignatius. *Geometrica inquisitio in parabolas numero & specie infinitas & iisdem congenitae hyperbolas ac præcipue in quadraturam hyperbolæ apollonianæ*. Antwerp: n.p., 1688.

De la Hire, Philippe. *Sectiones conicae in novem libros distributæ [...]*. Paris: Stephanum Michallet, 1685.

De Maeyer, Marcel. *Albrecht en Isabella en de Schilderkunst: Bijdrage tot de Geschiedenis van de 17e-eeuwse schilderkunst in de Zuidelijke Nederlanden*. Brussels: Koninklijke

Vlaamse Academie voor Wetenschappen, Letteren en Schone Kunsten van België, 1955.
De Meneses, Fernando. *Historiarum Lusitanarum ab anno 1640 usque ad 1657 libri decem, Vol. 2.* Lisbon: Joseph Anthony da Sylva, 1734.
De Oliveira, Natália, and Sezinando Menezes. "Ciência moderna em Portugal: A 'Aula Da Esfera' no Colégio de Santo Antão." *Acta scientiarum: Education* 39, no. 3 (2017): 243–53.
De Ribadeneyra, Pedro, and Philippe Alegambe. *Bibliotheca scriptorum Societatis Jesu: Post excusum anno M.DC.VIII catalogum R.P. Petri Ribadeneirae [...] Nunc [...] Ad Annum M.DC.XLII. [...] Concinnata.* Antwerp: Joannes Meursius, 1643.
De Ridder-Symoens, Hilde. "Het hoger onderwijs." In *België in de 17de Eeuw: De Spaanse Nederlanden en het Prinsbisdom Luik. Band I: Politiek*, edited by Paul Janssens, 84–96. Brussels: Dexia Bank-Snoeck, 2006.
De Ridder-Symoens, Hilde. "Het onderwijs te Antwerpen in de zeventiende eeuw." In *Antwerpen in de XVIIde Eeuw*. Antwerp: Genootschap voor Antwerpse Geschiedenis, 1989.
De Sarasa, Alfonso Antonio. *Solutio problematis a R.P. Marino Mersenno Minimo propositi. duo a proponente de hac propositione pronuntiantur: Unum quod fortisan longe difficiliorem quam ipsa quadratura solutionem requirat; Alterum quo quadratura circuli à R.P. Gregorio a S[anc]to V.* Antwerp: Ioannem & Iacobum Meursios, 1649.
De Smet, H. "Pieter Huyssens, architect van de Sint-Walburgakerk." In *Sint-Walburga: een Brugse kerk vol geschiedenis*, edited by Jozef van den Heuvel, 77–86. Bruges: Jong Kristen Onthaal voor Toerisme, 1982.
De Smet, Rudolf. "Simon Stevin en de paradox van het gefragmenteerde humanisme." In *Simon Stevin 1548–1620: De geboorte van een nieuwe wetenschap*, edited by Wouter Bracke, 27–34. Turnhout and Brussels: Brepols and KBR, 2004.
De Vries, Jan. *European Urbanization 1500–1800*. London: Methuen & Co., 1984.
Dear, Peter. *Discipline and Experience: The Mathematical Way in the Scientific Revolution.* Chicago and London: University of Chicago Press, 1995.
Dear, Peter. "Jesuit Mathematical Science and the Reconstitution of Experience in the Early Seventeenth Century." *Studies in the History and Philosophy of Science* 18 (1987): 133–75.
Dekoninck, Ralph, and Agnes Guiderdoni. "Knowledge in Transition: A Case of 'Scientific Emblematics' (Ciermans and Vaenius) at the Turn of the Seventeenth Century." In *Embattled Territory: The Circulation of Knowledge in the Spanish Netherlands*, edited by Sven Dupré, Bert de Munck, Werner Thomas, and Geert Vanpaemel, 279–96. Ghent: Academia Press, 2015.
Delattre, Pierre, S.J., ed. *Les établissements des jésuites en France depuis quatre siècles: Répertoire topo-bibliographique publié à l'occasion du quatrième centenaire de la*

fondation de la Compagnie de Jésus 1540–1940. Enghien: Institut supérieur de théologie, n.d.

Delée, Joseph. "Liste d'élèves du collège des pères jésuites à Anvers de 1575 à 1640." *De Schakel* 22 (1967): 1–94.

Della Faille, Joannes. *Theoremata de centro gravitatis partium circuli et ellipsis*. Antwerp: Ex officina typographica Joannis Meursii, 1632.

Derkennis, Ignatius. *De Deo Uno, Trino, Creatore*. Brussels: Franciscus Foppens, 1655.

Derkennis, Ignatius, and Andreas Aloysius Zamora. *Positiones sacræde avgvstiss.Mo Sacramento Evcharistiæ Rationibvs illvstratæ [...] Præside R. P. Ignatio Der-Kennis [...] tuebitur, ac explicabit Andreas Aloysivs de Zamora in Collegio Societatis Iesv lovanii*. Antwerp: Joannes Meursius, 1638.

Descartes, René. *Œuvres de Descartes, 2: Correspondance; Mars 1638–Décembre 1639*. Edited by Charles Adam and Paul Tannery. Paris: Cerf, 1898.

Descartes, René. *Œuvres de Descartes, 2: Correspondance; Mars 1638–Décembre 1639*. Edited by Charles Adam and Paul Tannery. Nouv. éd. Paris: Vrin, 1996.

Descotes, Dominique. "Pascal's Indivisibles." In *Seventeenth-Century Indivisibles Revisited*, edited by Vincent Jullien, 211–48. Cham: Springer International, 2015.

Descotes, Dominique. "Two Jesuits against the Indivisibles." In *Seventeenth-Century Indivisibles Revisited*, edited by Vincent Jullien, 249–73. Cham: Springer International, 2015.

Dhombres, Jean. "Is One Proof Enough? Travels with a Mathematician of the Baroque Period." *Educational Studies in Mathematics* 24 (1993): 401–19.

Dhombres, Jean. "L'innovation comme produit captif de la tradition: Entre Apollonius et Descartes, une théorie des courbes chez Grégoire de Saint-Vincent." In *Geometria, flussioni e differenziali*, edited by Marco Panza and Clara Silvia Roero, 17–102. Naples: La Città del Sole, 1995.

Dhombres, Jean, and Patricia Radelet-de-Grave. *Une mécanique donnée à voir: Les thèses illustrées défendues à Louvain en Juillet 1624 par Grégoire de Saint-Vincent, S.J.* Turnhout: Brepols, 2008.

Dhombres, Jean, and Jacques Sakarovitch. *Desargues en son temps*. Paris: A. Blanchard, 1994.

Dijksterhuis, Eduard Jan. *Clio's Stiefkind*. Edited by Karel van Berkel. Amsterdam: Bert Bakker, 1990.

Donnelly, John Patrick, s.j. "Padua, Louvain, and Paris: Three Case Studies of University–Jesuit Confrontation." *Louvain Studies* 15 (1990): 38–52.

Dooreman, Herman. "Andreas Tacquet *Cylindrica et annularia quinque libris comprehensa*." Master's thesis, Katholieke Universiteit Leuven, 1966.

Drake, Stillman. *Galileo Galilei: Operations of the Geometric and Military Compass 1606*. Washington, DC: Dibner Library Smithsonian Institution Press, 1978.

Droeshout, Charles. "Histoire du Collège d'Anvers." Manuscript, Kadoc Leuven, ABML 3279, c.1908.

Droeshout, Charles. "Histoire de la Compagnie de Jésus à Anvers." 6 Vols. Manuscript, Kadoc Leuven, ABML 3284–89.

Duerloo, Luc. *Dynasty and Piety: Archduke Albert (1598–1621) and Habsburg Political Culture in an Age of Religious Wars*. London and New York: Routledge, 2012.

Duffy, Christopher. *Siege Warfare: The Fortress in the Early Modern World 1494–1660*. London: Routledge & Kegan Paul, 1996.

Durand, Jacques, and Joannes de Swienthohliwicz Kamiensky. *Problema mathematicum ex architectonia militari de moenibus inferioribus sive falsabraga [...]*. Graz: Apud haeredes Ernesti Widmanstadij, 1636.

Dürer, Albrecht. *Underweysung der Messung*. Nuremberg: n.p., 1525.

Elazar, Michael, and Rivka Feldhay. "Jesuit Conceptions of Impetus after Galileo: Honoré Fabri, Paolo Casati, and Francesco Eschinardi." In *Emergence and Expansion of Pre-classical Mechanics*, edited by Rivka Feldhay, Jürgen Renn, Matthias Schemmel, and Matteo Valleriani, 285–324. Cham: Springer International, 2018.

Elliott, John H. *The Count-Duke of Olivares: The Statesman in an Age of Decline*. New Haven—London: Yale University Press, 1986.

Elliott, John H. *The Revolt of the Catalans: A Study in the Decline of Spain*. Cambridge: Cambridge University Press, 1984.

Evans, Robert John Weston. *Rudolf II and His World: A Study in Intellectual History 1576–1612*. Oxford: Clarendon Press, 1984.

Eves, Howard. *College Geometry*. Boston and London: Jones & Bartlett, 1990.

Eytzinger, Michael. *De Leone Belgico, ejusque topographica atque historica descriptione liber*. Cologne: Gerardus Campensis, 1583.

Eytzinger, Michael. *De Leone Belgico, ejusque topographica atque historica descriptione liber*. Cologne: Gerardus Campensis, 1585.

Fabri, Ria. "Light and Measurement: A Theoretical Approach of the Interior of the Jesuit Church in Antwerp." In *Innovation and Experience in the Early Baroque in the Southern Netherlands: The Case of the Jesuit Church in Antwerp*, edited by Piet Lombaerde, 125–40. Turnhout: Brepols, 2008.

Feingold, Mordechai. "The Grounds for Conflict: Grienberger, Grassi, Galileo, and Posterity." In *The New Science and Jesuit Science: Seventeenth-Century Perspectives*, edited by Mordechai Feingold, 121–57. Dordrecht: Springer Netherlands, 2003.

Feingold, Mordechai, ed. *The New Science and Jesuit Science: Seventeenth-Century Perspectives*. Dordrecht: Kluwer Academic, 2003.

Feldhay, Rivka. "Knowledge and Salvation in Jesuit Culture." *Science in Context* 1 (1987): 195–213.

Feldhay, Rivka. "On Wonderful Machines: The Transmission of Mechanical Knowledge by Jesuits." *Science & Education* 15, no. 2 (2006): 151–72.

Feldhay, Rivka, and Ayelet Even-Ezra. "Heaviness, Lightness, and Impetus in the Seventeenth Century: A Jesuit Perspective." In *Emergence and Expansion of Preclassical Mechanics*, edited by Rivka Feldhay, Jürgen Renn, Matthias Schemmel, and Matteo Valleriani, 255–84. Cham: Springer International, 2018.

Fermat, Pierre de. *Oeuvres: [Suivi de] observations sur Diophante/de Fermat tome premier, oeuvres mathématiques diverses*. Edited by Paul Tannery and Charles Henry. Paris: Gauthier-Villars et fils, 1891.

Field, Judith V., and Jeremy J. Gray. *The Geometrical Work of Girard Desargues*. New York: Springer Verlag, 1987.

Fiolhais, Carlos, and José Eduardo Franco. "Portuguese Jesuits and Science: Continuities and Ruptures (16th–18th Centuries)." *Antiguos jesuitas en Iberoamérica* 5, no. 1 (2017): 163–78.

Fladt, Kuno. *Geschichte und Theorie der Kegelschnitte und der Flächen zweiten Grades*. Stuttgart: E. Klett Verlag, 1967.

Fletcher, John Edward. *Athanasius Kircher und seine Beziehungen zum Gelehrten Europa seiner Zeit*. Wiesbaden: Harrassowitz, 1988.

Fletcher, John Edward. *A Study of the Life and Works of Athanasius Kircher, "Germanus Incredibilis": With a Selection of His Unpublished Correspondence and an Annotated Translation of His Autobiography*. Leiden: Brill, 2011.

Folkerts, Menso. "Die Entwicklung und Bedeutung der Visierkunst als Beispiel der praktischen Mathematik der frühen Neuzeit." *Humanismus und Technik* 18, no. 1 (1974): 1–41.

Foppens, Jan Frans. *Bibliotheca Belgica, sive virorum in Belgio vita, scriptisque illustrium catalogus, librorumque nomenclatura*. Brussels: Petrus Foppens, 1739.

Fried, Michael N., and Sabetai Unguru. *Apollonius of Perga's Conics: Text, Context, Subtext*. Leiden: Brill, 2001.

García Villoslada, Ricardo. *Storia del Collegio Romano dal suo inizio (1551) alla soppressione della Compagnia di Gesù (1773)*. Rome: Università Gregoriana, 1954.

Garstein, Oskar. *Rome and the Counter-Reformation in Scandinavia: Jesuit Educational Strategy*. Leiden: Brill, 1992.

Gatto, Romano. "Christoph Clavius's *Ordo servandus in addiscendis disciplinis mathematicis* and the Teaching of Mathematics in Jesuit Colleges at the Beginning of the Modern Era." *Science and Education* 15 (2006): 235–58.

Gessner, Samuel. "The Conception of a Mathematical Instrument and Its Distance from the Material World: The 'Pantometra' in Lisbon, 1638." *Studium* 4 (2011): 210–27.

Giard, Luce. "Les collèges jésuites des anciens Pays-Bas et l'élaboration de la *Ratio studiorum*." In *The Jesuits of the Low Countries: Identity and Impact*, edited by Rob Faesen and Leo Kenis, 83–108. Leuven, Paris, and Walpole, MA: Peeters, 2012.

Gilissen, John. "Le Père Guillaume Hesius, architecte du XVIIe siècle." *Annales de La Société Royale d'Archéologie de Bruxelles: Mémoires, rapports et documents* 42 (1938): 216–55.

Golvers, Noël. "F. Verbiest's Mathematical Formation: Some Observations on Post-Clavian Jesuit Mathematics in Mid-17th Century Europe." *Archives Internationales d'Histoire des Sciences* 54 (2004): 29–47.

Golvers, Noël. "Ferdinand Verbiest's 1668 Observation of an Unidentified Celestial Phenomenon in Peking, Its Lost Description, and Some Parallel Observations, Especially in Korea." *Almagest: International Journal for the History of Scientific Ideas* 5, no. 1 (2014): 33–51.

Golvers, Noël. "The Instrument Collection of Prince-Bishop Ernest of Bavaria (Liège, 1612), Nicolas Trigault & Johann Terrentius, and the Chinese Destiny of Some of Its Items." *Archives Internationales d'Histoire des Sciences* 62, no. 169 (2012): 691–701.

Golvers, Noël. "De recruteringstocht van M. Martini, S.J. door de Lage Landen in 1654: Over geomantische kompassen, Chinese verzamelingen, lichtbeelden en R.P. Wilhelm van Aelst, S.J." *De Zeventiende Eeuw* 10 (1994): 331–44.

Golvers, Noël. "The XVIIth-Century Jesuit Mission in China and Its 'Antwerp Connections.'" *De Gulden Passer* 74 (1996): 157–88.

Golvers, Noël, and Ulrich Libbrecht. *Astronoom van de Keizer: Ferdinand Verbiest en zijn Europese sterrenkunde*. Leuven: Davidsfonds, 1988.

Goris, Jan-Albert. *Lof van Antwerpen: hoe reizigers Antwerpen zagen, van de XVe tot de XXe Eeuw*. Brussels: Standaard, 1940.

Gorman, Michael John. "From 'The eyes of all' to 'Usefull quarries in philosophy and good literature': Consuming Jesuit Science, 1600–1655." In *The Jesuits: Cultures, Sciences, and the Arts 1540–1773*, edited by John W. O'Malley, s.j., Gauvin Alexander Bailey, Steven J. Harris, and T. Frank Kennedy, 170–89. Toronto, Buffalo, and London: University of Toronto Press, 1999.

Gorman, Michael John. "Mathematics and Modesty in the Society of Jesus: The Problems of Christoph Grienberger." In *The New Science and Jesuit Science: Seventeenth-Century Perspectives*, edited by Mordechai Feingold, 1–120. Dordrecht: Kluwer Academic, 2003.

Gray, Howard, s.j. "The Experience of Ignatius Loyola." In *The Jesuit* Ratio studiorum *of 1599: 400th Anniversary Perspectives*, edited by Vincent J. Duminuco, 1–21. New York: Fordham University Press, 2000.

Grendler, Paul F. "The Culture of the Jesuit Teacher 1548–1773." *Journal of Jesuit Studies* 3, no. 1 (2016): 17–41.

Grendler, Paul F. "Jesuit Schools in Europe: A Historiographical Essay." *Journal of Jesuit Studies* 1, no. 1 (2014): 7–25.

Grendler, Paul F. *The Jesuits and Italian Universities, 1548–1773*. Washington, DC: Catholic University of America Press, 2017.

Grendler, Paul F. "Laínez and the Schools in Europe." In *Diego Laínez (1512–1565) and His Generalate: Jesuit with Jewish Roots, Close Confidant of Ignatius of Loyola, Preeminent Theologian of the Council of Trent*, edited by Paul Oberholzer, s.j., 649–78. Rome: Institutum Historicum Societatis Iesu, 2015.

Grendler, Paul F. *The Universities of the Italian Renaissance*. Baltimore, MD: Johns Hopkins University Press, 2002.

Grienberger, Christopher. *Euclidis sex primi Elementorum geometricorum libri cum parte undecimi ex majoribus Clavii commentariis [...]: Contracti per P. Christophorum Grienbergum e Societate Iesu; Brevis trigonometria planorum, cum tabulis sinuum, tangentium, et secantium, ad pantes*. Graz: Haeredes Ernesti Widmanstadii, 1636.

Guicciardini, Lodovico. *Antwerpen, Mechelen, Lier e.a. (vertaald door Kiliaan)*. Deurne: Soethoudt, 1979.

Haeger, Barbara. "The Façade of the Jesuit Church in Antwerp: Representing the Church Militant and Triumphant." In *Innovation and Experience in the Early Baroque in the Southern Netherlands: The Case of the Jesuit Church in Antwerp*, edited by Piet Lombaerde, 97–124. Turnhout: Brepols, 2008.

Harris, Steven J. "Transposing the Merton Thesis: Apostolic Spirituality and the Establishment of the Jesuit Scientific Tradition." *Science in Context* 3 (1989): 29–65.

Heath, Thomas L. *Apollonius of Perga: Treatise on Conic Sections*. Cambridge: W. Heffer, 1961.

Heath, Thomas L. *Euclid: The Thirteen Books of the Elements (3 Vols.)*. New York: Dover Publications, 1956.

Heilbron, J. L. [John Lewis]. *The Sun in the Church: Cathedrals as Solar Observatories*. Cambridge, MA, and London: Harvard University Press, 1999.

Hellyer, Marcus. *Catholic Physics: Jesuit Natural Philosophy in Early Modern Germany*. Notre Dame, IN: University of Notre Dame Press, 2005.

Hengst, Karl. *Jesuiten an Universitäten und Jesuitenuniversitäten: Zur Geschichte der Universitäten in der oberdeutschen und rheinischen Provinz der Gesellschaft Jesu im Zeitalter der konfessionellen Auseinandersetzung*. Paderborn: Schöningh, 1981.

Heredia Herrera, Antonia. *Catálogo de las consultas del Consejo de Indias, 8 Vols*. Seville: Diputación Provincial Sevilla, n.d.

Herman, J. B. *La pédagogie des jésuites au 16e siècle: Ses sources, ses caractéristiques*. Leuven: Bureaux de recueil, 1914.

Hermans, Michel, and Jean-François Stoffel, eds. *Le Père Henri Bosmans SJ, (1852–1928): Historien des mathématiques*. Brussels: Académie Royale de Belgique, 2010.

Heron of Alexandria. *The Mechanics of Heron of Alexandria*. Edited by Jutta Miller, 1999. http://www.faculty.umb.edu/gary_zabel/Courses/Bodies,%5C (accessed February 14, 2020).

Hesius, Willem. "Emblemata de fide, spe, charitate." Manuscript, Erfgoedbibliotheek Hendrik Conscience, Antwerp, B 129141, 1624.

Hesius, Willem, s.j. *Gvilielmi HesI* [...] *Emblemata sacra de fide, spe, charitate*. Antwerp: Ex officina Plantiniana Balthasaris Moreti, 1636.

Hoffmann, Hermann. "Der Breslauer Mathematiker Theodor Moretus, S.J." *Jahresbericht der Schlesischer Gesellschaft für Vaterländische Cultur* 107 (1934): 118–55.

Hofmann, Joseph E. *Das Opus geometricum der Gregorius à Sancto Vincentio und seine Einwirkung auf Leibniz*. Vol. 13. Abhandlungen der Preußischer Akademie der Wissenschaften. Berlin: Verlag der Akademie der Wissenschaften, 1941.

Hogendijk, Jan P. "Desargues's *Brouillon Project* and the *Conics* of Apollonius." *Centaurus* 34 (1991): 1–43.

Homann, Frederick A. "Christopher Clavius and the Renaissance of Euclidean Geometry." *Archivum historicum Societatis Iesu* 52 (1983): 233–46.

Homann, Frederick A. "Introduction." In *Church, Culture, and Curriculum: Theology and Mathematics in the Jesuit* Ratio studiorum, edited by László Lukács, s.j., and Guiseppe Cosentino, 1–16. Philadelphia: Saint Joseph's University Press, 1999.

Hufton, Olwen. "Every Tub on Its Own Bottom: Funding a Jesuit College in Early Modern Europe." In *The Jesuits II: Cultures, Sciences, and the Arts 1540–1773*, edited by John W. O'Malley, s.j., Gauvin Alexander Bailey, Steven J. Harris, and T. Frank Kennedy, 5–23. Toronto, Buffalo, and London: University of Toronto Press, 2006.

Huygens, Christiaan. *Oeuvres complètes: Tome I; Correspondance 1638–1656*. Edited by David Bierens de Haan. The Hague: Martinus Nijhoff, 1888.

Huygens, Christiaan. *Oeuvres complètes: Tome II; Correspondance 1657–1659*. Edited by David Bierens de Haan. The Hague: Martinus Nijhoff, 1889.

Imhof, Dirk. *Jan Moretus and the Continuation of the Plantin Press: A Bibliography of the Works Published and Printed by Jan Moretus I in Antwerp (1589–1610)*. Leiden: Brill, 2014.

Israel, Jonathan I. *The Dutch Republic*. Oxford: Clarendon Press, 1998.

Jaffe, Michael. "Rubens and Optics: Some Fresh Evidence." *Journal of the Warburg and Courtauld Institutes* 34 (1971): 362–66.

Janssens, Paul, ed. *België in de 17de Eeuw: De Spaanse Nederlanden en het Prinsbisdom Luik; Band I; Politiek*. Brussels: Dexia Bank-Snoeck, 2006.

Janssens, Paul, ed. *België in de 17de Eeuw: De Spaanse Nederlanden en het Prinsbisdom Luik; Band II; Cultuur en leefwereld*. Brussels: Dexia Bank-Snoeck, 2006.

Jullien, Vincent. "Archimedes and Indivisibles." In *Seventeenth-Century Indivisibles Revisited*, edited by Vincent Jullien, 451–57. Basel: Birkhäuser, 2015.

Jullien, Vincent. "Roberval's Indivisibles." In *Seventeenth-Century Indivisibles Revisited*, edited by Vincent Jullien, 177–210. Basel: Birkhäuser, 2015.

Kemp, Martin. *The Science of Artisans*. New Haven and London: Yale University Press, 1990.

Kepler, Johannes. *Außzug Außder Uralten MesseKunst Archimedis* [...]. Linz: Hansen Blancken, 1616.

Kepler, Johannes. *Briefe 1612–1620*. Gesammelte Werke/Kepler, Johannes 17. Munich: Beck, 1955.

Kircher, Athanasius. *Ars magna lucis et umbræ in decem libros digesta*. Rome: Hermanni Scheus, 1646.

Knaap, Anna C. "Meditation, Ministry, and Visual Rhetoric in Peter Paul Rubens's Program for the Jesuit Church in Antwerp." In *The Jesuits II: Cultures, Sciences, and the Arts 1540–1773*, edited by John W. O'Malley, s.j., Gauvin Alexander Bailey, Steven J. Harris, and T. Frank Kennedy, s.j., 157–81. Toronto, Buffalo, and London: University of Toronto Press, 2006.

Krayer, Albert. *Mathematik im Studienplan der Jesuiten: Die Vorlesung von Otto Cattenius an der Universität Mainz (1610/11)*. Stuttgart: Franz Steiner Verlag, 1991.

Lamalle, Edmond, s.j. "La propagande du P. Nicolas Trigault en faveur des missions de Chine (1616)." *Archivum historicum Societatis Iesu* 9 (1940): 49–120.

Lattis, James M. *Between Copernicus and Galileo: Christoph Clavius and the Collapse of Ptolemaic Cosmology*. Chicago: University of Chicago Press, 1994.

Leitão, Enrique. "Entering Dangerous Ground: Jesuits Teaching Astrology and Chiromancy in Lisbon." In *The Jesuits II: Cultures, Sciences, and the Arts 1540–1773*, edited by John W. O'Malley, s.j., Gauvin Alexander Bailey, Steven J. Harris, and T. Frank Kennedy, s.j., 371–84. Toronto, Buffalo, and London: University of Toronto Press, 2006.

Leitão, Enrique. "Jesuit Mathematical Practice in Portugal 1540–1759." In *The New Science and Jesuit Science: Seventeenth-Century Perspectives*, edited by Mordechai Feingold, 229–47. Dordrecht: Springer Netherlands, 2003.

Lennon, Thomas M. "The Significance of the Barrovian Case." *Studies in History and Philosophy of Science* 38, no. 1 (2007): 36–55.

Lesger, C. [Clé]. *Handel in Amsterdam Ten Tijde van de Opstand*. Hilversum: Verloren, 2001.

Leyder, Dirk. "L'éclosion scolaire." *Paedagogica historica* 36, no. 3 (2000): 1003–51.

Lipenio, Martino. *M. Martini Lipenii bibliotheca realis philosophica omnium materiarum, rerum, & titulorum, in vniverso totivs philosophiæ ambitu occurrentium: Ordine alphabetico sic disposita, ut primo statim aspectu titvli, et sub titulis autores ordinata velut acie dispo*. Frankfurt am Main: Fridericus, 1682.

Loach, Judi. "Revolutionary Pedagogues? How Jesuits Used Education to Change Society." In *The Jesuits II: Cultures, Sciences, and the Arts 1540–1773*, edited by John W. O'Malley, s.j., Gauvin Alexander Bailey, Steven J. Harris, and T. Frank Kennedy, s.j., 66–85. Toronto, Buffalo, and London: University of Toronto Press, 2006.

Lombaerde, Piet. "The Façade and the Towers of the Jesuit Church in the Urban Landscape of Antwerp during the Seventeenth Century." In *Innovation and Experience in the Early Baroque in the Southern Netherlands: The Case of the Jesuit Church in Antwerp*, edited by Piet Lombaerde, 77–96. Turnhout: Brepols, 2008.

Lucca, Denis De. *Jesuits and Fortifications: The Contribution of the Jesuits to Military Architecture in the Baroque Age*. Leiden: Brill, 2012.

Lukács, László, s.j. "A History of the Jesuit *Ratio studiorum*." In *Church, Culture, and Curriculum: Theology and Mathematics in the Jesuit* Ratio studiorum, edited by László Lukács, s.j., and Guiseppe Cosentino, 17–46. Philadelphia: Saint Joseph's University Press, 1999.

Lukács, László, s.j., ed. *Monumenta pædagogica Societatis Iesu 7 Vols*. Rome: Institutum Historicum Societatis Iesu, n.d.

M. D. [Jean Charles Joseph de Vegiano (seigneur de Hovel)]. *Nobiliaire des Pays-Bas, et du comté de Bourgogne, Volume 2*. Mechelen: P. J. Hanicq, 1779.

MacDonnell, Joseph F., s.j. *Jesuit Geometers: A Study of Fifty-Six Prominent Jesuit Geometers during the First Two Centuries of Jesuit History*. St. Louis, MO, and Vatican City: Institute of Jesuit Sources, 1989.

Magruder, Kerry V. "Jesuit Science after Galileo: The Cosmology of Gabriele Beati." *Centaurus* 51 (2009): 189–212.

Mahoney, Michael Sean. *The Mathematical Career of Pierre de Fermat*. Princeton: Princeton University Press, 1973.

Mannaerts, Rudi. *Sint-Carolus Borromeus: De Antwerpse Jezuïetenkerk, een Openbaring*. Antwerp: Vzw Maria-Elisabeth Belpaire, vzw Toerismepastoraal Antwerpen, 2011.

Marinus, Marie Juliette. *De Contrareformatie te Antwerpen (1585–1676): Kerkelijk leven in een grootstad*. Brussels: Paleis der Academiën, 1995.

Marinus, Marie Juliette. "De financiering van de contrareformatie te Antwerpen (1585–1700)." In *Geloven in het Verleden: Studies over het godsdienstig leven in de Vroegmoderne Tijd, aangeboden aan Michel Cloet*, edited by Eddy Put Marinus, Marie Juliette, Hans Storme, 239–52. Leuven: Universitaire Pers, 1996.

Marinus, Marie Juliette. *Laevinus Torrentius als tweede bisschop van Antwerpen (1587–1595)*. Brussels: Paleis der Academiën, 1989.

Marnef, Guido. *Antwerp in the Age of Reformation: Underground Protestantism in a Commercial Metropolis*. Baltimore, MD: Johns Hopkins University Press, 1996.

Marnef, Guido. *Antwerpen in de tijd van de Reformatie: Ondergronds protestantisme in een handelsmetropool 1550–1577*. Antwerp: Kritak, 1996.

Marnef, Guido. "Protestant Conversions in an Age of Catholic Reformation: The Case of Sixteenth-Century Antwerp." In *The Low Countries as a Crossroads of Religious Belief*, edited by Arie J. Gelderblom, Jan L. de Jong, and Marc van Vaeck, 33–48. Leiden: Brill, 2004.

Martin, John R. *The Ceiling Paintings for the Jesuit Church in Antwerp*. Brussels: Arcade Press, 1968.

Maryks, Robert A. *The Jesuit Order as a Synagogue of Jews: Jesuits of Jewish Ancestry and Purity-of-Blood Laws in the Early Society of Jesus*. Leiden: Brill, 2009.

Meeus, Hubert. "Peeter Heyns, a 'French Schoolmaster.'" In *Grammaire et enseignement du Français 1500–1700*, edited by Jan de Clercq, Nico Lioce, and Pierre Swiggers, 301–16. Leuven: Peeters, 2000.

Meeus, Hubert. "De *Spieghel der Werelt* als spiegel van Peeter Heyns." In *Ortelius's Spieghel der Werelt: A Facsimile for Francine de Nave*, edited by Elly Cockx-Indestege, Dirk Imhof, Hubert Meeus, and Norbert Moermans, 31–47. Antwerp: Vereniging van Antwerpse Bibliofielen, 2009.

Memorial histórico Español. Madrid: Academia Real de la Historia, 1851.

Mennher, Valentin. *Practique pour brievement apprendre a ciffrer, & tenir livre de domptes*. Anvers: Aegidius Diest [Coppens van Diest, Gillis I], 1565.

Mersenne, Marin. *F. Marini Mersenni minimi cogitata physico-mathematica, Volumes 1–3*. Paris: Antoni Bertier, 1644.

Meskens, Ad. *Familia universalis: Coignet*. Antwerp: Koninklijk Museum voor Schone Kunsten, 1998.

Meskens, Ad. *Joannes della Faille, S.J.: Mathematics, Modesty, and Missed Opportunities*. Brussels and Rome: Belgisch Historisch Instituut te Rome, 2005.

Meskens, Ad. "Die Mathematikschule der Antwerpener Jesuiten." In *Die Entwicklung der Mathematik in der Frühen Neuzeit*, edited by Rainer Gebhardt. Annaberg-Buchholz: Adam-Ries-Bund, 2020.

Meskens, Ad. "Michiel Coignet's Contribution to the Development of the Sector." *Annals of Science* 54 (1997): 143–60.

Meskens, Ad. "Peter Heyns: Leaving the Market Place of the World." In *Rechenmeister und Mathematiker der frühen Neuzeit*, edited by Rainer Gebhardt, 287–96. Annaberg-Buchholz: Adam-Ries-Bundes, 2017.

Meskens, Ad. "The Portrait of Jan-Karel Della Faille by Anthony van Dijck." *Koninklijk Museum voor Schone Kunsten Jaarboek* 39 (1999): 124–37.

Meskens, Ad. *Practical Mathematics in a Commercial Metropolis: Mathematical Life in Late 16th-Century Antwerp*. Dordrecht: Springer Science & Business Media B.V., 2013.

Meskens, Ad. "Some New Biographical Data about Michiel Coignet." *Nuncius* 17 (2002): 447–54.

Meskens, Ad. "Veralgemening van de stelling van Pythagoras." *Wiskunde en Onderwijs* 45, no. 177 (2019): 26–29.

Meskens, Ad. *Wiskunde tussen Renaissance en Barok, aspecten van wiskunde-beoefening te Antwerpen 1550–1620*. Antwerp: Stadsbibliotheek Antwerpen, 1994.

Meskens, Ad, and Godelieve van Hemeldonck. "Quignet, Quingetti, Cognget, Coignet: An Antwerp Family of Goldsmiths, Some Painters, One Mathematician, and a Lot of Merchants." *Antwerp Royal Museum Annual 2015–2016* (2018): 75–138.

Meskens, Ad, and Paul Tytgat. *Exploring Classical Greek Construction Problems with Interactive Geometry Software*. Compact Textbooks in Mathematics. Basel: Birkhäuser Verlag, 2017.

Moehlig, Marianne. "Het Jezuïetencollege te Antwerpen in de 17de en 18de eeuw." Master's thesis, Katholieke Universiteit Leuven, 1988.

Möllmann, Heinz-Helmut. *Über Beweise und Beweisarten bei Wilhelm Ockham*. Amsterdam and Philadelphia: John Benjamins, 2013.

Moretus, Theodorus. "Exercitationes mathematicae, poetice atque sermones." Manuscript, Národní Knihovna VI B 12, Prague, n.d.

Moretus, Theodorus. "Praelecyiones naturalis Theodori Moreti, S.J." Manuscript, Národní Knihovna XIV G7, Prague, n.d.

Moretus, Theodorus. *Tractatus physico-mathematicus de Aestu Maris*. Antwerp: Apud Jacobum Meursium, 1665.

Moretus, Theodorus. *Tractatus physico-mathematicus de Aestu Maris*. Vienna: Ignatius Dominicus Voigt, 1719.

Moretus, Theodorus, and Ferdinand Ernest L. B. de Bukau. *Virgini Matri fonti sapientiae e terra in caelos exilienti hoc de fontibus problema mathematicum dicabat illustrimus D. Ferdinandus Ernestus L.B. de Buckaw philosophiae et matheseos auditor*. Prague: Typis academicis, 1641.

Moretus, Theodorus, Georg Franz Czesch, and Georg Becke. *Propositiones mathematicae ex optica, de imagine visionis, demonstratio prooemialis matutina tubum opticum conficere ex lentibus sphaericis aeriis, quibus intra aquam [...] Eo modo imagines rerum repraesentantur, sicut passim crystallinis*. Wrocław: Grunder, 1661.

Moretus, Theodorus, and Gottfried Fibig. *Propositiones mathematicae ex harmonica, de soni magnitudine/Propositae [...] a nobili & erudito D. Godefrido Fibig, Silesio Vratislaviensi [...] Anno 1664: Die 4. Septembris, horis pomeridianis, in gymnasio Caesarei Regiiq[Ue] Collegii Vratislaviensis in B*. Wrocław: Baumannische Druckerey Johann Christoph Jacob, 1664.

Moretus, Theodorus, and Friedrich Füssel. "De mathematica et scientiis naturalibus varii: Praecepta medicinalia." Manuscript, Knihovna kláštera premonstrátu Teplá d23, 1635.

Moretus, Theodorus, and Caspar Knittel. *Propositiones mathematicae ex astronomia de luna paschali et solis motu*. Wrocław: Jacobi, 1666.

Moretus, Theodorus, and Josephus Nicotius. *Propositiones: Mathematicae; Ex hydro-statica de prima; Suppositione; Archimedis*. Brzeg: Christoph Tschorn, 1667.

Moretus, Theodorus, and Johann Heinrich Joseph von Schenckendorff. *Propositiones mathematicae ex geographia de Aestu Maris/A [...] D. Joanne Henrico Josepho a Schenckendorff & Milgast, philosophiae & matheseos auditore: Die & anno DeCIMo QVInto IULII in gymnasio Caesarei Regiiq[Ue] Collegii Vratislaviensis in Burgo Socie*. Wrocław: Baumannische Druckerey Johann Christoph Jacob, 1665.

Moretus, Theodorus, Joannes Ignatius Stephan, and Balthasar Kirstenio. *Propositiones mathematicae, ex statica de raro et denso, demonstratio prooemialis matutina, de aequilibri in liquido natatantium difficultate & modo, propositio a D. Joanne Ignatio Stephan Wansoviensi Silesio, physicae & matheseos auditore: Demonstratio P.* Wrocław: Baumannische Druckerey Johann Christoph Jacob, 1660.

Muller, Jeffrey. "Jesuit Uses of Art in the Province of Flanders." In *The Jesuits II: Cultures, Sciences, and the Arts 1540–1773*, edited by John W. O'Malley, S.J., Gauvin Alexander Bailey, Steven J. Harris, and T. Frank Kennedy, S.J., 113–56. Toronto, Buffalo, and London: University of Toronto Press, 2006.

Napolitani, Pier Daniele. "Between Myth and Mathematics: The Vicissitudes of Archimedes and His Work." *Lettera matematica* 1, no. 3 (2013): 105–12.

Narratio de Christina Svecorum serenissima Regina Gvstavi Adolphi Regis. Liège: Joannes Mathias Hovius, 1656.

Naux, Charles. "Grégoire de St. Vincent et les propriétés logarithmiques de l'hyperbole équilatère." *Revue des questions scientifiques* 143, no. 2 (1972): 209–21.

Navarro-Loidi, Juan, and José Llombart. "The Introduction of Logarithms into Spain." *Historia mathematica* 35 (2008): 83–101.

Noyes, Ruth Sargent. "'Per modum compendii a Leonardo Damerio Leodiensi in lucem editum': Odo van Maelcote and Leonard Damery's *astrolabium aequinoctiale* and the Catholic Reformation Converting (Im)print." In *Winning Back with Books and Print: At the Heart of the Catholic Reformation in the Low Countries (16th–17th Centuries)*, edited by Renaud Adam, Rosa De Marco, Malcolm Walsby. Library of the Written Word Series. Leiden: Brill, forthcoming 2021.

Nuchelmans, Gabriel. "A 17th-Century Debate on the *Consequentia mirabilis*." *History and Philosophy of Logic* 13 (1992): 43–58.

Nuyts, C. J. [Charles Joseph]. *Philippe Nutius à la cour de Suède*. Brussels: J. Vandereydt, 1836.

Oberholzer, Paul, S.J., ed. *Diego Laínez (1512–1565) and His Generalate: Jesuit with Jewish Roots, Close Confidant of Ignatius of Loyola, Preeminent Theologian of the Council of Trent*. Rome: Institutum Historicum Societatis Iesu, 2015.

O'Malley, John W., S.J. *The First Jesuits*. Cambridge, MA: Harvard University Press, 1993.

O'Malley, John W., S.J. "How the First Jesuits Became Involved in Education." In *The Jesuit Ratio studiorum of 1599: 400th Anniversary Perspectives*, edited by Vincent J. Duminuco, 56–74. New York: Fordham University Press, 2000.

Ostermann, Alexander, and Gerhard Wanner. *Geometry by Its History*. Dordrecht: Springer Verlag, 2012.

Ottenheym, Koen A. "De Correspondentie Tussen Rubens en Huygens over Architectuur." *Bulletin KNOB* 96 (1997): 1–11.

Ottenheym, Konrad. "Peter Paul Rubens's 'Palazzi di Genova' and Its Influence on Architecture in the Netherlands." In *The Reception of P. P. Rubens's "Palazzi di*

Genova" during the 17th Century in Europe: Questions and Problems, edited by Piet Lombaerde, 81–98. Turnhout: Brepols, 2002.

Ottenheym, Konrad, and Krista de Jonge. "Civic Prestige: Building the City 1580–1700." In *Unity and Discontinuity: Architectural Relationships between the Southern and Northern Low Countries (1530–1700)*, edited by Krista de Jonge and Konrad Ottenheym, 209–50. Turnhout: Brepols, 2007.

Paar, Edwin. "Jan Ciermans: Een Bossche vestingbouwkundige in Portugal." *De Brabantse Leeuw* (2000): 201–16.

Pappus of Alexandria. *La collection mathématique*. Edited by Paul ver Eecke. Bruges: Desclée De Brouwer, 1933.

Parker, Geoffrey. *The Army of Flanders and the Spanish Road*. Cambridge: Cambridge University Press, 1972.

Parker, Geoffrey. *Spain and the Netherlands: Ten Studies*. Glasgow: Fontana Press, 1990.

Parker, Geoffrey. *The Thirty Years' War*. London: Routledge, 1993.

Parkhurst, Charles. "Aguilonius's *Optics* and Rubens's Colour." *Nederlands Kunsthistorisch Jaarboek* 12 (1961): 35–49.

Pavur, Claude, s.J., ed. *The* Ratio studiorum: *The Official Plan for Jesuit Education*. St. Louis, MO: Institute of Jesuit Sources, 2005.

Piñeiro, Mariano Esteban, and Mauricio Jalón. "Juan de Herrera and the Royal Academy of Mathematics." In *Scientific Instruments in the Sixteenth Century: The Spanish Court and the Louvain School*, edited by Jacques van Damme, Koenraad van Cleempoel, and Gérard L'Estrange Turner, 33–42. Madrid: Fundación Carlos de Amberes, 1997.

Poncelet, Alfred, s.J. *Histoire de la Compagnie de Jésus dans les anciens Pays-Bas: Établissement de la Compagnie de Jésus en Belgique et ses développements jusqu'à la fin du règne d'Albert et d'Isabelle*. Brussels: Marcel Hayez, n.d.

Price, Audrey. "Pure and Applied: Christopher Clavius's Unifying Approach to Jesuit Mathematics Pedagogy." PhD diss., University of California at San Diego, 2017.

Prims, Floris. "Letterkundigen, Geleerden en Kunstenaars." *Verslagen en Mededeelingen van de Koninklijke Vlaamsche Academie voor Taal- en Letterkunde* (March 1931): 171–221.

Put, Eddy. *De Cleijne Scholen: Het volksonderwijs in het Hertogdom Brabant tussen Katholieke Reformatie en Verlichting (eind 16de eeuw-1795)*. Leuven: Universitaire Pers, 1990.

Put, Eddy, Marie-Juliette Marinus, and Hans Storme, eds. *Geloven in Het Verleden*. Leuven: Universitaire Pers Leuven, 1996.

Put, Eddy, and Maurice Wynants, eds. *De Jezuïeten in de Nederlanden en het Prinsbisdom Luik (1542–1773)*. Brussels: Algemeen Rijksarchief, 1991.

Put, Erik [Puteanus, Erycius]. *De cometa anni M.DC.XVIII: Novo mundi spectaculo, libri dvo. Paradoxologia*. Leuven: Apud Bernardinum Masium, 1619.

Radelet-de Grave, Patricia. "Guarini et la structure de l'Univers." *Nexus Network Journal* 11 (2009): 393–414.

Radelet-de Grave, Patricia. "Kepler, Cavalieri, Guldin: Polemics with the Departed." In *Seventeenth-Century Indivisibles Revisited*, edited by Vincent Jullien, 57–86. Basel: Birkhäuser, 2015.

Radelet-de Grave, Patricia. "Matematica, architettura e meccanica nella scuola di François d'Aguillon e di Grégoire de Saint-Vincent." In *Matematica, arte e tecnica nella storia, in memoria di Tullio Viola*, edited by Livia Giacardi and Clara Silvia Roero, 275–92. Turin: Ken Williams Books, 2006.

Radelet-de Grave, Patricia, and Edoardo Benvenuto, eds. *Entre mécanique et architecture/Between Mechanics and Architecture*. Basel: Birkhäuser, 1995.

Raeymaekers, Dries. *One Foot in the Palace: The Habsburg Court of Brussels and the Politics of Access in the Reign of Albert and Isabella, 1598–1621*. Leuven: Leuven University Press, 2013.

Raphael, Renee. "Teaching Sunspots: Disciplinary Identity and Scholarly Practice in the Collegio Romano." *History of Science* 52, no. 2 (2014): 130–52.

Riccioli, Giovanni Battista, s.j. *Almagestum novum astronomiam veterem novamque complectens observationibus aliorum, et propiis novisque theorematibus, problematibus, ac tabulis promotam, in tres tomos distributam quorum argumentum sequens pagina explicabit*. Bologna: Ex typographia haeredis Victorii Benatii, 1651.

Richard, Claude. *Conicorvm libri IV: Cvm commentariis*. Antwerp: Apud Hieronymum & Joannem Bapt. Verdussen, 1655–61.

Rocca, Giovanni Antonio, and Cosimo Gaetano Rocca. *Lettere d'uomini illustri del secolo XVII: A Giannantonio Rocca; Filosofo, e matematico reggiano*. Modena: Presso la Societa' Tipografica, 1785.

Rochhaus, Peter. "Adam Ries und die Annaberger Rechenmeister zwischen 1500 und 1604." In *Rechenmeister und Cossisten der Frühen Neuzeit*, edited by Rainer Gebhardt, 95–106. Annaberg: Technische Universität Bergakademie, 1996.

Roegiers, Jan. "Awkward Neighbours: The Leuven Faculty of Theology and the Jesuit College (1542–1773)." In *The Jesuits of the Low Countries: Identity and Impact*, edited by Rob Faesen and Leo Kenis, 153–76. Leuven, Paris, and Walpole, MA: Peeters, 2012.

Romano, Antonella. "Les jésuites et les mathématiques: Le cas des collèges français de la Compagnie de Jésus (1580–1640)." In *Christoph Clavius e attività scientifica dei gesuiti nell'età di Galileo*, edited by Ugo Baldini, 243–82. Rome: Bulzoni Editore, 1995.

Romano, Antonella. *La Contre-Réforme mathématique: Constitution et diffusion d'une culture mathématique jésuite à la Renaissance (1540–1640)*. Rome: École française de Rome, 1999.

Romano, Antonella. "Réflexions sur la construction d'un champ disciplinaire: Les mathématiques dans l'institution jésuite à la Renaissance." *Paedagogica historica* 40, no. 3 (2004): 245–59.

Roosens, Ben H. "Habsburgse defensiepolitiek en vestingbouw in de Nederlanden (1520–1560)." PhD diss., University of Leiden, 2005.

Rooses, Max, *Le Musée Plantin Moretus*. Antwerp: Zazzarini, 1913.

Rubens [illustrator], Peter Paul, Jacob de Wit [illustrator], and Jan Punt [engraver]. *De Plafonds, of Gallerystukken Uit de Kerk der Eerw. P.P. Jesuiten Te Antwerpen*. Amsterdam: Jan Punt, 1751.

Rubens, Peter Paul, and Nicolaes Ryckemans. *Palazzi Di Genova*. Antwerp: n.p., 1622.

Russo, Lucio. "Archimedes: Between Legend and Fact." *Lettera matematica* 1, no. 3 (2013): 91–95.

Sadler, James R. "Family in Revolt: The Van Der Meulen and Della Faille Families in the Dutch Revolt." PhD diss., University of California at Los Angeles, 2015.

Saito, Ken. "Archimedes and Double Contradiction Proof." *Lettera matematica* 1, no. 3 (2013): 97–104.

Sale, Giovanni, s.j. "Architectural Simplicity and Jesuit Architecture." In *The Jesuits and the Arts 1540–1773*, edited by John W. O'Malley, s.j., and Gauvin Alexander Bailey, 27–44. Philadelphia: Saint Joseph's University Press, 2005.

San Vicente, Gregorio a. *Opus geometricum posthumum ad mesolabium per rationum proportionalium novas Proprietates*. Ghent: Balduinus Manilius, 1668.

San Vicente, Gregorio a. *Problema Avstriacvm plvs vltra qvadratvra circvli*. Antwerp: Joannes et Jacobus Meursius, 1647.

San Vicente, Gregorio a, and Walter van Aelst. *Theoremata mathematica [...] Defenda ac demonstranda in Collegio Societatis Iesv Iouanij [...] Die 29. Iulij ante mtridiem*, [!] *Anno 1624*. Leuven: Henrici Hastenii, 1624.

San Vicente, Gregorio a, and Jan Ciermans. *Theoremata mathematica [...] Defenda ac demonstranda in Collegio Societatis Iesv Louanij [...] Die 29. Iulij ante mtridiem*, [!] *Anno 1624*. Leuven: Henrici Hastenii, 1624.

Sauvenier-Goffin, Elisabeth. "Les manuscrits de Grégoire de Saint-Vincent." *Bulletin de la Société royale des sciences de Liège* 20 (1951): 413–738.

Schmitz, Yves. *Les Della Faille*. Brussels: F. Van Buggenhoudt, n.d.

Schneider, Ivo. *Der Propotionalzirkel, Ein universelles Analoginstrument der Vergangenheit*. Munich: Deutsches Museum, 1970.

Schoffer, Ivo, Herman van der Wee, and Johannes A. Bornewasser. *De Lage Landen 1500–1780*. Amsterdam: Agon, 1988.

Schott, Caspar. *Magia universalis naturae et artis, sive recondita naturalium [...], Volume 1*. Bamberg: Joh. Martini Schönwetteri, 1677.

Schuppener, Georg. *Jesuitische Mathematik in Prag im 16. und 17. Jahrhundert (1556–1654)*. Leipzig: Leipziger Universitätsverlag, 1999.

Schuppener, Georg. "Theodor Moretus (1602–1667): Ein Prager Jesuiten-Mathematiker." In *Bohemia Jesuitica 1556–2006*, edited by Petronilla Cemus, 2:661–75. Prague and Würzburg: Nakladatelství Karolinum and Echter Verlag, 2010.

Sefrin-Weis, Heike. *Pappus of Alexandria: Book 4 of the* Collection. London: Springer, 2010.

Serlio, Sebastiano. *Des antiquites, le troisiesme livre translaté d'italien en Franchois*. Antwerp: Pieter Coecke van Aelst, 1550.

Serlio, Sebastiano. *Den Eersten Boeck van Architecturen [...] Tracterende van Geometrye*. Antwerp: Mayken Verhulst, 1553.

Serlio, Sebastiano. *Reglen van Metselrijen, Op de Vijve Manieren van Edificien, Te Wetene, Thuscana, Dorica, Jonica, Corintha, Ende Composita*. Antwerp: Peter Coecke van Aelst, 1549.

Serlio, Sebastiano. *Den Tweeden Boeck van Architecturen S. Serlii, Tracterende van Perspectyven, Dat Is, Het Insien Duer Toercorten*. Edited by Peter van Aelst. Antwerp: Mayken Verhulst, wwe Peter Coecke van Aelst, 1553.

Serlio, Sebastiano. *Den Vijfsten Boeck van Architecturen [...] Inden Welcken van Diversche Formen van Templen Getracteert Wordt*. Edited by Peter van Aelst. Antwerp: Mayken Verhulst, wwe Peter Coecke van Aelst, 1553.

Sherry, David. "The Jesuits and the Method of Indivisibles." *Foundations of Science* 23, no. 2 (2018): 367–92.

Shipley, Thorne, Kenneth Neil Ogle, and Samuel C. Rawlings. "The *Nonius horopter*: I; History and Theory." *Vision Research* 10, no. 11 (1970): 1225–62.

Šípek, Richard. *Die Jauerer Schlossbibliothek Ottos des Jüngeren von Nostitz Teil 1 und 2*. Frankfurt am Main: Peter Lang GmbH, 2014.

Smith, A. Mark, ed. and trans. "Alhacen on the Principles of Reflection: A Critical Edition, with English Translation and Commentary, of Books 4 and 5 of Alhacen's *De aspectibus*, the Medieval Latin Version of Ibn Al-Haytham's *Kitāb Al-Manāzir*; Volume One; Introduction and Latin." *Transactions of the American Philosophical Society* 96, no. 2 (2006): i–288.

Smith, A. Mark, ed. and trans. "Alhacen on the Principles of Reflection: A Critical Edition, with English Translation and Commentary, of Books 4 and 5 of Alhacen's *De aspectibus*, the Medieval Latin Version of Ibn al-Haytham's *Kitāb al-Manāzir*; Volume Two; English Translation." *Transactions of the American Philosophical Society* 96, no. 3 (2006): 289–697.

Smith, A. Mark, ed. and trans. "Alhacen on Image-Formation and Distortion in Mirrors: A Critical Edition, with English Translation and Commentary, of Book 6 of Alhacen's *De aspectibus*, the Medieval Latin Version of Ibn al-Haytham's *Kitāb al-Manāzir*; Volume One; Introduction A." *Transactions of the American Philosophical Society* 98, no. 1 (2008): i–153.

Smith, A. Mark, ed. and trans. "Alhacen on Image-Formation and Distortion in Mirrors: A Critical Edition, with English Translation and Commentary, of Book 6 of Alhacen's *De aspectibus*, the Medieval Latin Version of Ibn al-Haytham's *Kitāb al-Manāzir*;

Volume Two; English Transl." *Transactions of the American Philosophical Society* 98, no. 1 (2008): 155–393.

Smith, A. Mark, ed. and trans. "Alhacen on Refraction: A Critical Edition, with English Translation and Commentary, of Book 7 of Alhacen's *De aspectibus*, the Medieval Latin Version of Ibn al-Haytham's *Kitāb al-Manāzir*; Volume One; Introduction and Latin Text." *Transactions of the American Philosophical Society* 100, no. 3 (2010): i–212.

Smith, A. Mark, ed. and trans. "Alhacen's Approach to 'Alhacen's Problem.'" *Arabic Sciences and Philosophy* 18, no. 2 (2008): 143–63.

Smolarski, Dennis Chester. "The Jesuit *Ratio studiorum*, Christopher Clavius, and the Study of Mathematical Sciences in Universities." *Science in Context* 15, no. 3 (2002): 447–57.

Smolarski, Dennis Chester. "Jesuits on the Moon: Seeking God in All Things; Even Mathematics!" Seminar on Jesuit Spirituality, St. Louis, MO, 2005.

Smolarski, Dennis Chester. "Teaching Mathematics in the Seventeenth and Twenty-First Centuries." *Mathematics Magazine* 75, no. 4 (2002): 256–62.

Smolka, Josef, and René Zandbergen. "Athanasius Kircher und seine ersten Prager Korrespondenten." In *Bohemia Jesuitica 1556–2006*, edited by Petronilla Cemus, 2:677–704. Prague and Würzburg: Nakladatelství Karolinum and Echter Verlag, 2010.

Snaet, Joris. "De bouwprojecten voor de Antwerpse jezuïetenkerk." In *Bellissimi ingegni, grandissimo splendore*, edited by Krista de Jonge, Annemie de Vos, and Joris Snaet, 43–66. Leuven: Universitaire Pers Leuven, 2000.

Snaet, Joris, and Krista de Jonge. "The Architecture of the Jesuits in the Southern Low Countries: A State of the Art." In *La arquitectura jesuitica: Actas del simposio internacional*, edited by Isabel Álvaro-Zamora, Javier Ibanez Fernáñdez, and Jesús Criado Maínar, 239–76. Zaragoza: IFC, 2012.

Sober, Elliott. *Ockham's Razors: A User's Manual*. Cambridge: Cambridge University Press, 2015.

Sollier, Jean-Baptiste, Jean Pien, Willem Cuypers, and Pieter van den Bosch, eds. *Acta sanctorum Julii Tomus Septimus*. Antwerp: Jacob du Moulin, 1731.

Soly, Hugo. "De schepenregisters als bron voor de conjunctuurgeschiedenis van Zuid- en Noordnederlandse steden in het Ancien Régime: Een concreet voorbeeld; De Antwerpse immobiliënmarkt in de zestiende eeuw." *Tijdschrift Voor Geschiedenis* 87 (1974): 521–44.

Soly, Hugo. *Urbanisme en kapitalisme te Antwerpen in de 16de eeuw: De stedebouwkundige en industriële ondernemingen van Gilbert van Schoonbeke*. Brussels: Gemeentekrediet van België, 1977.

Stansel, Valentin. *Legatus Uranicus ex Orbe Novo in veterem, hoc est observationes Americanae cometarum factae, conscriptae ac in Europam missae*. Prague: Typis Universitatis Carolo-Ferdinandi, 1683.

Staubermann, Klaus. "Making Stars: Projection Culture in Nineteenth-Century German Astronomy." *British Journal for the History of Science* 34, no. 123 (2001): 439–51.

Stevin, Simon. *De Beghinselen des Waterwichts*. Leiden: Christoffel Plantijn, by Françoys van Raphelinghen, 1586.

Stevin, Simon. *De Beghinselen der Weeghconst*. Leiden: Christoffel Plantijn, by Françoys van Raphelinghen, 1586.

Stevin, Simon. *Hypomnemata mathematica, hoc est eruditus ille pulvis, in quo se exercuit [...] Maurits Princeps Auraicus*. Leiden: Jan Paets Jacobszoon, 1608.

Stevin, Simon. *Wisconstige Gedachtenissen: Inhoudende t'ghene Daer Hem in Gheoeffent Heeft*. Leiden: Jan Bouwensz, 1608.

Stradling, Robert A. *Europe and the Decline of Spain: A Study of the Spanish System, 1580–1720*. London: Allen & Unwin, 1981.

Stradling, Robert A. *Philip IV and the Government of Spain 1621–1665*. Cambridge: Cambridge University Press, 1988.

Swedenborg, Emanuel. *The* Principia: *Or, The First Principles of Natural Things*. London: W. Newbury, 1846.

Tacquet, Andreas. *Cylindricorum et annularium liber quintus addendus ad quatuor priores anno 1651 editus*. Antwerp: Jacobus Meursius, 1659.

Tacquet, Andreas. *Elementa Geometriae planae ac solidae: Quibus accedunt selecta ex Archimede theoremata*. Antwerp: Jacobus Meursium, 1654.

Tacquet, Andreas. *Opera mathematica*. Antwerp: Apud Henricum & Cornelium Verdussen, 1707.

Tacquet, Andreas, and Simon Laurent. *Illustrissimo D. Simoni Laurentio veterani baccalaureorum Jurisprudentiae fisco disputationibus mathematicis teste quae tota in capitibus suis adfuit et applausit Lovaniensi academia felicissime peractis singularem doctrinae et ingenii laudem consecuto Gr*. n.p.: n.p., n.d.

Tacquet, Andreas, and Philippe Eugène d'Hornes et d'Herlies. *Dissertatio physico-mathematica de motv circvli et sphœra*. Leuven: Corn. Coenestenii, 1650.

Tacquet, Andreas, and Theodore d'Imerselle. *Theodorvs d'Imerselle comes de Bovchove et S. Rom. Imp. defendet in Collegio Soc. Iesv lovanii Præside R.P. Andrea Tacquet eiusdem Soc. Math: Cos prof; E A° M DC LII Tertiâ Sept*. Antwerp, 1652.

Tacquet, Andreas, and Theodore d'Imerselle. *Theses mathematicæ [...] Quas serenissimo Archiduci Leopoldo Wilhelmo Dicatas [...] Demonstrabit illustrissimus Dominus Theodorus d'Imerselle [...] In Collegio Societatis Jesu lovanii [] Septembris*. Antwerp: Andreæ Bouveti, 1652.

Tacquet, Andreas, Philippe Eugène d'Hornes et d'Herlies (Comte de), Peter Dannoot [illustrator], and Antoon Sallaert [designer]. "Thesis Print." n.p.: n.p., 1650–51.

Tadisi, Jacopo Antonio. *Memorie della vita di Monsignore Giovanni Caramuel di Lobkowitz*. Venice: Giovanni Tevernin, 1760.

Thijs, Alfons K. L. "De Contrareformatie en het economisch transformatieproces te Antwerpen na 1585." *Bijdragen Tot de Geschiedenis* 70 (1987): 97–124.

Thijs, Alfons K. L. *Van Geuzenstad tot katholiek bolwerk: Antwerpen en de Contrareformatie*. Turnhout: Brepols, 1990.

Thomas, Werner, ed. *De Val van Het Nieuwe Troje: Het Beleg van Oostende, 1601–1604*. Leuven and Ostend: Davidsfonds, 2004.

Thomas, Werner, and Luc Duerloo, eds. *Albert and Isabella 1598–1621: Essays*. Turnhout: Brepols, 1998.

Thomas, Werner, and Luc Duerloo, eds. "Andromeda Unbound: The Reign of Albert and Isabella in the Southern Netherlands, 1598–1621." In *Albert and Isabella 1598–1621: Essays*, edited by Werner Thomas and Luc Duerloo, 1–14. Turnhout: Brepols, 1998.

Timmermans, Bert. "The Chapel of the Houtappel Family and the Privatisation of the Church in Seventeenth-Century Antwerp." In *Innovation and Experience in the Early Baroque in the Southern Netherlands: The Case of the Jesuit Church in Antwerp*, edited by Piet Lombaerde, 175–86. Turnhout: Brepols, 2008.

Treweek, A. P. [Athanasius Pryor]. "Pappus of Alexandria, The Manuscript Tradition of the *Collectio Mathematica*." *Scriptorium* 11, no. 2 (1957): 195–233.

Udías, Agustín. "Los libros y manuscritos de los profesores de matemáticas del Colegio Imperial de Madrid, 1627–1767." *Archivum historicum Societatis Iesu* 74, no. 148 (2005): 369–448.

Unguru, Sabetai. "Witelo and Thirteenth-Century Mathematics: An Assessment of His Contributions." *Isis* 63, no. 4 (1972): 496–508.

Uppenkamp, Barbara. "Gilles Coignet: A Migrant Painter from Antwerp and His Hamburg Career." *De Zeventiende Eeuw* 31 (2015): 55–77.

Van Ceulen, Ludolph. *De circulo et adscriptis liber*. Leiden: Colster, 1619.

Van Couwerven, Norbert Everard. *Sermoon Ter Eeren Vanden H. Ignatius, Fondateur Vande Societeyt Iesu [...] Tot Antwerpen, Op Den XXXI. Iulij, M.DC.LVI*. Antwerp: Balthasar Moretus, 1656.

Van de Vyver, Omer, s.j. "L'école de mathématiques des jésuites de la province Flandro-Belge au XVIIe siècle." *Archivum historicum Societatis Iesu* 49, no. 97 (1980): 265–71.

Van de Vyver, Omer, s.j. *Jan Ciermans (Pascasio Cosmander) 1602–1648, Wiskundige en vestingbouwer*. Leuven: Katholieke Universiteit Leuven, Departement wiskunde, 1975.

Van de Vyver, Omer, s.j. "Lettres de Jean Charles della Faille, S.J., cosmographe du roi à Madrid à M. Fl. van Langren, cosmographe du roi à Bruxelles." *Archivum historicum Societatis Iesu* 46, no. 91 (1977): 73–183.

Van den Gheyn, Joseph-Marie-Martin. *Catalogue des manuscrits de la Bibliothèque Royale de Belgique: 6; Histoire des ordres religieux et des églises particulières*. Brussels: Lamertin, 1906.

Van den Heuvel, Charles. *"Papiere Bolwercken": De introductie van de Italiaanse stede- en vestingbouw in de Nederlanden (1540–1609) en het gebruik van tekeningen.* Alphen aan den Rijn: Canaletto, 1991.

Van Eysinga, Willem J. M. *De wording van het Twaalfjarig Bestand van 9 april 1609.* Amsterdam: Noord-Hollandsche Uitgeversmij, 1959.

Van Goethem, Herman, Marie-Juliette Marinus, Piet Lenders, Carl Reyns, eds. *Antwerpen en de Jezuïeten, 1562–2002.* Antwerpen: UFSIA, 2002.

Van Looy, Herman. "De analogie tussen de bol en de ungula bij Gregorius a Sancto Vincentio." *Wiskunde En Onderwijs* 3 (1977): 56–62.

Van Looy, Herman. "Chronologie et analyse des manuscrits mathématiques de Grégoire de Saint Vincent (1584–1667)." *Archivum historicum Societatis Iesu* 49 (1980): 279–303.

Van Looy, Herman. "Chronologie en analyse van de mathematische handschriften van G. a. Sancto Vincentio (1584–1667)." PhD diss., Katholieke Universiteit Leuven, 1979.

Van Looy, Herman. "A Chronology and Historical Analysis of the Mathematical Manuscripts of Gregorius a Sancto Vincentio (1584–1667)." *Historia mathematica* 11 (1984): 57–75.

Van Nouhuys, Tabitta. *The Ages of Two-Faced Janus: The Comets of 1577 and 1618 and the Decline of the Aristotelian World View in the Netherlands.* Leiden: Brill, 1998.

Van Nouhuys, Tabitta. "Copernicus als randverschijnsel: De kometen van 1577 en 1618 en het wereldbeeld in de Nederlanden." *Scientiarum historia* 24 (1998): 17–38.

Van Roomen, Adriaan. *Ideæ mathematicae pars prima, sive methodus polygonorum.* Antwerp: Jan I van Keerberghen, 1593.

Van Sasse van Ysselt, A. [Alexander] F. O. "De transformatie der Illustre Lieve Vrouwe Broederschap te 's-Hertogenbosch." *Taxandria: Tijdschrift Voor Noordbrabantsche Geschiedenis en Volkskunde 2de R.* 13 (1906): 237–46.

Vanden Bosch, Gerrit. "Saving Souls in the Dutch Vineyard: The Missio Hollandica of the Jesuits (1592–1708)." In *The Jesuits of the Low Countries: Identity and Impact*, edited by Rob Faesen and Leo Kenis, 139–52. Leuven, Paris, and Walpole, MA: Peeters, 2012.

Vanderspeeten, Hubert P., S.J. *Le R.P. Jean-Charles della Faille de la Compagnie de Jésus dans la seconde moitié du XVIe siècle.* Brussels: Vromant, 1874.

Vanderspeeten, Hubert P., S.J. "Le R.P. Jean-Charles della Faille de la Compagnie de Jésus dans la seconde moitié du XVIe siècle." *Précis historiques 2e série* 3 (1874): 77–125.

Vanpaemel, Geert. *Echo's van een wetenschappelijke revolutie: de mechanistische natuurwetenschap aan de Leuvense Artesfaculteit (1650–1797).* Brussels: Koninklijke Academie voor Wetenschappen, Letteren en Schone Kunsten van België, 1986.

Vanpaemel, Geert. "Jesuit Mathematicians, Military Architecture, and the Transmission of Technical Knowledge." In *The Jesuits of the Low Countries: Identity and Impact*, edited by Rob Faesen and Leo Kenis, 109–28. Leuven, Paris, and Walpole, MA: Peeters, 2012.

Vanpaemel, Geert. "Mechanics and Mechanical Philosophy in Some Jesuit Mathematical Textbooks of the Early 17th Century." In *Mechanics and Natural Philosophy before the Scientific Revolution*, edited by Walter Roy Laird and Sophie Roux, 259–74. Dordrecht: Springer Netherlands, 2008.

Venard, Marc. "Y-a-t-il une 'Stratégie Scolaire' des jésuites en France au XVIe siècle?" In *L'Université de Pont-à-Mousson et les problèmes de son temps*, 67–85. Nancy: Université de Nancy 2, 1974.

Vermeir, René. *In Staat van Oorlog: Filips IV en de Zuidelijke Nederlanden 1629–1648*. Maastricht: Shaker, 2001.

Vermeir, René. "'Oorloghsvloeck en Vredens Zegen': Madrid, Brussel en de Zuid-Nederlandse Staten over oorlog en vrede met de Republiek, 1621–1648." *Bijdragen en Mededelingen betreffende de Geschiedenis der Nederlanden* 115 (2000): 1–32.

Vicente Maroto, M. Isabel, and Mariano Esteban Piñeiro. *Aspectos de la ciencia aplicada en la España del Siglo de Oro*. Salamanca: Junta de Castilla y Leòn, Consejería de Cultura y Turismo, 1991.

Wallart, Joannes Franciscus. *Illvstrissimo Domino D. Phillipo Evgenio, Comiti de Hornes & de Herlies &c., Domino svo dum publicam de optica, statica, architectura militari disputationem instituit, clypevm gentilitivm, [...] felici omine adumbrantem [...]*. n.p., 1677.

Wepster, Steven. "Hoe van Ceulen π Insloot." *Pythagoras* 49, no. 3 (2010): 26–28.

Wepster, Steven. "In de Ban van de Cirkel." *Euclides* 85, no. 3 (2009): 98–100.

Wepster, Steven. "Ludolph van Ceulen (1540–1610), Meester der Rekenmeesters." *Pythagoras* 49, no. 3 (2010): 12–15.

Wepster, Steven, and Marjanne de Nijs. *Meester Ludolphs Koordenvierhoek*. Utrecht: Epsilon Uitgaven, 2010.

Westra, Frans. *Nederlandse ingenieurs en de fortificatiewerken in het eerste tijdperk van de Tachtigjarige Oorlog, 1573–1604*. Alphen aan den Rijn: Uitg. Vis, 1992.

Wydra, Stanisław. *Historia matheseos in Bohemia et Moravia cultae*. Prague: Ioannem Adamum Hagen, 1778.

Yeomans, Donald K. *Comets: A Chronological History of Observation, Science, Myth, and Folklore*. Wiley Science Editions. New York: Wiley, 1991.

Ziggelaar, August, S.J. *François de Aguilón, S.J. (1567–1617): Scientist and Architect*. Rome: Institutum Historicum S.I., 1983.

Ziggelaar, August, S.J. "Peter Paul Rubens and François de Aguilón." In *Innovation and Experience in the Early Baroque in the Southern Netherlands: The Case of the Jesuit Church in Antwerp*, edited by Piet Lombaerde, 31–40. Turnhout: Brepols, 2008.

Ziggelaar, August, S.J. "The Sine Law of Refraction Derived from the Principle of Fermat: Prior to Fermat? The Theses of Wilhelm Boelmans, S.J., in 1634." *Centaurus* 24 (1980): 246–62.

Index

Acquaviva, Claudio   23, 30, 222, 225
Aguilón, Franciscus de   10, 82
   biography   62–63
   as an architect
      Saint Charles Borromeo
         Church   224–31
   as an architect   63, 223–31
   as confessor   26
   as teacher at mathematics school   26
   Catoptrics (manuscript)   68–80
   erection of mathematics school   30
   experimental philosophy   64–65
   influence on della Faille   172
   influence on San Vicente   30, 47, 68, 71, 80, 87, 94
   influenced by painters   65, 67
   on conics
      properties in an ellipse   76–79
   on optics   64–80, 244
      attenuation of light   66
      camera obscura   66–67
      colour theory   65
      harmonic pencils   72–73
      harmonic quadruples and harmonic conjugates   70–75, 78
      horopter concept   65–66, 67
      neusis-like constructions   68–70
      on Alhazen's problem   75–76
      on mirrors   75–76, 78–79
      on projections and perspective   67, 78, 237
      on trisection of an angle   68–70, 94, 98, 101
   Opticorum libri sex   64–67
Aguilón, Juan de   62n2
Aguilón, Pedro de   62n1
Albert, Ferdinand   30
Alegambe, Antonius   31, 190
Alegambe, Philippe   63
Alexander, Amir   ix, 58, 59, 60, 247
Alhazen   68, 71, 75, 76, 97, 101
   Alhazen's problem   75n40, 101 fig5.10
Álvarez de Toledo, Fernando, Duke of Alba   7n21, 25n79

America   4
Amsterdam   ix, 2, 3n3, 187
Andalusia   5
Annaberg   11n2
Anselmo, Antonio   2n3, 37n150
Antwerp   ix, x, 17, 30, 33, 41n158, 63
   churches   10, 184, 192, 208, 211, 222–32
      Chapel of Jesuit College   235–37
      Saint Charles Borromeo Church   see church, Charles Borromeo
   Counter-Reformation   8
   fortifications   63, 184, 203
   Gregorio a San Vicente in Antwerp   30–33, 82, 87, 94–115, 132
   Ignatius of Loyola and Antwerp   62n2
   Jesuit domus professa   29, 186, 187, 208, 220, 233n, 60
   Jesuits in Antwerp   6, 7, 8fo.2,9, 188
   merchants and trade   ix, 2, 2n2, 4, 63, 232
   plague epidemic   63
   printers   5n12, 20n50, 33, 39, 64, 172, 182, 184, 192, 198,
   Protestant Revolt in Antwerp   1, 7, 8
   Reconciliation   1, 8, 25,
   schools   11, 12, 49
      Jesuit College   see Jesuit College, Antwerp
Archdukes Albert and Isabella   2, 3, 5, 10, 29, 232, 239
Aristotle   13, 14, 172, 193
Army of Flanders   5, 218
Arriaga, Rodrigo de   85
astronomy
   Brahe, Tycho   21, 193n12
   Cassini, Giovanni Domenico
      eclipse of Jupiter   218
   comet   21, 31, 32, 218n33
   Copernican system   187n19, 196, 217
   Peurbach, Georg
      Planetary theory   14
   Ptolemaic system   37, 196, 217
   transit of Venus   21n54, 202–3
   Tychonic system   187n19, 196
Aynscombe, Thomas   10n37

Badajoz 240, 241
Barvoets, Alexander 34
Beati, Gabriele 21n50
Bellarmine, Robert, Cardinal 22
Besson, Jacques 35n147
Beys, Gilles 1
   printer 192
Bierthe, Thomas de 187
Blocq, Jacques du 223, 224
Boechelion, Karl 187
Boelmans, Willem 8, 33, 246
   biography 206
   and the ductus-method 87, 140
   as teacher at mathematics school 34, 199, 211
   manuscripts by Boelmans 87, 119–21, 140
   on optics 80
   sine law 207–8
Bohemia 33, 193, 232
Bosmans, Henri xii, 246, 249
Braikenridge, William 47
   MacLaurin-Braikenridge theorem 47, 104, 175
Breda
   reduction of 5, 239
Breslau 39, 194, 195, 197
Briggs, Henry
   logarithms 220
Broetius, Paschasius 198n40
Bruges 6, 81, 223, 255
   youth of San Vicente in Bruges 81
Brussels 6, 27, 62, 82, 189–90, 208, 218
   churches 223–24, 234
Bürgi, Jost 35n147

Candone, Guiseppi 218n33
Carafa, Vincenzo 189
Carpentier, Gillis 82
Casati, Paolo 21n50, 186, 187n17
Cassini, Giovanni Domenico 218
Catalonia 241
   Revolt (1650–52) 5, 171
Cavalieri, Bonaventura 47, 169, 213
   correspondence with Moretus 86n33, 195
   correspondence with San Vicente 85f4.2, 86
   Geometria Indivisibilibus 89, 157
   Jesuit response to the concept of indivisibles 58–61
   San Vicente's criticism 128–31
   Tacquet's criticism 214–15
   on indivisibiles 57–58, 246–47
   possible influence on San Vicente 109, 132, 140, 161
Charles V, Emperor xi, 6, 23
Chifflet, Laurent 40
Christina, Queen of Sweden 186
   Casati and de Malines's mission 186
   conversion and abdication 186–89
   de Manderscheidt's account 186
   Francken as intermediary 187–88
   Macedo as confessor 186
   Philip Nutius's mission 186–88
   sojourn in the Southern Netherlands 188–89
church
   Charles Borromeo, Antwerp 28f1.3, 224
   architectural features 228–30
   debts incurred to build 232
   designs by Aguilón 225–28
   interior 230–32
   Collégiale Notre-Dame-de-la-Crypte, Cassel 234, 235f14.6
   Discalced Carmelites' Church, Brussels 224
   Eglise Saint Loup, Namur 234
   Gesu, Rome 227
   Mons 223
   Saint Rumbold's Cathedral, Mechelen 234
   Scherpenheuvel 234–35, 236f14.7
   Sint-Michielskerk, Leuven
      design by Hesius 233–34
   Tournai
      cooperation du Blocq and Aguilón 224
Ciermans, Joannes 20n50, 33
   biography 198–99
   and mechanics 33, 51, 91–93, 210, 246
   as military architect 238–41
   as teacher at mathematics school 31, 34–35, 199
   correspondence with Descartes on optics 199, 201

Disciplinae Mathematicae 201–3
  in Portuguese military service 205
  in Spanish military service 205–6
  logarithms 139n9
  supervision of theses 203
Clavius, Christopher 83, 205
  and astronomy 21
  and mathematical curricula 16–17
  and Michiel Coignet 35–38
  and San Vicente 20–21, 81–82
  and the invention of the sector 35, 36f1.5
  and the Ratio Studiorum 17
  as mathematics teacher at the Roman College
    Academy of mathematics 18, 20–21
  as mathematics teacher at the Roman College 16, 18
  cited by San Vicente 94
Cobergher, Wenceslaus
  as architect 224
Coignet (family) 2n2
Coignet, Gillis ix, 2n3
Coignet, Michiel 35
  and astronomy 36–37
  and the sector 35–41, 202,
  as surveyor 63
Colégio de Santo Antão 205
Colegio Imperiale 41, 168
Cologne 2n2
Commandino, Federico 21
  editions of books 45, 46, 94n2, 99
Consejo de Indias 169, 241
Coster, Jacob de 41
Cox, Joannes 31, 191
Cuellar, Juan de 62n2

de Gottignies, Gilles-François 20, 21n50, 198, 246
  biography 218–20
  as astronomer 218–19
  as consultant for building plans 223
  Elementa geometriae planae 213, 219
  on logistics 219
de Gottignies, Lancelot 218
del Monte, Guidobaldo 35n147
Delange, Boudewijn 18n38, 24n76
Delgado, João 205
della Faille, family 2n2

della Faille, Jan (van) Karel sr. 165
della Faille, Joannes 5n12, 20n50, 39, 84–85, 193, 205, 246
  biography 165–71
  as senior cosmographer 169
  at the college in Barcelona 171
  at the college in Dôle 166
  in Spanish military service 169, 241, 243
  in the retinue of don Juan 170–71
  and San Vicente x, 31, 33, 84, 165–66, 168–69, 172, 174, 193, 246
  as student at mathematics school 31, 33, 165
  as teacher at mathematics school 33–34, 168
  De Centro Gravitatis 172, 182–83
  influenced by painters 172, 175, 177f10.8,
  on architecture 172, 236–38, 241, 243
  On conic sections 173–79
    pole and polar 173–74, 176–77, 179
  on optics 80
  Theses Mechanicae 166, 169
  use of the sector 35, 40–41
Den Bosch   see 's-Hertogenbosch
Derkennis, Ignatius 31, 34, 238
  biography 189–90
  De uno, trino, creatore 190
Desargues, Girard 47, 181
Descartes, René 103, 104, 219, 246
  correspondence with Ciermans 199, 201
  on optics 199, 201
  sine law 207
Drebbel, Cornelis 201
Dudith, Andreas 15n24,
Dunkirk 1, 6
Durandus, Jacobus 85n29
  biography 190–91
  on fortification 191, 238
Dürer, Albrecht
  Underweysung der Messung
    on conic sections 44, 45f2.2
Dutch Republic ix, xi, 1, 3, 5, 21n53, 206n71, 239
d'Yllan, García 188

Élvas 205, 240
Engelgrave, Johannes Baptiste 187–88
England ix, 1, 9

Errard, Jean   205
Eucharist   59

Fallon Simão   240
Fardella, Michelangelo   219
Farnese, Alexander, Duke of Parma   1, 7–9, 23, 25
Fayd'herbe, Lucas   234–35
Ferdinand II, Emperor   84, 193
Fermat, Pierre de   35
   Fermat's principle   208, 246
   possible influence of San Vicente   158, 161
   Treatise on Quadrature
      higher hyperbola   158
   Treatise on Quadrature   158–59
Fernemont, Count   194
Finé, Oronce
   Arithmetica practica   14
Flüske, Franz   197
France   ix, 5, 6n14, 168, 171n38, 248, 249
Francken, Gottfried   187–88
Frankfurt   2n3
   book fair   37, 53
Franq, Gaspar Alexius   197
Frederik Hendrik, stadtholder   239
Frias, Ferdinand de   25
Füssel, Friedrich   196

Galileï, Galileo   35n147
Galle, Theodoor
   engraver   64, 66f3.3
Galucci, Paolo   35n147
Gechauff, Thomas   45
Ghetaldi, Marino   35–36
Ghisoni, Stephanus   85
Giattini, Giovanni Battista   21n50
Giornale di letterati
Girard, Albert   51, 176f10.7
Goa   218n33
Granvelle, Antoine Perrenot de, Cardinal   62n1, 62n3
Grassi, Orazio   20, 32
Graz   85, 190
Greek mathematicians
   Apollonius of Perga   42–44, 70, 71, 87, 103
      biography   42–43
      Apollonius's theorem   111

      Apollonius's problem   46, 46 fig.2.3
   Konika   42–44, 78, 97n5, 98, 106, 114
   Konika (edition by Commandino)   45, 99n10
   Konika (edition by Richard)   45
   Konika (first edition by Maurolico)   45
   Work on conic sections   42–43, 102
   Archimedes   42, 97
      determination of the center of gravity   50, 52
      geometry   213
      infinitesimal calculus   50, 52, 121, 149
      On conoids and spheroids   45, 110
      On the quadrature of the parabola   42, 43, 116, 120
      squaring of the circle   47–48, 54
      trisection of an angle   68
   Euclid
      Elements   14, 16, 18, 19f1.1, 66, 94n2, 97n5, 122–23, 173n53, 190, 212n11, 213
   Heron of Alexandria   99, 101–2
   Hippias of Elis   68
   Nicomedes   68, 100
   Pappus of Alexandria   46, 68, 70, 71, 79, 93f4.7, 94, 149, 197, 213
      Collection   46, 97, 99, 100, 102
   Philon   102
Grenzing, Christoph   85, 232
Grienberger, Christoph   20, 21, 22n56, 190
   and della Faille   172, 182
   and San Vicente   33
   as architect   233
   judgment on San Vicente's quadrature   83–85, 152, 161, 245
   mathematics professor   20
Guldin, Paul   20, 82, 169
   criticism on indivisibles   57, 61
   Pappus Guldin theorem   213

Haarlem   2n3
Habsburg, family   3
Hamburg   2n2, 2n3
Happaert, Remigius   165
Hay, John   37, 38
Hesius, Willem   33

INDEX

biography 208–9
 as architect 233–35, 236fi4.7, 237fi4.8
 Emblemata de fide, spe, charitate 208–9
Hevelius, Johannes
 correspondent of Moretus 195
Heyns, Peter 2n3
Hoens, Jacobus 203
Hoeymaker, Hendrik
 as architect 223–24, 234
Holy Roman Empire 4
Houtappel sisters 10, 217, 246
Huygens, Christiaan 158, 161
 Huygens, Christiaan correspondence with Tacquet 212–13
 critcism on San Vicente's quadrature 86, 89
 visit to San Vicente 86
Huyssens, Peter
 biography 225
 as architect
  Church of Charles Borromeo 225–28
 as architect 29n107, 224, 234

ibn al-Haytham, Abū ʿAlī al-Ḥasan ibn al-Ḥasan   see Alhazen
Iconoclastic Revolt 1, 4
India 205n62
Indies 4
Inquisition 1, 81n2
Italy ix, 4, 9, 58, 62n3, 81n2, 85, 171, 231, 249

Jesuit College
 Aalst 208, 210
 Antwerp 23–35, 63, 82, 165, 224, 235
  Academy of Church History 28–29, 30–32, 34
  Engels Huys (also Hof van Liere) 26n88, 27, 29, 234, 235
  erection 23–25
  Huys van Aecken 24, 25n78, 27, 28fi.3, 29
  mathematics school 27–28
  professed house 29
  re-erection 25–26
 Breslau (Wrocław) 194–96
 Breznice 194
 Bruges 190, 193, 209, 211

Brussels 82, 184, 218
Cambrai 6
Cassel 199, 234
Collegio Romano 15, 16, 36, 83, 85, 188n31
 Academy of Mathematics 20–21
 and astronomy 32, 60n36, 81, 203, 218, 223, 249
Cologne 27
Dinant 6, 23
Dôle 166
Douai 22, 27, 59, 62, 63, 81
Dunkirk 210
Ghent 184, 190, 206, 208, 218, 220, 234
Glogau (Głogów) 194
Graz 85, 190
Jihlava 194
Klatovy 194
Klementinum, Prague 84, 184, 193–94, 232
Kortrijk 62, 82, 209, 234
Leuven 6, 7, 22–23, 30, 59, 82, 189, 192, 199, 206, 208, 211, 217, 233
 mathematics school 33–34, 51, 168, 244
Liège 27
Lier 168
Maastricht 7, 191, 206
Mechelen 165, 189, 234
Messina 14, 15
Molsheim 27
Münster 194
Neisse 194
Olomouc 193
Oudenaarde 210
Paris 62
's-Hertogenbosch 82, 198–99, 210
Saint Omer 7
Tournai 6, 23
Trier 27
Ypres 189, 199, 209
Znojmo 194
João IV, King of Portugal 186
Jonghe, Ignatius de 34–35, 128, 244
 Geometrica inquisitio
  application of San Vicente's methods 159–60
Juan José of Austria 40, 170–71

Kamiensky, Joannes de Swienthohliwicz   191
Kepler, Johannes   38, 61
  infinitesimals   53–57, 58
  Messekunst Archimedis   53
  Nova stereometria   53–54, 58
Kinner, Godefred   195
Kircher, Athanasius   20n50, 65f3.2, 190n51, 195, 196, 217n25

Laínez, Diego   6n18, 13
Langres, Nicolas de   241
Lassart, Charles   241
Le Vray, Valentin   33, 210
Leibniz, Gottfried Wilhelm   158, 161, 221, 247
  influenced by San Vicente   89, 246
Leopold Wilhelm, Archduke   193, 208
Leuven   6, 11, 34, 82
  Jesuit College   see Jesuit College, Leuven
  University   22, 27, 28n101, 218
Lier   33
Lille   6
Lipsius, Justus   15n24
Lobkowitz, Juan Caramuel   189
Losson brothers   235
Low Countries   ix, 1–7, 9, 23, 33, 39, 62, 82, 168–69, 203, 207, 217, 227, 238–40, 248, 249
Loyola, Ignatius of   6, 62n2
Lübeck   188

Macedo, Antonio
  and Queen Christina   186–87
MacLaurin, Colin   47
Madrid   xi, 41, 84, 85, 168–69, 172, 184, 193, 206, 243
Malines, Franciscus de   186, 187n17
Manare, Olivier   26, 27, 225n28
Manderscheidt, Charles   186
Marolois, Samuel   201, 205
Martini, Martino   217
Mary of Hungary   6
Maselli, Ludovico   20
mathematical instruments
  astrolabe   14, 67
  sector   35–41
  sundial   44n10, 67, 196

mathematics
  see also San Vicente, Gregorio a, and della Faille, Joannes
  Alhazen's problem   75–76, 101, 156–60
  Apollonius' s theorem   111
  Apollonius's problem   46, 46 fig.2.3
  conic section   42–47, 103–16, 132–40, 173–83
  curve (other than conic section)
    cissoid   101
    quadratrix   101
  horopter   65–66, 67
  infinitesimal calculus   49–56
    Cavalieri's indivisibles   57–61, 128–31, 213
    criticism on indivisibles   58–61, 214
    exhaustion method   121–24
    quadrature of the hyperbola   132–40
    San Vicente's infinitesimals   121–28, 132–52, 158–61, 213–16
  logarithms   108, 132–40, 220–21
  Mystic Hexagram Theorem   47, 174–75
  neusis-like constructions   44, 68–69, 94–95
  perspective   67, 70, 175–76, 177f10.8
  projective geometry   47, 70–75, 173–81
  Pythagoras' theorem   89
  squaring of the circle   47–49, 82–89, 124, 126, 161–64, 182–83
  trisection of an angle   47–48, 54, 75n40, 101f5.10
Matthias, Archduke of Austria   8
Maurolyci, Franciscus   45
Mechelen   235
  novitiate   165, 184, 199, 218
Mennher, Valentin   49–50
Mercurian, Everard   6n18, 20, 24
Mersenne, Marin   128, 130–31, 220, 247
Missions
  China   18, 34, 82, 205, 217n25
  Holland Mission   9
  Missio castrensis   9, 168n20
  Missio Navalis   210
Montano, Benito Arias   192
Montmorency, Floris de   84
Mordente, Fabrizio   35n147
Moretus, Balthasar I   64
Moretus, Balthasar II   5n12, 20n50, 33, 39

INDEX

Moretus, Jan   64, 192
Moretus, Theodorus   80, 155n2, 172, 192, 245, 246
   biography   192–94
   as a teacher in Bohemia   195–98
   as an author   196–98
   as collaborator of San Vicente   84, 85, 86n33, 87, 90, 117
Mydorge, Claude   46

Nadal, Hieronimo   14, 16, 23
Naples   5, 171, 219
Netherlands (Nederlanden)   ix, x, 4, 6, 24, 85, 223–24, 238–39
Netherlands, Southern   see Netherlands, Spanish
Netherlands, Spanish   ix x, 2, 4, 12n5, 22, 24, 86, 172, 184, 187, 206n71, 248
Nickel, Goswin   188
Nieuwpoort   220
Noyelle, Charles de   234
Nutius, Philip   see Nuyts, Philip
Nuyts, Philip   10, 31, 165
   biography   184–85
Nutius, Philip
   mission to Stockholm to convert Queen Christina   185–89
   Templars, Philip (pseud.)   187
Nuyts, Martin   184

Oliva, Giovanni Paolo   86, 234
Olivença   206, 240
Ortelius, Abraham
   Theatrum Orbis terrarum   197
Ostend   238

Palermo   171
Papebrochius, Daniel   203n58
Paris   192, 198n40
Pascal, Blaise   47, 181, 213n16, 246
   Mystic Hexagram Theorem   47, 174–75
   Treatise on conics   47
Pels, Anna   62
Pels, Clara   62n3
Philip Eugene, Count of Hornes et d'Herlies   211–12
Philip II, King of Spain   xi, 1–3, 6, 7n21, 62
Philip IV, King of Spain   40, 169–71, 232, 241

philosophy
   al Ghazali's wheel   60, 61f2.12
   Aristotle's wheel   60, 61f2.12, 212
   Ockham's razor   59
physics
   Aristotelian   15, 16, 21, 32, 60, 61, 121
   Aristotle
      Achilles and the tortoise   121
      Aristotle's wheel   60, 61f2.12, 212
      On the heavens   18
      Physics   60, 121
   energy   91
   energy, kinetic   91
   free fall   91
   inertia   91
   optics   68–79, 94–101, 172, 198, 199–201, 207–8
   statics   31, 33, 85, 91, 182, 197–99, 201, 212, 244–46
Pinto Pereira, Joseph   186
Plantijn, Christoffel
   printer   192
Polanco, Juan Alfonso de   12n7
Portugal   5, 9, 35, 205, 240
   war of independence   169, 205, 240, 241, 249
Prague   xi, 39n155, 84, 87, 155, 157, 162, 166, 184, 193–94, 245, 248
   Sack of
      by Saxons   85, 87, 88, 91, 193, 195n22
      by Swedes   193
Prince, Henri de   217
Puteanus, Erycius   32n124
Pynappel, Reynier   33, 210

Ratio Studiorum   17–18
   Messina program   14–15
Regiomontanus   45
Requesens, Luis   24
Ribadeneyra, Pedro de   6
Ricci, Matteo   195
Richard, Claude   45
Roberval, Gilles Personne de   216
Roermond   218
Rosis Giovanni de
   as consultant for building plans.   223
Rotterdam   2
Rouen   2n2

Rozdrażewski, Jakub Hieronim   203
Rozdrażewski, Krzysztof Aleksander   203
Rubens, Peter Paul
   and de Aguilon   64–66, 70
   Saint Charles Borromeo Church   224, 227, 228, 231

's-Hertogenbosch   82, 187n25, 198–99, 210, 239
Saenredam, Pieter   70f2.8, 71
Salamanca
   University   63
San Vicente, Gregorio a   ix–xii, 205
   biography   81–87
   in Antwerp   30–33, 82, 87, 94–115, 132
   and Aguilon
      influence   30, 47, 68, 71, 80, 87, 94
   and astronomy
      comets   21
      transit of Venus   21n54, 203
   and Cavalieri
      correspondence with San Vicente   85f4.2, 86
      possible influence on San Vicente   109, 132, 140, 161
      San Vicente's criticism on Cavalieri's indivisibles   128–31
   and Clavius   20, 21, 81–82
      cited by San Vicente   94
   and de Gottignies   218
   and de Jonghe
      application of San Vicente's methods   159–60
   and de Sarasa
      defense of San Vicente's methods   131, 220–21
   and Fermat
      possible influence on Fermat   158, 161
   and Grienberger   83, 161
   and Huygens   86
   and mechanics   91, 92f4.6, 93f4.7
   and sector   39–40
   and Tacquet   211
      use of San Vicente's methods   213–16
   Antwerp mathematics school   30–34
   as an architect   232–34
   at the Roman College   18, 20
   classic Greek problems
      squaring of the circle   47, 82–84
      trisection of an angle   94–98, 117–18

conic sections   47
   as an aid to construct mean proportionals   102–3, 106–8, 113
   construction by transformations   104–5, 108–9, 111–12, 115, 150–51, 156–57, 174
   inscribed polygons   115–16
   logarithmic properties of a hyperbola   108, 137
   properties   103, 146
   properties in an ellipse   105, 109–11, 113
   properties of a hyperbola   111, 132–40
   properties of a parabola   111
construction of mean proportionals   98–103, 106–8, 113, 133–36
existence of the orthocenter in a triangle   90–91
infinitesimal calculus   60–61
   analagy between the area of a parabola and a spiral   157
   analogy between an ungula and the sphere   147–48
   calculation of lateral surface areas   147–49
   calculation of volumes using the ductus method   143–46, 151–52
   ductus method   82, 140–43, 153–55
   exhaustion method   121–24, 145–46, 161
   expanding a geometric sequence by insertion   126–28
   influence of Stevin   51
   use of infinitesimals   124–28
influence of   244–48
law of cosines   90
manuscripts sent to Rome   153–55
manuscripts written in Rome and Prague   155–57
mathematical manuscripts   87–88
Opus geometricum   86
Problema Austriacum   57, 87–89, 121
Pythagoras's theorem   89–90
quadrature of the circle   82–86, 88, 89, 124, 146, 152, 154, 156, 161–64, 245
sequences and series   117
   Achilles and the tortoise   121
   convergence of   117–21
sojourn in Prague   84

sojourn in Rome   84, 245
students of San Vicente
   Boelmans   206
   Ciermans   199
   della Faille   x, 31, 33, 84, 165–66,
      168–69, 172, 174, 181–82, 193, 246
   Derkennis   189
   Moretus   192–93, 196
   other students of San Vicente   209–10
   youth of San Vicente in Bruges   81
Sapieha, Leo Carolus   203
Sarasa, Alphonsus Antonius de   131, 140
   biography   220
   and logarithms   220
   defense of San Vicente's methods   131,
      220–21
Schall von Bell, Adam   18
Schott, Kaspar   213
Scribani, Carlo   11, 26–29, 83, 224
Seventeen Provinces   xi, 1, 17
Sicily   5, 81, 171
Snel van Royen, Willebrord   207
Spain   ix, 1, 4, 5, 8, 9, 35, 40, 41, 63, 168–71,
   236, 240, 249
Spinola, Ambrogio   5, 239
Stade   2n3
Stafford, Ignace   35, 205
Stansel, Valentin   195
Steenvoorde   1
Stevin, Simon   2n3, 15n24, 50, 51, 53, 93f4.7,
   201
   infinitesimal calculus   51, 122
Stockholm   xi, 186–88
Stöffler, Johannes   14
Succa, Jacob de   31n116

Tacquet, Andreas   32n121, 34, 61, 89, 160, 190,
   206, 217–19, 238, 242f14.11
   biography   211
   criticism on indivisibles   128n22, 214
   infinitesimal calculus   213–16
Tartaglia, Nicolo   45
Teodósio, Prince of Portugal   205
Teplá   196
Thirty Years' War   4, 5
Torrentius, Laevinus   11, 23
Torres, Baltasar de   16, 20
Torricelli, Evangelista   128–30, 158n10
Tournai   6, 11n2, 62, 225

Tristano, Giovanni   222
Tucher, Jan Antoon   203–4
Turin   186
Twelve Years' Truce   2, 4, 21n53

University of Douai
   Collège d'Anchin   22
Uppsala   193
Uwens, Hendrik   205

Valerio, Luca   52–53
van Aelst, Walter   33, 91, 199, 246
   biography   210
van Ceulen, Ludolff
   squaring of the circle   48–49
van de Vyver S.J, Omer   xii, 249
van Etten, Hendrik, mayor of Antwerp   29
van Langren, Arnold-Florent   166, 169n32,
   241
van Maelcote, Odo   20–22
van Moerbeke, Willem
   Archimedes translation   45
van Raphelingen, François
   printer   192
van Rasseghem, Jacob   10n37, 33
   biography   209
van Roomen, Adriaan   36, 37
   solution of Apollonius's problem
      46 f2.3
   squaring of the circle   48–49
van Schoonhoven, Jan   24
van Schooten, Frans   47
Venice   2n2
Verbiest, Ferdinand   18, 34, 218
Vienna   85, 195n22, 198
Viète, Francois   36, 46, 94, 100
Ville, Antoine de   205
Vitelleschi, Muzio   82, 84, 85, 169n24, 193,
   205, 225n26, 232, 245
von Bukau, Ferdinand   197
Vredeman de Vries, Hans   70, 226

Wales   9
Wallis, John   47
Westphalia, Peace of   5, 231
Whiston, William   213
Wroclaw   see Breslau

Zucchi, Nicolo   193

Printed in the United States
by Baker & Taylor Publisher Services